World Sustainability Series

Due to its scope and nature, sustainable development is a matter which is very interdisciplinary, and draws from knowledge and inputs from the social sciences and environmental sciences on the one hand, but also from physical sciences and arts on the other. As such, there is a perceived need to foster integrative approaches, whereby the combination of inputs from various fields may contribute to a better understanding of what sustainability is, and means to people. But despite the need for and the relevance of integrative approaches towards sustainable development, there is a paucity of literature which address matters related to sustainability in an integrated way.

Notes on the quality assurance and peer review of this publication

Prior to publication, the works published in this book are initially assessed and reviewed by an in-house editor. If suitable for publication, manuscripts are sent for further review, which includes a combined effort by the editorial board and appointed subject experts, who provide independent peer-review. The feedback obtained in this way was communicated to authors, and with manuscripts checked upon return before finally accepted. The peer-reviewed nature of the books in the "World Sustainability Series" means that contributions to them have, over many years, been officially accepted for tenure and promotion purposes.

Kavindra Kumar Kesari · Vishnu D. Rajput ·
Chander Prakash
Editors

Smart Sustainable Materials Advancing Re-manufacturing and Technology

Smart and Sustainable Development in Agricultural Technology: Ecological and Practical Applications (Volume IV)

Springer

Editors
Kavindra Kumar Kesari
University Center for Research
and Development
Chandigarh University
Mohali, Punjab, India

Vishnu D. Rajput
Academy of Biology and Biotechnology
Southern Federal University
Rostov-on-Don, Russia

Chander Prakash
School of Engineering
Geeta University
Panipat, India

ISSN 2199-7373 ISSN 2199-7381 (electronic)
World Sustainability Series
ISBN 978-3-032-17906-7 ISBN 978-3-032-17907-4 (eBook)
https://doi.org/10.1007/978-3-032-17907-4

This Springer imprint is published by the registered company Springer Nature Switzerland AG
The registered company address is: Gewerbestrasse 11, 6330 Cham, Switzerland

Contents

Conversion of Agricultural Waste into Functional Nanomaterials for Environmental Applications

Niranjan Patra and Ştefan Ţălu

Abstract The conversion of agricultural waste into functional nanomaterials offers a sustainable pathway to address environmental challenges while promoting circular economy principles. This chapter explores recent advances in the valorization of crop residues, fruit peels, husks, shells, and other agro-wastes into carbon-based, metal/metal oxide, biopolymer, and hybrid nanostructures. Green synthesis strategies, including plant extract-mediated, sol–gel, hydrothermal, and precipitation methods, are highlighted for their low environmental footprint. Applications in water treatment, air purification, soil remediation, and energy–environment synergies are critically reviewed, demonstrating the versatility of these materials in pollutant removal, catalytic degradation, nutrient recovery, and CO_2 capture. Challenges such as feedstock variability, scalability, material stability, and economic feasibility are discussed alongside future perspectives for industrial integration. By linking waste management with nanotechnology innovation, this work underscores the potential of agricultural residues as renewable resources for next-generation environmental solutions.

Keywords Agricultural waste valorization · Functional nanomaterials · Green synthesis · Environmental remediation · Circular economy · Sustainable nanotechnology

1 Introduction

The intensification of global agricultural activity—driven by population growth, urbanization, and rising food demand—has led to an unprecedented accumulation of agricultural residues and by-products. These wastes, which include crop residues

N. Patra (✉)
Department of Chemistry, Koneru Lakshmaiah Education Foundation, Greenfield, Vaddeswaram 522502, Andhra Pradesh, India
e-mail: patraji@gmail.com

Ş. Ţălu
The Directorate of Research, Development and Innovation Management (DMCDI), Technical University of Cluj-Napoca, 15 Constantin Daicoviciu St., Cluj-Napoca, Cluj 400020, Romania

K. K. Kesari et al. (eds.), *Smart Sustainable Materials Advancing Re-manufacturing and Technology*, World Sustainability Series, https://doi.org/10.1007/978-3-032-17907-4_1

such as rice husks, wheat straw, corn cobs, and sugarcane bagasse, as well as fruit peels, shells, and livestock-derived materials, are generated in vast quantities every year across both developed and developing economies (Pandey et al. 2025). In many cases, such residues are either left unutilized or are disposed off through open burning, landfilling, or uncontrolled dumping. These practices not only secures potentially valuable resources but also aggravate environmental degradation by contributing to greenhouse gas emissions, air pollution, eutrophication, and soil quality deterioration (Abubakar et al. 2022). Addressing the environmental challenges posed by agricultural waste requires innovative, sustainable approaches that align with the principles of the circular economy and global sustainability goals, such as the United Nations Sustainable Development Goals (Islam et al. 2024).

One promising strategy lies in the conversion of agricultural waste into high-value functional nanomaterials. Agricultural residues are rich in biogenic components such as cellulose, hemicellulose, lignin, silica, proteins, chitin, and various bioactive molecules, which can serve as ideal precursors for the synthesis of nanostructures. These constituents act not only as raw material sources but also as reducing, capping, and stabilizing agents during nanoparticle formation, reducing the need for toxic reagents and enabling greener synthesis routes (Sanoja-López et al. 2025). Over the past decade, the emerging discipline of green nanotechnology has provided a platform for exploiting these intrinsic properties, offering environmentally benign, cost-effective, and scalable alternatives to conventional nanomaterial production methods (Osman et al. 2024).

Nanomaterials derived from agricultural waste have demonstrated remarkable potential in environmental remediation applications. Carbon-based nanostructures, such as biochar, activated carbon, graphene derivatives, and carbon nanotubes obtained from biomass precursors, have been successfully employed for heavy metal adsorption, dye removal, oil spill clean-up, and photocatalytic degradation of organic contaminants (Rao et al. 2021). Similarly, metal and metal oxide nanoparticles—including ZnO, TiO_2, Fe_3O_4, and CaO—synthesized from waste-rich sources like eggshells, banana peels, and rice husks have shown excellent catalytic and antimicrobial activities for water and air purification (Almeida-Naranjo et al. 2025). Biopolymer-based nanomaterials, such as cellulose nanofibers, nanocrystals, and chitosan nanoparticles, also offer sustainable solutions for soil remediation, nutrient delivery, and environmental sensing (Patra et al. 2024). Moreover, the versatility of these waste-derived nanomaterials extends beyond remediation, encompassing applications in energy storage devices, photocatalytic hydrogen production, and carbon dioxide capture, thereby linking waste valorization to both environmental and energy sustainability (Elezz et al. 2025; Ganesan and Kadhar Mohamed 2025).

Despite these advances, several challenges remain. Variability in feedstock composition, seasonal availability, and processing constraints can influence yield and material performance. Scaling up synthesis processes while maintaining quality and ensuring economic feasibility continues to be a critical bottleneck. Furthermore, the environmental, health, and safety implications of deploying nanomaterials—regardless of their green origin—require comprehensive assessment before large-scale adoption (Pierrat and García-Triñanes 2024; Hemavathy et al. 2025).

Nevertheless, ongoing research integrating biotechnology, material science, and environmental engineering offers promising avenues to overcome these barriers. With supportive policies, industry–academia collaboration, and continued innovation, agricultural waste can transition from an environmental liability to a strategic resource for sustainable nanomaterial production.

This chapter explores the scientific and technological foundations of converting agricultural waste into functional nanomaterials for environmental applications. It examines the nature and composition of various agricultural residues, discusses the synthesis routes and mechanisms for nanomaterial formation, categorizes the resultant nanostructures, and analyzes their roles in environmental remediation. It further addresses the challenges of large-scale implementation and envisions future directions that align with global sustainability priorities. By integrating concepts from green chemistry, waste valorization, and nanotechnology, the chapter aims to provide a comprehensive understanding of how agricultural waste can be transformed into high-value functional materials that contribute to environmental protection and sustainable development.

2 Agricultural Waste: Sources, Composition, and Challenges

Agricultural waste encompasses a diverse range of materials generated during farming, harvesting, processing, and agro-industrial operations. These wastes can broadly be categorized into **primary production residues**, which arise directly from cultivation and harvesting, and **secondary or processing residues**, which are generated during post-harvest handling, processing, and value addition (Riseh et al. 2024). Additionally, livestock and aquaculture activities contribute organic wastes of animal origin. The types of agricultural waste vary considerably in composition, moisture content, and potential for valorization, depending on the crop type, processing method, and local agricultural practices as shown in Table 1.

Crop residues: Crop residues are the largest component of agricultural waste globally and include **straws, stalks, husks, and leaves** left in the field after harvesting. Common examples are rice straw, wheat straw, maize stalks, sorghum stover, sugarcane trash, and cotton stalks (Riseh et al. 2024; Sharma et al. 2018). Rice husk, generated during milling, is particularly abundant in rice-producing nations and is rich in silica, lignin, and cellulose, making it suitable for nanomaterial synthesis such as silica nanoparticles and activated carbon (Nzereogu et al. 2023; Hincapié Rojas et al. 2019). These residues are often burnt in situ to clear fields for the next cropping cycle, contributing significantly to seasonal air pollution episodes, such as stubble burning in Northern India and Southeast Asia (Lin and Begho 2022).

Table 1 Major agricultural waste types, their chemical composition, potential nanomaterial products

Waste type	Major constituents	Potential nanomaterial products
Rice husk	Cellulose (~35–45%), lignin (~20–25%), hemicellulose (~15–20%), silica (~15–20%)	Silica nanoparticles, carbon nanostructures, mesoporous silica
Sugarcane bagasse	Cellulose (~40–50%), hemicellulose (~20–25%), lignin (~18–24%), ash (~1–3%)	Cellulose nanofibers, carbon quantum dots, activated carbon
Coconut husk	Lignin (~40–45%), cellulose (~35–40%), hemicellulose (~15–20%), minerals (~2–4%)	Activated carbon, lignin-derived nanoparticles, carbon nanotubes
Wheat straw	Cellulose (~38–45%), hemicellulose (~27–30%), lignin (~15–20%), ash (~5–8%)	Cellulose nanocrystals, silica nanoparticles, biochar
Banana peel	Pectin (~20–30%), cellulose (~15–20%), hemicellulose (~10–15%), lignin (~6–12%), minerals (~5–7%)	Metal oxide nanoparticles (ZnO, Ag), biochar, carbon quantum dots
Corn cob	Cellulose (~33–42%), hemicellulose (~35–40%), lignin (~10–15%), ash (~1–2%)	Carbon nanofibers, cellulose nanofibers, activated carbon
Eggshell	Calcium carbonate (~94–97%), magnesium carbonate (~1%), organic matrix (~2–4%)	CaO nanoparticles, hydroxyapatite, calcium phosphate nanomaterials

Processing by-products from agro-industries: Agro-industrial activities, such as milling, juicing, oil extraction, and canning, generate substantial amounts of by-products. Examples include **fruit peels (orange, banana, mango), seed shells (peanut, coconut, walnut), bagasse from sugarcane processing, and oil cakes from soybean or sunflower seed pressing**. These materials often contain high levels of carbohydrates, proteins, phenolics, and minerals, which can be used as feedstocks for producing carbon-based nanomaterials, metal oxide nanoparticles, and biopolymer nanostructures (Hincapié Rojas et al. 2019; Jafarzadeh et al. 2024; Panesar et al. 2016). For instance, coconut husk contains lignocellulose and silica, while banana peel is rich in potassium and polyphenols—both have been exploited in green nanoparticle synthesis (Ansari et al. 2024; Machado et al. 2022).

Animal husbandry and aquaculture waste: Livestock and aquaculture sectors contribute **manure, feathers, eggshells, bones, and fish scales** as agricultural waste streams. Manure contains nitrogen, phosphorus, and organic matter, making it suitable for composting and biochar production. Feathers are a rich source of keratin protein, which has potential as a precursor for nanofibers and biopolymer films (Chilakamarry et al. 2021; Lazarus et al. 2021). Eggshells, composed mainly of calcium carbonate, have been widely utilized for the synthesis of CaO nanoparticles and as bio-adsorbents in water treatment (Zia et al. 2025). Similarly, fish

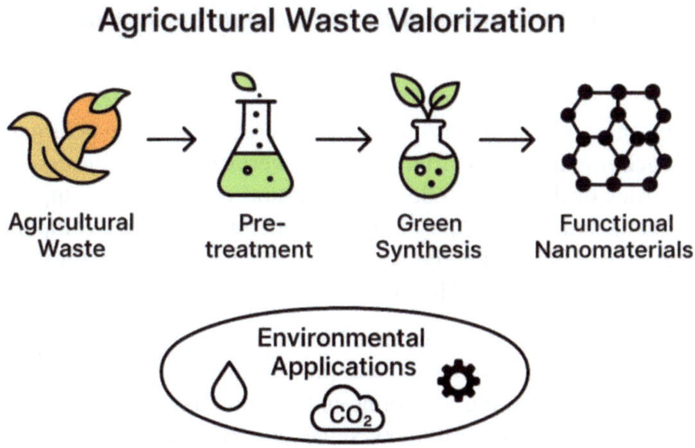

Fig. 1 Overview of the valorization pathway for agricultural waste into functional nanomaterials, including collection, pre-treatment, green synthesis, and environmental applications

scales and bones are sources of hydroxyapatite nanocrystals, useful in environmental remediation and biomedical applications (Zia et al. 2025).

Horticultural and greenhouse waste: The cultivation of fruits, vegetables, and ornamentals in open fields or greenhouses generates **trimmings, pruned branches, spoiled produce, and discarded flowers** (Fig. 1). These materials are highly biodegradable and typically rich in cellulose, hemicellulose, sugars, and essential oils (Patra et al. 2024). Floral waste from large-scale horticulture and religious ceremonies has also been explored as a precursor for carbon nanostructures and natural dye recovery (Srivastav and Kumar 2021).

Emerging waste streams from modern agriculture: Precision and protected farming systems, including hydroponics and vertical farming, produce **nutrient-rich plant roots, spent growth media, and residual nutrient solutions**. Although currently small in volume compared to traditional residues, these waste streams are expected to grow with the adoption of high-tech agriculture. They offer potential as specialized sources for nanomaterial precursors, especially due to their controlled nutrient profiles. The diversity in agricultural waste types underscores the importance of **tailoring valorization strategies** based on the physicochemical composition of each waste stream. For nanomaterial production, factors such as lignocellulosic content, mineral profile, and extractable bioactive compounds directly influence the choice of conversion pathway and the properties of the resultant nanomaterials.

2.1 *Chemical Composition Relevant to Nanomaterial Synthesis*

The suitability of agricultural waste for nanomaterial production is largely dictated by its **physicochemical composition**, which governs both the conversion pathway and the properties of the final product. Agricultural residues are complex matrices containing organic macromolecules, minerals, and trace bioactive compounds, all of which can serve as functional precursors for nanoscale materials (Jayakrishnan and Bora 2024; Zhang et al. 2025). The major components are outlined below.

Cellulose: Cellulose is the most abundant biopolymer on Earth and a key structural component of plant cell walls. It is a linear polysaccharide composed of β-(1 → 4)-linked D-glucose units, forming crystalline and amorphous regions. Cellulose derived from waste such as rice husk, corn stalks, sugarcane bagasse, and cotton residues can be processed into **cellulose nanofibers (CNFs)** or **cellulose nanocrystals (CNCs)** via mechanical, chemical, or enzymatic treatment (Ansari et al. 2024). These nanostructures exhibit exceptional tensile strength, biodegradability, and surface reactivity, making them useful in adsorption, membrane fabrication, and as supports for catalytic nanoparticles.

Hemicellulose: Hemicelluloses are branched heteropolysaccharides consisting of xylose, mannose, galactose, arabinose, and uronic acids. Although less crystalline than cellulose, hemicelluloses can be hydrolyzed into sugars for microbial fermentation or chemically functionalized for nanomaterial synthesis. Their amphiphilic nature enables them to act as stabilizers in green nanoparticle formation (Chibrikov et al. 2024).

Lignin: Lignin is a three-dimensional phenolic polymer composed mainly of p-coumaryl, coniferyl, and sinapyl alcohol units. Abundant in woody and grassy residues such as wheat straw and coconut husk, lignin is a valuable precursor for **carbon-based nanomaterials** like activated carbon, graphene-like sheets, and carbon dots. Its aromatic structure also imparts antioxidant and UV-absorbing properties to composite nanomaterials (Mariana et al. 2021).

Proteins: Certain agricultural wastes, particularly from animal by-products (feathers, milk residues, legumes), are rich in proteins such as keratin, casein, and albumin. Proteins contain functional groups ($-NH_2$, $-COOH$, $-SH$) capable of reducing metal salts and stabilizing nanoparticles in aqueous media. Keratin from poultry feathers has been used to produce nanofibers, while casein has served as a biotemplate for calcium phosphate nanomaterials (Lazarus et al. 2021).

Polysaccharides other than Cellulose: Wastes from crustaceans, insect farming, or fungal cultivation often contain chitin—a linear polymer of N-acetylglucosamine—which can be converted into **chitosan nanoparticles** with excellent adsorption and antimicrobial properties. Similarly, pectin from fruit peel waste acts as a natural capping agent in green nanoparticle synthesis (Muthu et al. 2021).

Minerals and inorganic components: Many plant-derived wastes contain significant inorganic fractions. Rice husk is rich in amorphous silica, which can be extracted and used to produce silica nanoparticles. Eggshells and mollusk shells are predominantly calcium carbonate, which can be thermally converted to calcium oxide nanoparticles. Banana peels and sugarcane bagasse contain potassium, magnesium, and trace metals that can contribute to catalytic activity in the resulting nanomaterials (Pandey et al. 2025).

Phenolic compounds and secondary metabolites: Agricultural residues such as tea waste, grape pomace, and citrus peels are rich in polyphenols, flavonoids, tannins, and essential oils. These compounds possess reducing and chelating abilities, enabling them to mediate the formation of metal nanoparticles without external chemical reducing agents (Rudrapal et al. 2024; Rodríguez-Félix et al. 2022).

Moisture and ash content: Although often overlooked, the moisture and ash content of agricultural residues affect both storage stability and processing efficiency. High moisture content can lead to microbial degradation before utilization, while high ash content can influence the purity and morphology of nanomaterials produced through thermal routes (Pandey et al. 2025). In essence, the **diversity of chemical constituents in agricultural waste directly shapes its valorization potential**. Carbon-rich lignocellulosic materials are particularly suited for producing carbon nanostructures, while mineral-rich wastes are ideal for inorganic nanoparticles. Understanding the composition–structure–property relationship is therefore critical to selecting optimal synthesis strategies and achieving targeted functionality in environmental applications.

2.2 Challenges in Utilizing Agricultural Waste for Nanomaterial Production

Despite the abundance and promising composition of agricultural residues, their direct utilization for nanomaterial synthesis faces multiple technical, economic, and regulatory challenges. These barriers must be addressed to achieve consistent quality, scalability, and environmental safety in practical applications. One of the primary issues is **feedstock variability**. The chemical composition of agricultural waste can fluctuate considerably due to crop variety, cultivation practices, soil quality, climatic conditions, and harvest time (Le et al. 2023; Bhatia and Sindhu 2024). For instance, the silica content in rice husk or the lignin fraction in wheat straw can vary widely across regions and seasons, leading to inconsistencies in yield and properties of the derived nanomaterials. This variability necessitates pre-processing, characterization, and sometimes blending of different batches to maintain uniformity in production. **Seasonal availability and logistics** present additional hurdles. Many residues are generated in large quantities during specific harvest windows, resulting in surplus material that requires storage or rapid processing to prevent degradation.

High moisture content and susceptibility to microbial activity can reduce the usability of wastes such as fruit peels, floral residues, and green biomass if not promptly stabilized (Nirmal et al. 2023; Usman et al. 2025; Ganesh et al. 2022). Furthermore, transporting low-density agricultural residues over long distances can be economically unviable unless densification or localized processing facilities are established. From a technical perspective, **pre-treatment requirements** often add complexity and cost. Many conversion processes for nanomaterial synthesis require size reduction, drying, removal of impurities, or chemical modification of the feedstock to enhance reactivity. For lignocellulosic wastes, delignification or alkaline treatment may be necessary, while mineral-rich wastes may require calcination or acid washing before use. These steps can be energy-intensive and, if not optimized, may offset some of the environmental benefits of using waste-derived precursors (Saadan et al. 2024; Jain and Kumar 2024). Another important concern is **process scalability**. While numerous laboratory-scale studies have demonstrated the feasibility of producing high-performance nanomaterials from agricultural residues, translating these processes to industrial-scale operations remains challenging. Issues include reactor design, process control, continuous operation, and maintaining nanoparticle size and morphology at large volumes. The absence of standardized protocols also makes it difficult to compare results across studies or to benchmark against conventional nanomaterials. **Environmental, health, and safety (EHS) considerations** are increasingly critical in this domain. Although agricultural waste is a renewable resource, the nanomaterials produced from it may still present risks if they are released into the environment or enter biological systems. Limited data exists on the ecotoxicity, biodegradability, and long-term environmental fate of many waste-derived nanomaterials, especially in complex ecosystems. This knowledge gap can hinder regulatory approvals and public acceptance. Finally, **economic feasibility and market integration** remain decisive factors. Even when production processes are technically viable, their cost competitiveness with conventional nanomaterials depends on factors such as raw material collection, processing efficiency, energy consumption, and product performance. Without supportive policies, subsidies, or market incentives, industries may be reluctant to invest in scaling up agricultural waste valorization technologies.

3 Conversion Pathways for Nanomaterial Synthesis

The transformation of agricultural waste into functional nanomaterials can be achieved through several conversion pathways, each exploiting the inherent chemical and structural characteristics of the feedstock. These routes may be broadly classified into **thermochemical processes**, **biological or biochemical processes**, **green chemical synthesis methods**, and **hybrid approaches** that integrate multiple techniques for enhanced efficiency and tailored properties. The selection of a particular pathway depends on the nature of the waste, desired nanomaterial properties, resource availability, and economic and environmental considerations (Abady et al. 2025; Gupta et al. 2023).

3.1 Thermochemical Methods

Thermochemical processes involve the use of high temperatures to induce chemical transformations in agricultural residues, converting organic and inorganic constituents into nanoscale structures. These methods are particularly suited for carbonaceous and mineral-rich wastes. **Pyrolysis**: Pyrolysis is the thermal decomposition of biomass in an oxygen-limited environment, producing biochar, bio-oil, and syngas. Biochar derived from rice husks, coconut shells, or sugarcane bagasse can be further activated (physically or chemically) to yield activated carbon nanoparticles with high surface areas and porosity, suitable for adsorption and catalysis (Jayakrishnan and Bora 2024). Pyrolysis temperatures and residence times influence pore structure and particle size. **Gasification**: Partial oxidation of biomass at elevated temperatures converts organic material into syngas (CO and H_2) and ash. The ash fraction from certain residues, such as rice husk, can be rich in amorphous silica, which can be processed into silica nanoparticles (Chand et al. 2025). **Calcination**: Mineral-rich wastes like eggshells, mollusk shells, and bones can be subjected to calcination at 800–1000 °C to produce calcium oxide (CaO) or hydroxyapatite nanoparticles. These materials have demonstrated utility in water treatment (phosphate removal, heavy metal adsorption) and as catalysts (Hemmami et al. 2024; Das Lala et al. 2021). **Hydrothermal Carbonization (HTC)**: HTC uses subcritical water (typically 180–250 °C) to convert wet biomass into hydrochar without the need for drying. Hydrochar can then be functionalized to produce carbon dots or doped carbon nanoparticles with fluorescent or catalytic properties (Hemmami et al. 2024; Das Lala et al. 2021).

3.2 Biological and Biochemical Methods

Biological approaches harness microorganisms or biomolecules to synthesize nanomaterials under mild, environmentally benign conditions. **Microbial Synthesis**: Bacteria, fungi, and algae can act as biofactories for metal and metal oxide nanoparticles. Agricultural waste extracts can serve as nutrient sources or electron donors in these processes. For example, fungi grown on fruit peel hydrolysates have been shown to reduce silver ions into Ag nanoparticles with antimicrobial properties (Khursheed et al. 2023; Ahamed et al. 2024). **Enzymatic Biotransformation**: Specific enzymes such as laccases, peroxidases, and reductases can catalyze the synthesis of nanoparticles from metal salts. Enzyme-rich extracts derived from waste fermentation (e.g., from wheat bran) can mediate nanoparticle formation without toxic reducing agents (Seth and Meena 2024). **Fermentation-Derived Precursors**:

Biomass-derived sugars and organic acids produced during fermentation can be used to synthesize carbon nanostructures or as stabilizers for inorganic nanoparticles (Chauhan et al. 2025).

3.3 Green Chemical Synthesis Approaches

Green chemical synthesis focuses on producing nanomaterials with minimal environmental impact by reducing hazardous chemical usage, lowering energy requirements, and employing renewable feedstocks in line with green chemistry principles. A widely adopted route is **plant extract-mediated synthesis**, where bioactive compounds such as polyphenols, flavonoids, terpenoids, and proteins present in fruit peels, leaves, seeds, and husks serve as natural reducing and stabilizing agents for metal and metal oxide nanoparticles. For example, banana peel extract has been effectively used to synthesize ZnO nanoparticles with high photocatalytic efficiency for dye degradation (Ruangtong et al. 2020).

Another eco-friendly method is the **sol–gel process**, particularly suitable for silica-rich wastes like rice husk ash. This approach operates at relatively low temperatures, enabling precise control over particle size, morphology, and surface chemistry to produce monodisperse silica nanoparticles (Hincapié Rojas et al. 2019). **Precipitation and co-precipitation techniques** are also employed, where calcium carbonate from eggshells is dissolved in mild acids and subsequently precipitated into nanoparticles or converted into hydroxyapatite, which shows strong potential for pollutant adsorption and removal (Das Lala et al. 2021).

Hydrothermal and solvothermal synthesis methods provide an energy-efficient route for crystallizing nanomaterials from biomass-derived precursors under controlled temperature and pressure. By incorporating carbon derived from biomass or doping with heteroatoms, the catalytic and adsorption capabilities of the resulting materials can be significantly enhanced, making these approaches highly relevant for environmental remediation applications.

3.4 Hybrid and Integrated Approaches

Hybrid methods combine thermochemical, biological, and green chemical techniques to optimize efficiency, reduce costs, and tailor nanomaterial properties. For instance, hydrothermal carbonization of fruit waste can produce hydrochar, which is subsequently activated via chemical treatment using biomass-derived acids to yield high-performance activated carbon nanoparticles (Pasipanodya et al. 2025; Inkoua et al. 2023). Similarly, calcination of eggshells followed by green chemical functionalization with plant extracts can yield metal oxide nanoparticles with improved dispersibility and catalytic activity. Integrated biorefinery concepts—where multiple products such as bioenergy, biofertilizers, and nanomaterials are generated from a single biomass stream—are increasingly being explored to maximize value recovery and minimize waste (Ahmed et al. 2023). These diverse conversion pathways highlight the adaptability of agricultural waste valorization strategies. Each method offers specific advantages in terms of scalability, cost, and product functionality, but also comes with its own limitations. The choice of method depends on the targeted application, feedstock composition, and desired physicochemical properties of the nanomaterial.

4 Types of Functional Nanomaterials Derived from Agricultural Waste

4.1 Carbon-Based Nanomaterials

Agricultural waste is an abundant and renewable source of carbon, making it a promising feedstock for synthesizing carbon-based nanomaterials. These nanomaterials include **activated carbon nanoparticles**, **carbon nanotubes** (**CNTs**), **graphene and graphene oxide** (**GO**), and **carbon quantum dots** (**CQDs**). The high lignocellulosic content, inherent porosity, and presence of mineral impurities in agricultural residues enable diverse carbon architectures with tunable physicochemical properties (Fig. 2).

Activated Carbon Nanoparticles (**ACNPs**): Activated carbon nanoparticles are typically produced via pyrolysis or hydrothermal carbonization of biomass followed by physical or chemical activation. Agricultural wastes such as coconut shells, rice

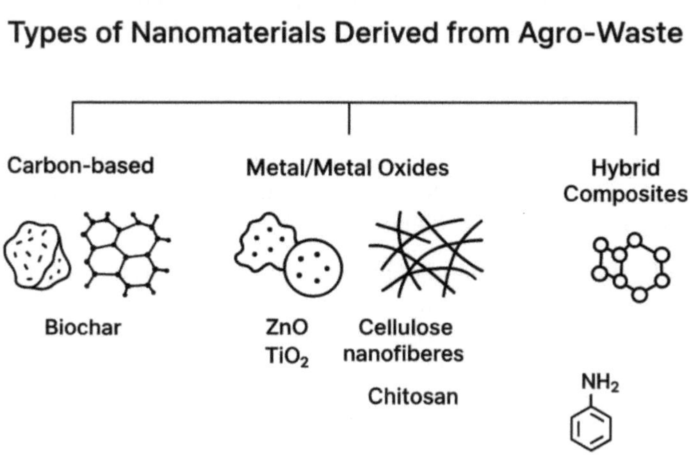

Fig. 2 Classification of nanomaterials from agro-waste

husks, corn cobs, and sugarcane bagasse have been extensively used to produce ACNPs with high surface area (500–2000 m^2/g), microporosity, and abundant surface functional groups (Laishram et al. 2025). These features make them effective adsorbents for heavy metals, dyes, and emerging contaminants in water treatment. Chemical activation agents like KOH, $ZnCl_2$, or phosphoric acid enhance pore development and control particle size (Laishram et al. 2025).

Carbon Nanotubes (CNTs): CNTs derived from agricultural wastes are generally synthesized through catalytic chemical vapor deposition (CVD), where carbon-rich volatiles from pyrolyzed biomass are deposited over metal catalysts (e.g., Fe, Ni, Co). Feedstocks such as palm kernel shells, corn stalks, and sawdust have been utilized as carbon precursors (Wang et al. 2021). CNTs possess high tensile strength, electrical conductivity, and large aspect ratios, making them suitable for applications in electrochemical energy storage, sensors, and pollutant removal. The sustainable production of CNTs from waste biomass significantly reduces the environmental footprint compared to petroleum-based precursors.

Graphene and Graphene Oxide (GO): Graphene and its derivatives can be obtained from agricultural wastes via two main routes: (i) conversion of biomass to graphite-like carbon followed by oxidation–exfoliation (Hummers' method or modified versions), and (ii) direct hydrothermal or microwave-assisted synthesis using biomass-derived carbon sources (Sheoran et al. 2022). For example, sugarcane bagasse and banana peels have been processed to yield GO sheets with high oxygen-containing functional groups, which enhance hydrophilicity and enable metal ion adsorption. Reduced graphene oxide (rGO) can also be obtained via green reduction using plant extracts, improving electrical conductivity for electrochemical applications.

Carbon Quantum Dots (CQDs): CQDs are quasi-spherical, photoluminescent nanoparticles (<10 nm) that can be synthesized from biomass via hydrothermal, microwave-assisted, or pyrolytic methods. Agricultural wastes such as orange peels, watermelon rinds, and coffee grounds have been used as carbon sources for CQD production (Aksu and Güzdemir 2025). CQDs exhibit unique optical properties, chemical stability, and biocompatibility, making them attractive for sensing, bioimaging, and photocatalysis. Biomass-derived CQDs often contain surface functional groups (–OH, –COOH, $-NH_2$) that facilitate functionalization and improve environmental remediation capabilities (Zhou et al. 2012).

Sustainability and Application Potential: Carbon-based nanomaterials from agricultural waste offer a dual advantage: they provide high-value products while mitigating environmental problems associated with open burning or landfill disposal of biomass. Their high adsorption capacity, tunable surface chemistry, electrical conductivity, and catalytic potential make them versatile for environmental applications such as heavy metal removal, dye degradation, capacitive deionization, photocatalysis, and pollutant sensing (Aksu and Güzdemir 2025; Zhou et al. 2012; Kasinathan et al. 2022).

4.2 Metal and Metal Oxide Nanoparticles

Agricultural wastes are rich in mineral components such as zinc, calcium, silica, iron, copper, and titanium precursors, making them suitable low-cost feedstocks for synthesizing metal and metal oxide nanoparticles. These nanomaterials possess excellent photocatalytic, magnetic, antimicrobial, and adsorption properties, enabling their deployment in diverse environmental applications. Biomass-derived synthesis approaches—often referred to as **green synthesis**—leverage plant extracts or ash as both a precursor and reducing/stabilizing agent, thereby reducing the need for hazardous chemicals (Kasinathan et al. 2022; Shah et al. 2025; Desmond et al. 2021).

Zinc Oxide (ZnO) Nanoparticles: ZnO nanoparticles are widely valued for their photocatalytic efficiency, antimicrobial activity, and UV-blocking properties. Agricultural wastes such as rice husk ash, corn husks, and banana peels can supply zinc precursors when combined with inexpensive zinc salts in green synthesis routes. For example, plant extract-mediated synthesis using banana peel or papaya leaf filtrates can produce crystalline ZnO nanoparticles with sizes below 50 nm, which exhibit efficient degradation of dyes and phenolic compounds under UV or sunlight irradiation (Kulis-Kapuscinska et al. 2023; Zainelabdin et al. 2012).

Titanium Dioxide (TiO_2) Nanoparticles: TiO_2 nanoparticles are renowned for their chemical stability, photocatalytic performance, and self-cleaning capabilities. Although titanium is less abundant in agricultural waste compared to other metals, certain residues such as ilmenite-rich rice husk ash or plant-based titanium adsorbates can serve as precursors. Green synthesis approaches using plant extracts from tea leaves or agricultural by-products as reducing agents have yielded anatase-phase TiO_2 nanoparticles with high photocatalytic activity for degrading persistent organic pollutants (Liu et al. 2021).

Magnetite (Fe_3O_4) Nanoparticles: Iron oxide nanoparticles—particularly Fe_3O_4—exhibit strong superparamagnetism, making them ideal for magnetic separation-based pollutant removal. Agricultural wastes such as sugarcane bagasse ash, banana peel ash, and corncob ash are rich in iron precursors that can be converted to Fe_3O_4 nanoparticles via co-precipitation or hydrothermal methods. These materials can be easily recovered from treated water using an external magnet, reducing secondary contamination risks (Trinh et al. n.d.).

Copper Oxide (CuO) Nanoparticles: CuO nanoparticles possess notable antimicrobial and catalytic properties. They can be synthesized from copper-rich agricultural residues, or from biomass-assisted reduction of copper salts. For example, onion peel extract or coffee husk filtrate has been used as both a reducing and capping agent to produce CuO nanoparticles with narrow size distributions. These particles are effective in catalyzing degradation of organic dyes and as antimicrobial agents against multidrug-resistant bacteria (Patra and Ravi 2024).

Calcium Oxide (CaO) Nanoparticles: CaO nanoparticles are valued for their strong alkalinity, adsorption potential, and use in wastewater treatment. Eggshells, a by-product of poultry farming, are one of the most abundant calcium-rich agricultural wastes, containing up to 94% $CaCO_3$. Upon calcination at elevated temperatures (800–900 °C), $CaCO_3$ is converted into reactive CaO nanoparticles. These have been applied in phosphorus removal, heavy metal precipitation, and as catalysts in biodiesel production (Hemmami et al. 2024).

Silicon Dioxide (SiO_2) Nanoparticles: Rice husks and sugarcane bagasse ash are among the richest natural sources of amorphous silica. Controlled combustion followed by sol–gel or precipitation methods can yield SiO_2 nanoparticles with high purity and surface area. These nanoparticles are used in adsorptive removal of dyes, immobilization of enzymes, and as carriers in nanocomposites for environmental remediation (Gupta et al. 2023). Using agricultural waste for metal and metal oxide nanoparticle synthesis offers a circular economy advantage by transforming low-value by-products into high-value functional materials. This approach reduces mining-related environmental impacts, lowers production costs, and aligns with green chemistry principles by minimizing the use of toxic precursors and solvents (Rahaman and Khan 2024; Aziz et al. 2025). Moreover, the adaptability of biomass-mediated synthesis allows for tuning nanoparticle size, morphology, and surface chemistry for targeted environmental applications.

4.3 Biopolymer-Based Nanomaterials

Agricultural waste is an abundant and renewable source of biopolymers such as cellulose, hemicellulose, starch, pectin, chitin, and proteins. These naturally occurring macromolecules can be converted into nanostructured materials—such as cellulose nanofibers, nanocrystals, and chitin/chitosan nanoparticles—offering biodegradability, biocompatibility, and tunable surface chemistry for environmental applications. The conversion of biomass-derived biopolymers into nanomaterials aligns with sustainable material engineering practices, as it minimizes waste, reduces dependence on petroleum-based polymers, and produces functional nanomaterials with minimal environmental footprint (Patra et al. 2024).

Cellulose Nanofibers (CNFs): Cellulose is the most abundant biopolymer in nature and is present in high concentrations in agricultural residues such as rice husks, wheat straw, sugarcane bagasse, banana pseudostems, and corn stalks. CNFs are long, flexible fibers with diameters typically below 100 nm, obtained through mechanical fibrillation, enzymatic treatment, or chemical pre-treatment. These nanofibers exhibit high tensile strength, low density, and a high aspect ratio, making them suitable as reinforcement agents in nanocomposites for water filtration, pollutant adsorption, and as support matrices for photocatalysts (Abualnaja et al. 2024). For example, CNF-based membranes derived from sugarcane bagasse have demonstrated high efficiency in removing dyes and heavy metals from wastewater.

Cellulose Nanocrystals (CNCs): CNCs are rod-like nanoparticles produced by the acid hydrolysis of cellulose-rich biomass, which removes the amorphous regions and leaves behind highly crystalline domains. CNCs from agricultural wastes such as pineapple leaves, cotton stalks, and rice husks exhibit crystallinities above 70% and surface areas exceeding 200 m^2/g (Singh et al. 2025). Their high stiffness, reactive hydroxyl groups, and potential for surface modification make them promising for adsorption of heavy metals, immobilization of enzymes, and as carriers for nanocatalysts in environmental remediation.

Chitin Nanofibers and Nanocrystals: Chitin is the second most abundant natural polymer after cellulose, primarily found in the exoskeletons of crustaceans and insects. Agricultural waste streams from seafood processing—such as shrimp and crab shells—are rich in chitin. Chitin nanofibers can be isolated via mechanical disintegration or chemical treatments, while chitin nanocrystals are obtained through partial acid hydrolysis. These nanostructures possess inherent antimicrobial properties, high surface charge, and excellent film-forming ability, making them suitable for antimicrobial filtration membranes and as bio-adsorbents for dyes and oil spills (Cao et al. 2023).

Chitosan Nanoparticles: Chitosan, derived from the deacetylation of chitin, is a cationic biopolymer with remarkable adsorption capacity for negatively charged pollutants, including dyes, phosphate, and heavy metals. Agricultural waste-based chitosan nanoparticles have been synthesized through ionic gelation, reverse micelle, and spray-drying methods (Muthu et al. 2021). Their nanoscale dimensions and abundant amino groups enhance adsorption kinetics and allow for facile regeneration, making them ideal candidates for wastewater treatment. Moreover, chitosan nanoparticles can act as stabilizing agents for green-synthesized metal nanoparticles, thereby creating multifunctional hybrid nanomaterials. The valorization of agricultural waste into biopolymer-based nanomaterials supports a circular economy by converting biodegradable resources into value-added products. The biodegradability of these nanomaterials ensures minimal long-term ecological burden, while their surface functionalization potential allows tailoring for specific environmental remediation tasks such as oil–water separation, flocculation, or photocatalysis enhancement (Pandey et al. 2025). Furthermore, biopolymer-based nanomaterials can replace synthetic polymers in filtration membranes and adsorbents, reducing reliance on fossil fuel-derived materials.

4.4 Composite and Hybrid Nanostructures

Composite and hybrid nanostructures derived from agricultural waste represent an advanced class of functional nanomaterials that integrate two or more components—such as carbon-based nanomaterials, metal/metal oxide nanoparticles, and biopolymer nanostructures—into a synergistic system. The objective of such integration is to combine the unique properties of each constituent, thereby enhancing overall performance in environmental remediation applications. By harnessing the multifunctionality of hybrid materials, these systems can address multiple environmental challenges simultaneously, such as pollutant adsorption, photocatalytic degradation, and antimicrobial action (Zhou et al. 2025).

Carbon–metal oxide composites: Agricultural wastes rich in lignocellulose, such as rice husks, coconut shells, or peanut shells, can be pyrolyzed or hydrothermally carbonized to yield porous carbon, which then serves as an excellent support matrix for metal oxide nanoparticles like ZnO, TiO_2, Fe_3O_4, and MnO_2. The high surface area and conductivity of bio-derived carbon facilitate efficient dispersion of nanoparticles and promote charge transfer in photocatalytic or electrochemical processes (Induja et al. 2025). For instance, rice husk-derived activated carbon decorated with ZnO nanoparticles has shown significant removal of organic dyes under UV light, owing to the synergistic adsorption–photocatalysis mechanism.

Biopolymer–nanoparticle hybrids: Biopolymer nanomaterials such as cellulose nanofibers, chitosan, and starch nanoparticles can be used as renewable templates or stabilizers for the green synthesis of metal and metal oxide nanoparticles. For example, chitosan's amine and hydroxyl groups can bind to metal ions and reduce them into nanoparticles, while also providing a biodegradable support matrix. Such hybrids exhibit enhanced adsorption capacity for heavy metals and pathogens in wastewater due to the combined effects of biopolymer surface chemistry and the catalytic/antimicrobial activity of the nanoparticles (Ganesan and Kadhar Mohamed 2025).

Carbon–biopolymer composites: Agricultural-waste-derived carbon materials can be blended with biopolymers to produce flexible and functional membranes or beads for water treatment. The carbon component contributes high adsorption capacity and, in some cases, electrical conductivity for electrochemical remediation, while the biopolymer ensures structural integrity and biodegradability. For example, activated carbon from coconut shells embedded in a cellulose nanofiber matrix has been reported to remove both organic and inorganic contaminants efficiently (Mariana et al. 2021; Kumari et al. 2025).

Ternary and multifunctional hybrids: Advanced approaches involve creating ternary hybrids—for example, carbon–metal oxide–biopolymer composites—that offer multiple functionalities in one material. Such systems can simultaneously adsorb pollutants, catalyze their degradation, and exhibit antimicrobial activity. A notable example is Fe_3O_4–biochar–chitosan composites derived from agricultural residues, which combine magnetic separability, high adsorption capacity, and easy regeneration for repeated use in wastewater treatment (Cheng et al. 2024). The integration of different nanostructures results in improved stability, higher reactivity, and broader applicability compared to individual components. By tailoring composition, morphology, and interfacial interactions, composite and hybrid nanomaterials can be engineered for specific remediation pathways, such as Fenton-like catalysis, solar-driven photocatalysis, capacitive deionization, and targeted adsorption of emerging pollutants (Jayakrishnan and Bora 2024). Furthermore, using agricultural waste as the primary feedstock ensures that these materials remain cost-effective, sustainable, and accessible, particularly in resource-limited settings.

5 Environmental Applications

Agricultural-waste-derived functional nanomaterials have shown significant promise in addressing critical environmental challenges, particularly in the domain of water treatment. Their unique physicochemical properties—high specific surface area, tunable surface chemistry, and the ability to host diverse functional groups—enable them to efficiently remove a wide range of contaminants from aqueous environments. Water pollution caused by heavy metals, synthetic dyes, petroleum hydrocarbons, and high salinity levels poses severe ecological and public health risks, making it imperative to develop low-cost, sustainable, and high-performance remediation materials (Fig. 3).

One of the most prominent applications of these nanomaterials is in **heavy metal adsorption**. Biochar, activated carbon, and carbon nanotubes derived from agricultural residues such as rice husks, coconut shells, and corn cobs have demonstrated exceptional adsorption capacities for toxic metals like Pb^{2+}, Cd^{2+}, Hg^{2+}, and Cr^{6+}. Functionalization with amine, thiol, or carboxyl groups further enhances their metal-binding affinity. For example, chitosan nanoparticles synthesized from shrimp shell waste have shown removal efficiencies exceeding 95% for Pb^{2+} in contaminated water, owing to strong chelation interactions between amino groups and metal ions (Zhou et al. 2025; Cheng et al. 2024). Similarly, cellulose nanofibers from crop

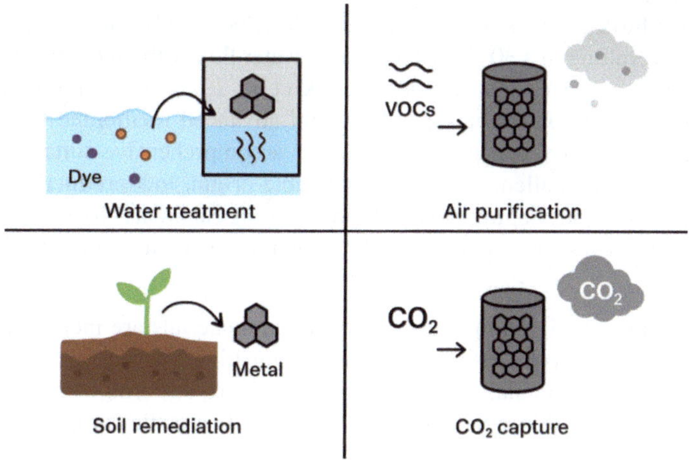

Fig. 3 Environmental applications of agro-waste-derived nanomaterials: **a** water treatment, **b** air purification, **c** soil remediation, and **d** CO_2 capture

residues, when combined with magnetic Fe_3O_4 nanoparticles, enable rapid adsorption followed by easy magnetic separation, offering a reusable solution for heavy metal removal.

Dye removal from textile and printing industry effluents represents another critical water treatment application. Photocatalytic nanomaterials such as ZnO, TiO_2, and CuO derived from agricultural waste sources (e.g., rice husk ash, banana peels, sugarcane bagasse) can effectively degrade synthetic dyes under UV or visible light irradiation. Biochar-supported TiO_2 composites not only adsorb dye molecules like methylene blue and rhodamine B but also catalyze their breakdown into non-toxic by-products, providing a dual adsorption–photodegradation mechanism (Abdu et al. 2025). Moreover, the use of green synthesis techniques imparts surface defects and dopant incorporation that enhance visible-light-driven photocatalytic activity.

For **oil spill cleanup**, hydrophobic and oleophilic carbon-based nanomaterials derived from biomass offer an efficient approach to recovering spilled petroleum products from aquatic environments. Superhydrophobic biochar sponges fabricated from sawdust or agricultural stalk waste have demonstrated high oil absorption capacities, while remaining buoyant and reusable over multiple cycles (Zubairu and Hambali 2025). By tuning pore structure and surface functional groups, these materials can selectively absorb hydrocarbons in the presence of water, making them suitable for marine spill response operations.

In **desalination and membrane-based treatment**, nanomaterials from agricultural waste are increasingly being incorporated into polymeric membranes to improve water permeability, salt rejection, and fouling resistance. Cellulose nanocrystals, chitosan, and silica nanoparticles from rice husks have been used as fillers in nanocomposite membranes for reverse osmosis and nanofiltration. These bio-derived additives enhance hydrophilicity, mechanical strength, and anti-biofouling properties, leading to improved long-term performance in desalination plants. For instance, rice-husk-derived SiO_2 nanoparticles have been embedded into polysulfone membranes, resulting in a 30–50% increase in water flux without compromising salt rejection efficiency (Heng et al. 2021; Sadek et al. 2023). By combining adsorption, photocatalysis, oil–water separation, and advanced membrane technologies, agricultural-waste-derived nanomaterials offer a comprehensive suite of solutions for water treatment challenges. Their sustainable origin, low production cost, and potential for large-scale deployment make them strong candidates for next-generation water purification systems aligned with the principles of green chemistry and circular economy.

Air Purification: Agricultural-waste-derived nanomaterials are increasingly being explored for air quality improvement due to their high surface reactivity, tunable porosity, and ability to catalytically degrade or adsorb airborne contaminants. Air pollution, comprising particulate matter (PM), volatile organic compounds (VOCs), nitrogen oxides (NO_x), sulfur oxides (SO_x), and ozone, is a major environmental and public health concern. Conventional filtration materials often suffer from limited efficiency, high replacement frequency, or the inability to break down harmful pollutants into benign products. In this context, functional nanomaterials synthesized from agricultural waste offer a sustainable, cost-effective, and multifunctional solution.

One significant application lies in the **catalytic degradation of gaseous pollutants**. Metal and metal oxide nanoparticles—such as TiO_2, ZnO, CuO, and Fe_3O_4—obtained from biomass precursors (e.g., rice husks, sugarcane bagasse, fruit peels) have been shown to effectively degrade toxic gaseous compounds via photocatalytic or thermocatalytic processes. For example, TiO_2 nanoparticles derived from rice husk ash exhibit strong photocatalytic oxidation of NO_x and formaldehyde under UV and visible light, converting them into less harmful nitrates or CO_2 and water (Heng et al. 2021; Sadek et al. 2023). Doping with non-metals such as nitrogen or carbon, achievable via bio-derived carbon sources, further enhances visible-light activity, making these materials more effective in indoor and outdoor applications.

For **VOCs removal**, activated carbons and biochars produced from agricultural residues like coconut shells, corn cobs, and sawdust demonstrate exceptional adsorption performance due to their large specific surface areas and abundant surface functional groups. VOCs such as benzene, toluene, ethylbenzene, and xylene (BTEX) can be effectively trapped via physisorption and chemisorption. Surface modification with amine or metal oxide coatings can further improve the affinity for polar VOC molecules (Li et al. 2022). Additionally, hybrid systems—where activated carbon is combined with photocatalytic nanoparticles—offer dual functionality, enabling simultaneous adsorption and catalytic decomposition of VOCs into non-toxic products.

In the domain of **filtration membranes and coatings**, cellulose nanofibers and chitin nanofibers extracted from agro-waste sources (e.g., crop residues, shrimp shells) have emerged as promising candidates for the fabrication of high-efficiency particulate air (HEPA)-grade filters. These biopolymer-based nanofibers can be electrospun or cast into thin mats with nanoscale pores capable of capturing ultrafine particles ($PM_{2.5}$ and $PM_{0.1}$) while maintaining high air permeability. Their biodegradability also addresses the end-of-life disposal concerns associated with synthetic polymer filters. Incorporating antimicrobial agents such as silver or copper nanoparticles, derived via green synthesis from agricultural biomass, imparts additional functionality by preventing microbial growth on filter surfaces.

Overall, the integration of catalytic, adsorptive, and filtration-based mechanisms within agricultural-waste-derived nanomaterials positions them as sustainable and versatile tools for combating air pollution. Their adaptability in form—from powders and coatings to membranes and sponges—allows deployment in diverse settings, including indoor air purifiers, industrial exhaust treatment units, and personal protective equipment.

Soil Remediation: The application of agricultural-waste-derived nanomaterials in soil remediation has gained significant attention due to their ability to immobilize contaminants, restore soil health, and enhance agricultural productivity. Contaminated soils, particularly those affected by heavy metals, pesticides, hydrocarbons, and excessive salinity, pose serious threats to food security, ecosystem stability, and human health. Traditional remediation strategies often involve expensive, energy-intensive processes or generate secondary waste, whereas nanomaterials synthesized from agricultural residues offer an environmentally friendly and cost-effective alternative.

Immobilization of Contaminants is one of the most prominent applications. Biochar, activated carbon, and nano-hydroxyapatite produced from crop residues (e.g., rice husks, sugarcane bagasse, corn stalks) can effectively immobilize heavy metals such as Pb^{2+}, Cd^{2+}, and As^{3+} in soil through adsorption, ion exchange, and precipitation mechanisms. For example, biochar derived from rice husk rich in silica and carbon functional groups can bind metal ions via surface complexation, reducing their bioavailability and mobility (Li et al. 2022). Similarly, Fe_3O_4 nanoparticles synthesized from agricultural biomass serve as both sorbents and magnetic agents, enabling recovery after treatment and minimizing long-term soil contamination.

Nutrient Recycling is another important aspect, particularly in degraded soils with depleted fertility. Biopolymer-based nanomaterials such as chitosan nanoparticles (from shrimp shell waste) and cellulose nanocrystals (from plant residues) can act as slow-release carriers for essential nutrients like nitrogen, phosphorus, and potassium. This controlled delivery improves nutrient use efficiency and reduces leaching into water bodies, thereby mitigating eutrophication risks. Additionally, agricultural waste-derived calcium oxide (CaO) nanoparticles, obtained from eggshells or mollusk shells, can serve as liming agents that release calcium and neutralize acidic soils while providing essential nutrients.

pH Adjustment is critical for maintaining optimal conditions for plant growth and microbial activity. Nano-CaO and nano-MgO derived from agro-waste sources not only adjust soil pH but also enhance the degradation of certain organic contaminants via alkaline hydrolysis. These nanomaterials, due to their high reactivity, can be applied in much smaller quantities than conventional liming materials, reducing transportation and application costs. The multifunctional nature of these nanomaterials allows them to simultaneously tackle multiple soil quality parameters—immobilizing contaminants, replenishing nutrients, and adjusting pH—while being derived from abundant, renewable agricultural residues. As such, they hold considerable promise for sustainable land management and restoration of contaminated agricultural lands.

Energy–environment synergy: Agricultural-waste-derived nanomaterials offer a unique platform for achieving an energy–environment synergy, wherein environmental remediation and sustainable energy generation/storage are addressed simultaneously. This dual functionality is particularly relevant in the context of circular economy and carbon neutrality goals, as these nanomaterials transform waste into high-value functional products capable of reducing environmental pollution while enabling renewable energy technologies.

Photocatalysis is one of the most promising applications in this domain. Biomass-derived carbon materials—such as activated carbon, carbon quantum dots, and graphitic carbon—can be combined with metal oxides (e.g., TiO_2, ZnO, Fe_2O_3) synthesized from agro-waste to form highly active photocatalysts. These hybrid materials can efficiently degrade organic pollutants in wastewater while simultaneously producing hydrogen via photocatalytic water splitting under solar irradiation. For instance, banana peel-derived carbon doped with ZnO nanoparticles has shown enhanced visible-light absorption and reduced electron–hole recombination, resulting in improved photocatalytic efficiency for both dye degradation and H_2 generation (Li et al. 2022).

Energy Storage applications, including supercapacitors and lithium-ion batteries, also benefit from agro-waste-derived nanomaterials. Porous activated carbon from coconut shells, date seeds, or rice husks offers high surface area and conductivity, making them excellent electrode materials for supercapacitors. When combined with metal oxides like MnO_2 or Fe_3O_4 derived from plant biomass ashes, these electrodes exhibit superior electrochemical performance due to synergistic pseudocapacitive and double-layer charge storage. Such devices not only store renewable energy but also utilize waste-derived materials, closing the sustainability loop.

CO_2 Capture and Utilization is another critical area where biomass-derived nanomaterials excel. Nanostructured biochars and functionalized carbon nanofibers from agricultural residues can be engineered with surface amine groups or metal–organic frameworks (MOFs) to selectively capture CO_2 from flue gases or the atmosphere. These captured CO_2 molecules can then be utilized in downstream processes, such as conversion into fuels or chemicals via photocatalytic or electrocatalytic pathways, thus integrating environmental cleanup with resource recovery. By integrating environmental remediation with renewable energy generation, storage, and carbon capture, agricultural-waste-derived nanomaterials offer a pathway toward a low-carbon and waste-free future. This multifunctional role underscores their significance not only in solving ressing environmental challenges but also in enabling a sustainable energy transition (Table 2).

6 Challenges, Limitations, and Future Perspectives

Agricultural-waste-derived functional nanomaterials hold strong potential for sustainable environmental remediation; however, several barriers must be addressed for large-scale adoption. One critical issue is **feedstock variability**. The proportions of lignin, cellulose, hemicellulose, and minerals vary with crop type, region, season, and cultivation practices, directly affecting nanomaterial yield, morphology, and functional properties. This heterogeneity can hinder reproducibility in industrial production, underscoring the need for standardized feedstock characterization and pre-treatment protocols.

Scalability remains a significant limitation. Many reported synthesis routes—such as solvothermal treatment, high-temperature pyrolysis, or acid/base extraction—are energy-intensive, batch-based, and optimized only at laboratory scale.

Table 2 Environmental applications of agro-waste-derived nanomaterials, their target pollutants, performance metrics

Nanomaterial type	Application (water/air/soil/energy)	Target pollutant	Performance metrics
Biochar from rice husk	Water treatment	Pb^{2+} ions	Adsorption capacity: 156 mg/g; >95% removal in 60 min
ZnO nanoparticles (banana peel)	Water treatment	Methylene blue dye	98% degradation in 120 min under UV light
Ag nanoparticles (tea waste)	Water treatment/ antibacterial	*E. coli*, *S. aureus*	99% bacterial reduction within 3 h
Fe_3O_4 nanoparticles (wheat straw)	Water treatment/ magnetic separation	Arsenic (As^{3+}, As^{5+})	94% removal efficiency; regeneration up to 5 cycles
Cellulose nanofibers (bagasse)	Air purification	Volatile organic compounds (VOCs)	85% removal of toluene vapors within 2 h
Hydroxyapatite (eggshell)	Soil remediation	Cd^{2+} ions	Immobilization efficiency: 92%; pH stabilization
Biochar/graphene hybrid	Energy–environment synergy	CO_2	Adsorption capacity: 3.2 mmol/g at 25 °C
TiO_2/carbon composite	Energy–environment synergy	Organic dyes	95% degradation in 90 min under visible light

While often described as "green," certain processes rely on concentrated acids, strong bases, or organic solvents, generating secondary wastes that require treatment. Additionally, fossil-fuel-driven high-temperature operations can offset environmental gains. From an application perspective, **material stability and reusability** require improvement. Adsorbents may lose efficiency after regeneration due to pore blockage, surface fouling, or nanoparticle leaching, while photocatalysts can deactivate via aggregation or surface poisoning. Environmental, health, and safety (EHS) aspects must also be considered, as nanoscale materials—regardless of origin—can present ecotoxicological risks, including bioaccumulation and inhalation hazards.

Economic feasibility depends not only on the low cost of feedstocks but also on total production costs, influenced by energy consumption, chemical usage, waste handling, and post-processing requirements. Without supportive policy incentives or integration into industrial symbiosis networks, large-scale deployment may remain limited. Looking ahead, advances in **green synthesis**—including microwave-assisted hydrothermal methods, enzymatic extraction, ionic liquid-mediated processing, and solar-driven pyrolysis—offer pathways to reduce energy demand, eliminate hazardous reagents, and enhance yields. The development of **hybrid nanostructures**, such as carbon nanomaterials combined with metal oxides or conducting polymers, can improve functional performance and stability. Furthermore, machine learning tools can accelerate optimization by linking feedstock characteristics to product properties.

Integrating agricultural waste valorization into **circular economy models**—where biofuels, platform chemicals, and nanomaterials are co-produced—will maximize resource efficiency and economic returns. Policy measures such as safety guidelines, environmental labeling, carbon credits, and pilot-scale demonstration projects will be crucial to bridging the gap between laboratory innovation and industrial adoption. Comprehensive life-cycle assessments (LCA) will ultimately determine the true sustainability of these technologies. With continued research, technological refinement, and enabling socio-economic frameworks, agricultural waste can transition from an underutilized residue to a key feedstock for next-generation, sustainable nanomaterials in environmental applications.

7 Conclusions

The transformation of agricultural waste into functional nanomaterials presents a sustainable and innovative solution to both waste management and environmental remediation challenges. Through advances in green synthesis methods—ranging from plant extract-mediated routes to hydrothermal and sol–gel processes—diverse nanostructures, including carbon-based, metal/metal oxide, biopolymer, and hybrid materials, can be derived from abundant agro-residues. These nanomaterials have demonstrated remarkable potential in water purification, air quality improvement, soil restoration, energy–environment applications, and CO_2 capture, thereby linking environmental protection with value-added material production. Despite the promise, challenges such as feedstock variability, process scalability, material stability, and economic feasibility must be addressed to enable widespread adoption. Integrating agricultural waste valorization into circular economy frameworks, supported by robust policy measures, industrial partnerships, and comprehensive life-cycle assessments, will be key to sustainable commercialization. With continued interdisciplinary research and technological refinement, agricultural residues can evolve from underutilized biomass into a cornerstone resource for next-generation environmental nanotechnologies.

References

Abady MM, Mohammed DM, Soliman TN, Shalaby RA, Sakr FA (2025) Sustainable synthesis of nanomaterials using different renewable sources. Bull Natl Res Cent 49. https://doi.org/10.1186/s42269-025-01316-4

Abdu M, Babaee S, Worku A, Diale P, Msagati TAM, Nure JF (2025) Synthesis and optimization of TiO_2-biochar composite for degradation of Basic Blue 41 dye in textile wastewater using response surface methodology. Inorg Chem Commun 174:114078. https://doi.org/10.1016/j.inoche.2025.114078

Abualnaja MM, Almotairy ARZ, Alorabi AQ, Alaysuy O, Almahri A, Alkhamis K, Alrefaee SH, El-Metwaly NM (2024) Optimizing cellulose nanofiber composite for Co^{2+} ions recovery from lithium-ion battery wastes and removal from wastewater: a green environmental solution. J. Water Process Eng 57:104621. https://doi.org/10.1016/j.jwpe.2023.104621

Abubakar IR, Maniruzzaman KM, Dano UL, AlShihri FS, AlShammari MS, Ahmed SMS, Al-Gehlani WAG, Alrawaf TI (2022) Environmental sustainability impacts of solid waste management practices in the global south. Int J Environ Res Public Health 19. https://doi.org/10.3390/ijerph191912717

Ahamed MS, Ali MS, Ahmed S, Sadia SI, Islam MR, Rahaman MA, Alam MA (2024) Synthesis of silver nanomaterials capping by fruit-mediated extracts and antimicrobial activity: a critical review. Int Res J Pure Appl Chem 25:45–60. https://doi.org/10.9734/irjpac/2024/v25i1844

Ahmed SF, Kabir M, Mehjabin A, Oishi FTZ, Ahmed S, Mannan S, Mofijur M, Almomani F, Badruddin IA, Kamangar S (2023) Waste biorefinery to produce renewable energy: bioconversion process and circular bioeconomy. Energy Rep 10:3073–3091. https://doi.org/10.1016/j.egyr.2023.09.137

Aksu M, Güzdemir Ö (2025) Food waste-derived carbon quantum dots and their applications in food technology: a critical review. Food Bioprocess Technol 18:6753–6778. https://doi.org/10.1007/s11947-025-03854-1

Almeida-Naranjo CE, Tejedor J, Villamar-Ayala CA, Vizuete G (2025) Transforming waste into solutions: raw and modified bioadsorbents for emerging contaminant removal. J Environ Chem Eng 13:116720. https://doi.org/10.1016/j.jece.2025.116720

Ansari MM, Heo Y, Do K, Ghosh M, Son YO (2024) Nanocellulose derived from agricultural biowaste by-products–sustainable synthesis, biocompatibility, biomedical applications, and future perspectives: a review. Carbohydr Polym Technol Appl 8:100529. https://doi.org/10.1016/j.carpta.2024.100529

Aziz T, Rohullah, Ullah A, Zeb U, Hussain M, Ali A, Haq F, Kiran M (2025) Advancements in cellulose nanocrystals: a review of functionalization, applications, and challenges. Int J Biol Macromol 315:144552. https://doi.org/10.1016/j.ijbiomac.2025.144552

Bhatia T, Sindhu SS (2024) Sustainable management of organic agricultural wastes: contributions in nutrients availability, pollution mitigation and crop production. Springer International Publishing. https://doi.org/10.1007/s44279-024-00147-7

Cao W, Zhang M, Ma W, Huang C (2023) Multifunctional electrospun nanofibrous membrane : an effective method for water purification. Sep Purif Technol 327:124952. https://doi.org/10.1016/j.seppur.2023.124952

Chand S, Kumar R, Thakur N, Kumar K, Umar A, Almas T, Baskoutas S (2025) Sustainable synthesis and multifunctional applications of biowaste-derived carbon nanomaterials and metal oxide composites: a review. Chemosphere 385:144540. https://doi.org/10.1016/j.chemosphere.2025.144540

Chauhan AK, Singh SP, Rajpoot AS, Chauhan S, Meraj A, Jawaid M, Varghese E, Kumar V, Awad SA (2025) Preparation and applications of nanoparticles from lignocellulosic biomass: a review. Int J Biol Macromol 321:146138. https://doi.org/10.1016/j.ijbiomac.2025.146138

Cheng Y, Li A, Shi W, Zhao L (2024) Magnetic chitosan-functionalized waste carton biochar composites for efficient adsorption of anionic and cationic dyes. Chem Eng J 481:148535. https://doi.org/10.1016/j.cej.2024.148535

Chibrikov V, Pieczywek PM, Cybulska J, Zdunek A (2024) The effect of hemicelluloses on biosynthesis, structure and mechanical performance of bacterial cellulose-hemicellulose hydrogels. Sci Rep 14:1–15. https://doi.org/10.1038/s41598-024-72513-w

Chilakamarry CR, Mahmood S, Saffe SNBM, Bin Arifin MA, Gupta A, Sikkandar MY, Begum SS, Narasaiah B (2021) Extraction and application of keratin from natural resources: a review. 3 Biotech 11:1–12. https://doi.org/10.1007/s13205-021-02734-7

Das Lala S, Barua E, Deb P, Deoghare AB (2021) Physico-chemical and biological behaviour of eggshell bio-waste derived nano-hydroxyapatite matured at different aging time. Mater Today Commun 27:102443. https://doi.org/10.1016/j.mtcomm.2021.102443

Desmond LJ, Phan AN, Gentile P (2021) Critical overview on the green synthesis of carbon quantum dots and their application for cancer therapy. Environ Sci Nano 8:848–862. https://doi.org/10.1039/d1en00017a

Elezz MA, Aboeleneen NM, Abd-ElMonem NM, Sorour FH (2025) A comprehensive review: functional nanomaterials for renewable energy: innovations, applications, and sustainable strategies. Next Mater 9:101001. https://doi.org/10.1016/j.nxmate.2025.101001

Ganesan V, Kadhar Mohamed MSB (2025) An eco-economic approach of valorization of plastic waste into functional materials for energy and environmental applications. INC. https://doi.org/10.1016/B978-0-443-23599-3.00008-3

Ganesh KS, Sridhar A, Vishali S (2022) Utilization of fruit and vegetable waste to produce value-added products: conventional utilization and emerging opportunities—a review. Chemosphere 287:132221. https://doi.org/10.1016/j.chemosphere.2021.132221

Gupta D, Boora A, Thakur A, Gupta TK (2023) Green and sustainable synthesis of nanomaterials: recent advancements and limitations. Environ Res 231:116316. https://doi.org/10.1016/j.envres.2023.116316

Hemavathy RV, Ragini YP, Shruthi S, Ranjani S, Subhashini S, Thamarai P (2025) Biofuel production from marine macroalgae: pathways, technologies, and sustainable energy solutions. Ind Crops Prod 224:120282. https://doi.org/10.1016/j.indcrop.2024.120282

Hemmami H, Zeghoud S, Ben Amor I, Alnazza Alhamad A, Tliba A, Alsalme A, Cornu D, Bechelany M, Barhoum A (2024) Green synthesis of CaO nanoparticles from chicken eggshells: antibacterial, antifungal, and heavy metal ($Pb2^+$, $Cr2^+$, $Cd2^+$ and $Hg2^+$) adsorption properties. Front Environ Sci 12:1–17. https://doi.org/10.3389/fenvs.2024.1450485

Heng ZW, Tan YY, Chong WC, Mahmoudi E, Mohammad AW, Teoh HC, Sim LC, Koo CH (2021) Preparation of a novel polysulfone membrane by incorporated with carbon dots grafted silica from rice husk for dye removal. J Water Process Eng 40:101805. https://doi.org/10.1016/j.jwpe.2020.101805

Hincapié Rojas DF, Pineda Gómez P, Rosales Rivera A (2019) Production and characterization of silica nanoparticles from rice husk. Adv Mater Lett 10:67–73. https://doi.org/10.5185/amlett.2019.2142

Induja M, Thyagarajan S, Mayildurai R, Vaithiyanathan R, Maruthavanan T, Sarangi PK, Singh S, Varshney D, Sivaprakash K, Mani D, Murphin Kumar PS (2025) Lignocellulose-derived carbon nanomaterials: unlocking sustainable solutions for waste management and water depollution. Biomass Bioenergy 201:108102. https://doi.org/10.1016/j.biombioe.2025.108102

Inkoua S, Li C, Shao Y, Lin H, Fan M, Zhang L, Zhang S, Hu X (2023) Co-hydrothermal carbonization of fruit peel with sugars or furfural impacts structural evolution of hydrochar. Ind Crops Prod 193:116221. https://doi.org/10.1016/j.indcrop.2022.116221

Islam NF, Gogoi B, Saikia R, Yousaf B, Narayan M, Sarma H (2024) Encouraging circular economy and sustainable environmental practices by addressing waste management and biomass energy production. Reg Sustain 5:100174. https://doi.org/10.1016/j.regsus.2024.100174

Jafarzadeh S, Nooshkam M, Zargar M, Garavand F, Ghosh S, Hadidi M, Forough M (2024) Green synthesis of nanomaterials for smart biopolymer packaging: challenges and outlooks. J Nanostruct Chem 14:113–136. https://doi.org/10.1007/s40097-023-00527-3

Jain S, Kumar S (2024) Advances and challenges in pretreatment technologies for bioethanol production: a comprehensive review. Sustain Chem Clim Action 5:100053. https://doi.org/10.1016/j.scca.2024.100053

Jayakrishnan U, Bora A (2024) Fabrication of functional nanomaterials from food waste. INC. https://doi.org/10.1016/B978-0-443-21428-8.00010-2

Kasinathan K, Samayanan S, Marimuthu K, Yim JH (2022) Green synthesis of multicolour fluorescence carbon quantum dots from sugarcane waste: investigation of mercury (II) ion sensing, and bio-imaging applications. Appl Surf Sci 601:154266. https://doi.org/10.1016/j.apsusc.2022.154266

Khursheed S, Dutta J, Ahmad I, Rather MA, Badroo IA, Bhat TA, Ahmad I, Amin A, Shah A, Qadri T, Habib H (2023) Biogenic silver nanoparticles: synthesis, applications and challenges in food sector with special emphasis on aquaculture. Food Chem X 20:101051. https://doi.org/10.1016/j.fochx.2023.101051

Kulis-Kapuscinska A, Kwoka M, Borysiewicz MA, Wojciechowski T, Licciardello N, Sgarzi M, Cuniberti G (2023) Photocatalytic degradation of methylene blue at nanostructured ZnO thin films. Nanotechnology 34. https://doi.org/10.1088/1361-6528/aca910

Kumari S, Thakur B, Sharma A, Gautam S, Kumar A (2025) Valorization of agricultural waste for wastewater treatment: a sustainable approach to environmental clean-up. Next Res 2:100590. https://doi.org/10.1016/j.nexres.2025.100590

Laishram D, Bin Kim S, Lee SY, Park SJ (2025) Advancements in Biochar as a sustainable adsorbent for water pollution mitigation. Adv Sci 12. https://doi.org/10.1002/advs.202410383

Lazarus BS, Chadha C, Velasco-Hogan A, Barbosa JDV, Jasiuk I, Meyers MA (2021) Engineering with keratin: a functional material and a source of bioinspiration. iScience 24:102798. https://doi.org/10.1016/j.isci.2021.102798

Le VS, Herrmann L, Bräu L, Lesueur D (2023) Sustainable green tea production through agroecological management and land conversion practices for restoring soil health, crop productivity and economic efficiency: evidence from Northern Vietnam. Soil Use Manag 39:1185–1204. https://doi.org/10.1111/sum.12885

Li Y, Lin X, Li Z, Liu J (2022) Highly-efficient and visible light photocatalytical degradation of organic pollutants using TiO_2-loaded on low-cost biomass husk. Materials (Basel) 15. https://doi.org/10.3390/ma15238671

Lin M, Begho T (2022) Crop residue burning in South Asia: a review of the scale, effect, and solutions with a focus on reducing reactive nitrogen losses. J Environ Manage 314:115104. https://doi.org/10.1016/j.jenvman.2022.115104

Liu D, Li H, Gao R, Zhao Q, Yang Z, Gao X, Wang Z, Zhang F, Wu W (2021) Enhanced visible light photoelectrocatalytic degradation of tetracycline hydrochloride by I and P co-doped TiO_2 photoelectrode. J Hazard Mater 406:124309. https://doi.org/10.1016/j.jhazmat.2020.124309

Machado TO, Grabow J, Sayer C, de Araújo PHH, Ehrenhard ML, Wurm FR (2022) Biopolymer-based nanocarriers for sustained release of agrochemicals: a review on materials and social science perspectives for a sustainable future of agri- and horticulture. Adv Colloid Interface Sci 303:102645. https://doi.org/10.1016/j.cis.2022.102645

Mariana M, Alfatah T, Abdul Khalil HPS, Yahya EB, Olaiya NG, Nuryawan A, Mistar EM, Abdullah CK, Abdulmadjid SN, Ismail H (2021) A current advancement on the role of lignin as sustainable reinforcement material in biopolymeric blends. J Mater Res Technol 15:2287–2316. https://doi.org/10.1016/j.jmrt.2021.08.139

Muthu M, Gopal J, Chun S, Devadoss AJP, Hasan N, Sivanesan I (2021) Crustacean waste-derived chitosan: antioxidant properties and future perspective. Antioxidants 10:1–27. https://doi.org/10.3390/antiox10020228

Nirmal NP, Khanashyam AC, Mundanat AS, Shah K, Babu KS, Thorakkattu P, Al-Asmari F, Pandiselvam R (2023) Valorization of fruit waste for bioactive compounds and their applications in the food industry. Foods 12:1–26. https://doi.org/10.3390/foods12030556

Nzereogu PU, Omah AD, Ezema FI, Iwuoha EI, Nwanya AC (2023) Silica extraction from rice husk: comprehensive review and applications. Hybrid Adv 4:100111. https://doi.org/10.1016/j.hybadv.2023.100111

Osman AI, Zhang Y, Farghali M, Rashwan AK, Eltaweil AS, Abd El-Monaem EM, Mohamed IMA, Badr MM, Ihara I, Rooney DW, Yap PS (2024) Synthesis of green nanoparticles for energy, biomedical, environmental, agricultural, and food applications: a review. Springer International Publishing. https://doi.org/10.1007/s10311-023-01682-3

Pandey AK, Thakur S, Mehra R, Kaler RSS, Paul M, Kumar A (2025) Transforming agri-food waste: innovative pathways toward a zero-waste circular economy. Food Chem X 28:102604. https://doi.org/10.1016/j.fochx.2025.102604

Panesar PS, Kaur R, Singla G, Sangwan RS (2016) Bio-processing of agro-industrial wastes for production of food-grade enzymes: progress and prospects. Appl Food Biotechnol 3:208–227. https://doi.org/10.22037/afb.v3i4.13458

Pasipanodya D, Seedat N, Patel B, Roopchund R (2025) Production of hydrochar from the hydrothermal carbonisation of food waste feedstock for use as an adsorbent in removal of heavy metals from water. Biomass Convers Biorefin 15:11819–11833. https://doi.org/10.1007/s13399-024-06097-w

Patra N, Ravi G (2024) Hydrothermal synthesis of CuO/WO_3 hybrid bifunctional composites with unprecedented dark adsorption and photocatalysis. ChemSel 9:1–5. https://doi.org/10.1002/slct.202400042

Patra N, Ramesh P, Donthu V, Ahmad A (2024) Biopolymer-based composites for sustainable energy storage: recent developments and future outlook. J Mater Sci Mater Eng. https://doi.org/10.1186/s40712-024-00181-9

Pierrat L, García-Triñanes P (2024) Optimising furfural production from lignocellulosic biomass: feedstock selection, process enhancement, and techno-economic and environmental viability. Chem Eng Res Des 212:261–280. https://doi.org/10.1016/j.cherd.2024.10.035

Rahaman MT, Khan MSH (2024) Applications of green nano textile materials for environmental sustainability and functional performance: past, present and future perspectives. Nano-Struct Nano-Objects 40:101332. https://doi.org/10.1016/j.nanoso.2024.101332

Rao N, Singh R, Bashambu L (2021) Carbon-based nanomaterials: Synthesis and prospective applications. Mater Today Proc 44:608–614. https://doi.org/10.1016/j.matpr.2020.10.593

Riseh RS, Vazvani MG, Hassanisaadi M, Thakur VK (2024) Agricultural wastes: a practical and potential source for the isolation and preparation of cellulose and application in agriculture and different industries. Ind Crops Prod 208:117904. https://doi.org/10.1016/j.indcrop.2023.117904

Rodríguez-Félix F, Graciano-Verdugo AZ, Moreno-Vásquez MJ, Lagarda-Díaz I, Barreras-Urbina CG, Armenta-Villegas L, Olguín-Moreno A, Tapia-Hernández JA (2022) Trends in sustainable green synthesis of silver nanoparticles using agri-food waste extracts and their applications in health. J Nanomater 2022. https://doi.org/10.1155/2022/8874003

Ruangtong J, Jiraroj T-T, T-Thienprasert NP (2020) Green synthesized ZnO nanosheets from banana peel extract possess anti-bacterial activity and anti-cancer activity. Mater Today Commun 24:101224. https://doi.org/10.1016/j.mtcomm.2020.101224

Rudrapal M, Rakshit G, Singh RP, Garse S, Khan J, Chakraborty S (2024) Dietary polyphenols: review on chemistry/sources, bioavailability/metabolism, antioxidant effects, and their role in disease management. Antioxidants 13. https://doi.org/10.3390/antiox13040429

Saadan R, Hachimi Alaoui C, Ihammi A, Chigr M, Fatimi A (2024) A brief overview of lignin extraction and isolation processes: from lignocellulosic biomass to added-value biomaterials. 3. https://doi.org/10.3390/eesp2024031003

Sadek AH, Abdel-Karim A, Mohsenpour S, Ismail SH, Bayoumy AM, Ibrahim M, Mohamed GG (2023) Polysulfone-based mixed matrix membranes loaded with a multifunctional hierarchical porous Ag-Cu dendrites@SiO_2 core-shell nanostructure for wastewater treatment. Process Saf Environ Prot 175:677–691. https://doi.org/10.1016/j.psep.2023.05.085

Sanoja-López KA, Cedeño-Muñoz JS, Rivadeneira-Mendoza BF, Vergara-Romero A, Luque R, Rodríguez-Díaz JM (2025) Agricultural residues to high-value nanomaterials: pathways to sustainability. Resour Conserv Recycl Adv 25. https://doi.org/10.1016/j.rcradv.2025.200243

Seth R, Meena A (2024) Enzymes-based nanomaterial synthesis: an eco-friendly and green synthesis approach. Clean Technol Environ Policy. https://doi.org/10.1007/s10098-024-02854-7

Shah KJ, Li J, Jiang A, You Z, Sun Y, Balu K, Ahn YH (2025) Green synthesis of the functional bio-macromolecular composite film based on cellulose material for the photocatalytic degradation of tetracycline: performance and mechanism. Int J Biol Macromol 309:143111. https://doi.org/10.1016/j.ijbiomac.2025.143111

Sharma IP, Kanta C, Gusain YS (2018) Crop residues utilization: wheat, paddy, cotton, sugarcane and groundnut. Int J Bot Stud 3:11–15. http://www.botanyjournals.com/archives/2018/vol3/issue3/3-3-14

Sheoran K, Kaur H, Siwal SS, Saini AK, Vo DVN, Thakur VK (2022) Recent advances of carbon-based nanomaterials (CBNMs) for wastewater treatment: synthesis and application. Chemosphere 299. https://doi.org/10.1016/j.chemosphere.2022.134364

Singh B, Kumar R, Sharma S, Ramachandran T, Raja VKB, Mahapatro A, Gupta D, Kedia A, Ismail AI, Kumar A (2025) Review on the synergistic mechanisms in harnessing rice residue-derived cellulose nanocrystals for sustainable water purification and wastewater treatment. Int J Biol Macromol 311:143225. https://doi.org/10.1016/j.ijbiomac.2025.143225

Srivastav AL, Kumar A (2021) An endeavor to achieve sustainable development goals through floral waste management: a short review. J Clean Prod 283:124669. https://doi.org/10.1016/j.jclepro.2020.124669

Trinh B, Le PTK, Werner D, Phuong NH, Le Luu T (n.d.) Rice husk biochars modified with magnetized iron oxides and nano zero valent iron for decolorization of dyeingwastewater

Usman I, Sana S, Afzaal M, Imran A, Jaffar HM, Sukhera S, Munir M, Asghar A, Urugo MM (2025) Phytochemicals in plant food waste: innovative insight and implications. J Agric Food Res 20:101779. https://doi.org/10.1016/j.jafr.2025.101779

Wang L, Wu J, Gao Y, Sha L, Ma H, Li X, Zhou J (2021) H_3PO_4-assisted preparation of precursor-derived porous carbon: construction of precursor/precursor-derived porous carbon texture properties relationship. Diam Relat Mater 119:108596. https://doi.org/10.1016/j.diamond.2021.108596

Zainelabdin A, Amin G, Zaman S, Nur O, Lu J, Hultman L, Willander M (2012) CuO/ZnO Nanocorals synthesis via hydrothermal technique: growth mechanism and their application as Humidity Sensor. J Mater Chem 22:11583–11590. https://doi.org/10.1039/c2jm16597j

Zhang Y, Zhang S, Dong H, Yin L (2025) Eco-friendly nanomaterials derived from lignocellulosic macromolecules for dual applications in geological waste and water treatment. Int J Biol Macromol 319. https://doi.org/10.1016/j.ijbiomac.2025.143485

Zhou J, Sheng Z, Han H, Zou M, Li C (2012) Facile synthesis of fluorescent carbon dots using watermelon peel as a carbon source. Mater Lett 66:222–224. https://doi.org/10.1016/j.matlet.2011.08.081

Zhou Y, Zhang Y, Nie Y, Sun D, Wu D, Ban L, Zhang H, Yang S, Chen J, Du H, Pan X (2025) Recent advances and perspectives in functional chitosan-based composites for environmental remediation, energy, and biomedical applications. Prog Mater Sci 152:101460. https://doi.org/10.1016/j.pmatsci.2025.101460

Zia J, Shringi AK, Riaz U (2025) Calcium oxide nanoparticles from eggshell waste: a green nanotechnological strategy for microwave-assisted environmental clean up. Clean Chem Eng 11:100182. https://doi.org/10.1016/j.clce.2025.100182

Zubairu A, Hambali HU (2025) Biomass-based sorbents for oil spill clean-up: a review. Malaysian J Catal 9:1–15. https://doi.org/10.11113/mjcat.v9n1.192

Rooted Futures: Cultivating Resilience in the Soil of Sustainability

Stanzin Angmo, Anjali Jha, Elham Vaghari, Vijay Kumar Bhatiya, Sonia Kumari Shishodia, Gaurav Mudgal, and Kavindra Kumar Kesari

Abstract As the global agricultural landscape faces unprecedented environmental, social, and technological challenges, *this chapter* shapes communities and supports planetary health. These have been framed and aligned with the United Nations Sustainable Development Goals (SDGs). This book revise agriculture industry as a living system that holds the power to heal both land and society. Spanning soil microbiomes to AI-driven precision farming, indigenous land ethics to circular bioeconomies, this chapter presents interdisciplinary insights that integrate science, policy, and grassroots innovation. Readers will discover how soil stewardship, water governance, agrobiodiversity, and ecological design intersect to shape resilient food systems. Emphasizing the climate, culture, and cultivation, the narrative draws on case studies, emerging technologies, and traditional knowledge to propose pathways for regenerative transformation. This chapter critically examines the power structures embedded in agricultural systems, offering frameworks for equitable finance, policy alignment, and community-led governance. It moves beyond sustainability as maintenance, envisioning a future where agriculture actively restores ecosystems, uplifts marginalized voices, and fosters a moral imagination for living systems. With its unique blend of science, and systems strategy, *the chapter* serves academics, practitioners, and change-makers seeking to cultivate agricultural models that are not only sustainable but also alive, and adaptive. It's both a roadmap and a rallying

S. Angmo · V. K. Bhatiya (✉) · S. K. Shishodia · G. Mudgal (✉)
University Institute of Biotechnology, Chandigarh University, Mohali, Punjab, India
e-mail: bhatiavijay1988@gmail.com

G. Mudgal
e-mail: garavmdgl@gmail.com

S. Angmo
e-mail: stanzinangmo727@gmail.com

A. Jha
ICAR-Indian Institute of Wheat and Barley Research, Karnal, Haryana, India

E. Vaghari
Islamic Azad University Ferdos Blvd, Tehran, Iran

K. K. Kesari
School of Engineering, Geeta University, Panipat, Haryana, India

K. K. Kesari et al. (eds.), *Smart Sustainable Materials Advancing Re-manufacturing and Technology*, World Sustainability Series, https://doi.org/10.1007/978-3-032-17907-4_2

call for those committed to seeding resilience in the soil beneath our feet—and in the societies we grow.

Keywords IOT · Artificial intelligence · Digital twin · Conservation · SDG · Water · Sensing · Precision · Irrigation · Satellite

1 Introduction

In the face of cascading global crises—climate disruption, food insecurity, soil degradation, biodiversity collapse, and escalating socio-economic inequality—agriculture sits at a historic crossroads. It holds a paradoxical position: both a significant contributor to environmental degradation and a potential linchpin for planetary regeneration. Agriculture feeds the world, yet the dominant systems we've relied upon are increasingly unsustainable, both ecologically and socially. The future of food—indeed, the future of the planet—depends on our ability to reimagine agriculture as more than a mechanized, extractive process. It must become a generative, living system rooted in resilience, justice, and ecological balance.

This chapter serves as the conceptual foundation that seeks to break new ground by blending science, technology, tradition, policy, and ethics to envision transformative pathways for sustainable agriculture. Here, we lay out the urgency of agricultural reform and introduce the systems-based approach that guides the structure of the work ahead. While the impacts of industrialized agriculture are now well-documented—greenhouse gas emissions, fertilizer runoff, deforestation, and reliance on monocultures—the deeper challenge lies in the worldview it represents: one of control over nature rather than collaboration with it. Soil, once revered as sacred and alive, has in many systems become merely a substrate for chemical inputs. Water is treated as an external utility rather than a shared life force. Farmers are often trapped between global commodity chains and climate volatility, disempowered by opaque policies and inequitable access to markets or technology.

But across the globe, new and old alternatives are taking root—regenerative practices that honor ecological cycles, recover indigenous wisdoms, and apply cutting-edge innovation not to dominate but to adapt, restore, and co-evolve. This chapter seeks to position those alternatives not at the margins, but at the center of our collective agricultural future. It proposes a shift from linear to circular thinking; from extractive to regenerative paradigms (Fig. 1). Rather than examining agricultural elements in isolation—soil, water, crops, labor, markets—we emphasize their interdependence within complex, adaptive systems. This systems thinking lens allows us to understand how feedback loops, thresholds, and leverage points shape both ecological health and human well-being. It also helps us navigate the trade-offs and synergies that emerge when integrating social justice, climate resilience, and food security goals.

We introduce a regenerative framework built on three primary pillars:

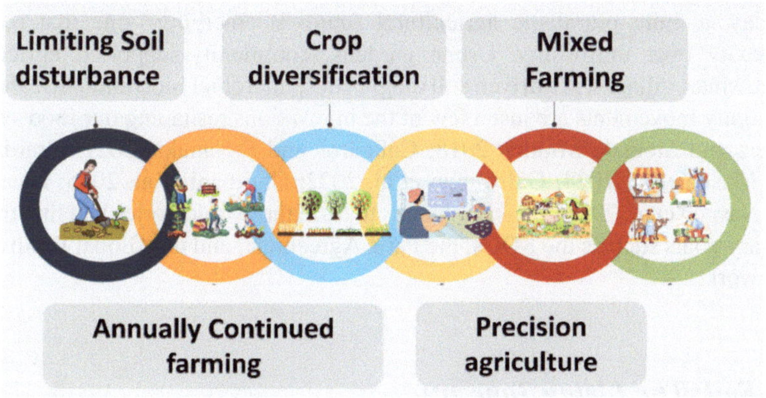

Fig. 1 Regenerative farming approaches

- **Ecological Resilience**—Grounded in biodiversity, soil health, and nature-based solutions;
- **Equity and Inclusion**—Recognizing farmers, especially smallholders and women, as knowledge holders and changemakers;
- **Technological and Institutional Innovation**—Applying AI, remote sensing, blockchain, and policy shifts to align incentives with sustainability.

Rather than idealizing tradition or fetishizing technology, this approach invites radical collaboration across domains—uniting data science with ancestral wisdom, policy with grassroots action, and science with storytelling (Shemshad et al. 2025; Zupancic 2023; Lidder et al. 2025; Addorisio et al. 2025; Gamage et al. 2024; Charatsari et al. 2022).

1.1 Agricultural Histories and Futures

The past holds powerful lessons for the future of agriculture. For millennia, indigenous and local communities cultivated highly resilient, biodiverse systems tailored to regional ecologies. Many of these practices—intercropping, agroforestry, seed saving, lunar planting—embodied an intimate understanding of ecosystems that modern models often overlook (Gamage et al. 2024; Birla et al. 2024; Bhattacharya 2024; Patel et al. 2020; Tripathi et al. 2024; Prasad et al. 2025; Akchaya et al. 2025; Mohamed Shaffril et al. 2024; Paut et al. 2024). Colonialism, industrialization, and the Green Revolution introduced high-yield varieties, mechanization, and petrochemical inputs, which, while increasing productivity in some regions, also reduced genetic diversity, degraded soils, and centralized power (Yu et al. n/a; Marambe et al. 2020; Peddyreddy 2023).

Today, a more pluralistic agricultural future is emerging—one that revalues complexity over uniformity. Urban gardens, community-supported agriculture, climate-smart villages, AI-driven soil diagnostics, microbial biostimulants, and seed sovereignty movements are just a few of the innovations reshaping our food systems (Walters and Stoelzle Midden 2018; Calheiros and Stefanakis 2021; Grard et al. 2017; Masha et al. 2024; Galiwango et al. 2022; Ghorbani et al. 2025; Pace et al. 2025; Naveed et al. 2024). This book aims to trace those trajectories and link them to planetary goals such as the SDGs, the Paris Agreement, and the Global Biodiversity Framework.

1.2 Soil: The Living Tapestry

Healthy soil is not dirt; it is a dynamic, living tapestry of microorganisms, minerals, organic matter, and root networks—capable of carbon sequestration, nutrient cycling, water retention, and disease suppression. It is the hidden infrastructure of resilience. Degraded soils, however, are symptomatic of broader disconnections in our food systems. Restoring them requires more than compost or cover crops—it demands a shift in mindset, from extract and replace to regenerate and reciprocate. Whether through biochar, vermiculture, deep-rooting perennials, or bioregional planning (Lorenz and Lal 2023; Okolo et al. 2019; Lastname et al. 2024; Kumar et al. 2025), the future depends on rebuilding our relationship with the soil, socially and scientifically.

1.3 Cultures of Cultivation

Agriculture is more than technique—it is culture. It shapes festivals, cuisines, myths, and identities. Yet modern agribusiness models have often eroded this cultural richness in favor of efficiency and scalability (Kusz 2014). In the name of progress, seed banks have replaced seed songs; pesticides replaced pollinators; local knowledge was dismissed in favor of standardized packages. The result is not only ecological collapse, but cultural homogenization and loss of community agency. This chapter argues for re-grounding agricultural knowledge in place-based traditions while innovating respectfully and collaboratively. By spotlighting success stories from around the world—from seed libraries in Odisha (Das and Mallick 2024; Panda et al. 2023) to aquaponics in Nairobi (Dijkgraaf et al. 2019) to satellite-supported agroecology in the Andes (Smoliak et al. 2022)—we advocate for diversified pathways rooted in local needs, local leadership, and planetary ethics.

1.4 A Call to Future-Rooted Leadership

This chapter presents a leadership framework which is not limited to CEOs, ministers, or academics, but a call for each actor in the food system to reimagine their roles. Farmers as scientists, youth as change, policy architects as community allies, AI as a tool for ecological literacy, and investors as stewards of long-term value. This chapter focusing on a dimension of sustainable agriculture: soil, water, energy, biodiversity, circularity, technology, economics, and policy. Each offers a blend of theoretical grounding, practical examples, and visionary provocations for transdisciplinary action. What holds them together is a common conviction: that agriculture, when re-rooted in ecological intelligence, cultural reverence, and collective will, can seed a world where abundance is shared, futures are regenerative, and food truly nourishes all.

2 Seeds of Change

2.1 The Agricultural Frontier in the Anthropocene

The Anthropocene—a proposed epoch signifying the profound and accelerating human impact on Earth's geology and ecosystems—has transformed agriculture into both a driver and casualty of planetary change (Lam and Rousselot 2024). Farming today occurs under conditions vastly different from those experienced by previous generations: rising global temperatures, altered precipitation patterns, degraded soils, and shifting biomes have redefined the ecological context of food production (Borsari 2021). As food systems increasingly intersect with climate change, migration, energy demand, and global trade, agriculture must evolve from a sectoral concern to a planetary priority (Abhilash et al. 2022). The frontier we now face is not geographic but systemic: how do we design food systems that operate within planetary boundaries while safeguarding ecological integrity and cultural resilience? This challenge calls for a new kind of stewardship—one that recognizes that growing food is inseparable from sustaining life itself. Emerging paradigms such as planetary health diets, ecosystem-based adaptation, and nature-based solutions reflect this shift, blending traditional knowledge systems with scientific innovation to restore biospheric vitality and ensure long-term food security (Mishra 2025).

2.2 From Green Revolution to Regenerative Renaissance

The mid-20th-century Green Revolution marked a pivotal moment in global agriculture, introducing high-yielding crop varieties, synthetic fertilizers, and irrigation technologies that dramatically increased food production, particularly in developing nations like India (John and Babu 2021; Eliazer Nelson et al. 2019; Gollin et al. 2021; Karunarathne et al. 2020; Kabir et al. 2023). While these innovations helped avert famine and boost productivity, they also triggered unintended consequences: soil degradation, reduced agrobiodiversity, and escalating dependence on chemical inputs that compromised both human and ecological health (Pingali 2012). Moreover, the Green Revolution's top-down implementation often marginalized indigenous farming knowledge and exacerbated socio-economic disparities (Tittonell et al. 2022). As a result, its legacy is now being critically re-evaluated through the lens of sustainability and equity (Trujillo Bilbao 2024; Chauhan and Wehrden 2025; Geels et al. 2015; Bjornlund and Bjornlund 2025).

In response, a Regenerative Renaissance is emerging—an agricultural paradigm shift that prioritizes ecological function, resilience, and holistic well-being over mere yield maximization (Garbisu et al. 2025). Practices such as agroecology, permaculture, silvopasture, and regenerative organic farming are gaining global traction. These approaches treat farms as dynamic ecosystems, integrating biological, social, and even spiritual dimensions (Tittonell et al. 2022; Giller et al. 2021). Rather than rejecting science, this movement embraces complexity, blending indigenous wisdom with cutting-edge technologies like precision agriculture, remote sensing, and soil carbon monitoring to restore degraded landscapes, enhance biodiversity, and promote food sovereignty. Regenerative agriculture thus represents not just a set of practices but a transformative narrative that reimagines the relationship between agriculture, nature, and society.

2.3 Systems Thinking in Food and Climate Policy

The intricate interdependence of agriculture, climate, and society demands a systems-thinking approach—one that moves beyond fragmented interventions and embraces complexity. Traditional policy responses that isolate subsidies, trade, or emissions have often failed to address the root causes of systemic vulnerability (Jagustović et al. 2019; Lalani et al. 2023). Transformative food policy must instead account for whole-system dynamics: how land use changes influence water cycles, how livestock contribute to methane emissions and diet-related health burdens, and how power asymmetries determine who benefits from innovation (Wopereis et al. 2024). Systems thinking reframes sustainability not as a checklist but as a web of relationships and feedback loops, inviting collaboration across disciplines—from farmers and technologists to economists and communities (Issac et al. 2025). This approach prioritizes adaptive governance, intersectional justice, and long-term scenario planning. As climate models become more granular and digital tools more accessible, there's a growing opportunity to embed resilience at every node of the agricultural value chain (Hassan et al. 2025). Ultimately, "Seeds of Change" is not a metaphor but a directive: to cultivate new narratives, new alliances, and new models of care for land, for people, and for the intricate systems that connect them (Tables 1 and 2).

Table 1 Key dimensions of regenerative agriculture and their core principles. This table outlines the foundational dimensions of regenerative agriculture discussed in the manuscript, detailing their core principles, key practices, innovative concepts, and supporting scientific literature citations

Dimension	Core principle/ focus	Key practices/innovations/ concepts	References
Soil	Soil parameters	Microbiomes, mycorrhizae and carbon cycles, soil microbiome, microbial carbon use efficiency (CUE), arbuscular mycorrhizal (AM) and ectomycorrhizal (ECM) systems, nitrogen fertilization, substrate stoichiometry, spatial heterogeneity	Kallenbach et al. (2019), Dang et al. (2024), Gavazov et al. (2022), Li and Qi (2025), Yu et al. (2025a, b)
	Soil health as the foundation of resilience	Soil functions as a living ecosystem, with physical, chemical, and biological attributes, crop yield, nitrogen use efficiency (NUE), conservation tillage, cover cropping, and organic amendments	Govindasamy et al. (2023), Ahmad et al. (2022), Ward et al. (2023), Chmelíková et al. (2021), You et al. (2023), Ankomah et al. (2024), Pandian et al. (2024), Ghiglieno et al. (2025), Omer et al. (2024), Jug et al. (2019), Carrara et al. (2021)
	Decolonizing land management practices	Integrating indigenous and local knowledge systems, relational understandings of land, stewardship, ecological reciprocity, pluriversal approaches, traditional soil classification, crop-livestock integration, community-led conservation, rematriating land	Pienaah et al. (2024), Gordon et al. (2023), Kimmerer (2011), Noel et al. (2023), Lucas (2025), Lal (2023), Ambus et al. (2023), Mudigere Sannegowda and Garkoti (2022), Farias et al. (2020), Manono and Gichana (2025), Bhagat et al. (2024), Hill et al. (2024), Dai et al. (2025)
Water	Agroecological hydrology: beyond irrigation	Ecological principles, landscape-level integration, socio-cultural dimensions, hydrological connectivity, soil–water-plant interactions, integrated watershed management, contour farming, vegetative buffers, decentralized water storage, agroforestry, hydro-ecological ethics	Sun et al. (2017), Douglas et al. (2019), Yu et al. (2024), Chen et al. (2024), Garrido-Baserba et al. (2024), Frimpong et al. (2023), Bhattacharyya et al. (2016), Wassie (2020), Estévez et al. (2022)

(continued)

Table 1 (continued)

Dimension	Core principle/focus	Key practices/innovations/concepts	References
	Rainwater harvesting and aquifer ethics	Rainwater harvesting (RWH), managed aquifer recharge (MAR), hydrogeological suitability, water quality, community consent, GIS-based multi-criteria decision analysis, check dams, percolation tanks, intergenerational equity, ecological integrity	Bonilla Valverde et al. (2016), Waghaye et al. (2023), Hiscock et al. (2024), Tahvili et al. (2021), Sengupta and Ghosh (2024), Hussein et al. (2024), Ali et al. (2024), Atta et al. (2022), Bhunia et al. (2018)
	Climate-smart water governance	Agility, inclusivity, ecological foresight, technological innovation, policy coherence, community engagement, IoT, AI, digital twins, real-time monitoring, predictive analytics, risk assessment, integrated strategies, participatory platforms, transdisciplinary research	Snapp et al. (2019), Qoseem et al. (2024), MacPherson et al. (2025), Musvoto et al. (2015), Jackson-Smith and Veisi (2023), Omrany et al. (2025), Bhanye (2025), Ahern (2025), Bibri et al. (2024)
The Sunlit equation	Renewable energy and solar farming	Solar farming, agrivoltaics, photovoltaic (PV) installations, land productivity, microclimatic conditions, PM-KUSUM scheme, solar pumps, bifacial solar panels, dynamic tracking systems	Widmer et al. (2024), Asa'a et al. (2024), Pal A 93–100, Mutum et al. (2024), Jouttijärvi et al. (2022), Zainol Abidin et al. (2021)
	Light-dependent biological pathways	Photosynthetically active radiation (PAR), photoreceptor-mediated pathways, anthocyanin biosynthesis, UVR8 photoreceptor, COP1, HY5, abiotic stress tolerance, carotenoid synthesis, circadian rhythms, photomorphogenesis, hormonal pathways	Zhai et al. (2020), Greef et al. (1983), Covington et al. (2008), Gangappa and Botto (2016), Yin et al. (2024), Kataria et al. (2014), Li et al. (n/d), Plunkett et al. (2019), Roeber et al. (2021), Bian et al. (2022)

(continued)

Table 1 (continued)

Dimension	Core principle/focus	Key practices/innovations/concepts	References
	Photosynthetic innovation and AI modeling	Synthetic biology, computational modeling, synthetic carbon fixation cycles, engineered Rubisco variants, CETCH cycle, AI modeling, machine learning algorithms, hyperspectral data, chlorophyll fluorescence data, digital twins, AI-driven phenotyping platforms	Varghese et al. (2023), Sharma et al. (2024), Kubis and Bar-Even (2019), Kumari et al. (2025), Lebrini and Ayerdi Gotor 2024)
Living biodiversity	Pollination agents	Pollinator pathways, habitat corridors, green corridors, pollinator foraging specialization, habitat heterogeneity, native forbs, floral scent dispersion, green overpasses	Kisliuk et al. (2023), Žák et al. (2020), Wang et al. (2021), Płaskonka et al. (2024)
	Crop rotation and genetic diversification	Crop rotation, genetic diversification, diverse rotations, legume inclusion, nitrogen fixation, landrace cultivation, varietal mixtures, intercropping, yield stability	Saikia et al. (2024), Cougnon et al. (2022), Ngwenya et al. (2024), Vasconcelos et al. (2020), Huang et al. (2024), Marin et al. (2025), McAlvay et al. (2022)

3 Soil—The Silent Superpower

3.1 Microbiomes, Mycorrhizae and Carbon Cycles

Soil is not merely a substrate for plant growth—it is a dynamic, living system teeming with microbial life that orchestrates critical biogeochemical processes. The soil microbiome, comprising bacteria, archaea, fungi, and protozoa, plays a pivotal role in nutrient cycling, organic matter decomposition, and carbon sequestration (Mzungu et al. 2024). Recent studies emphasize the functional diversity of microbial communities and their influence on soil carbon dynamics. For instance, microbial carbon use efficiency (CUE)—the proportion of assimilated carbon converted into biomass rather than respired—varies significantly between arbuscular mycorrhizal (AM) and ectomycorrhizal (ECM) systems (Fig. 2). ECM-dominated soils often exhibit lower CUE due to higher metabolic costs and fungal-bacterial competition 111. Mycorrhizal fungi, particularly AMF, form symbiotic relationships with over 90% of terrestrial plants, enhancing nutrient uptake and modulating carbon flow in the rhizosphere. Long-term isotope labeling experiments have revealed that AMF can sequester substantial amounts of photosynthetically derived carbon in their storage

Table 2 Transitioning from conventional to regenerative agriculture: challenges and pathways. This table compares key aspects of conventional and regenerative agriculture, highlighting the challenges inherent in traditional practices and outlining the pathways and solutions offered by regenerative approaches, supported by examples and scientific literature from the manuscript

Aspect of comparison	Conventional agriculture (challenges/ characteristics)	Regenerative agriculture (pathways/ solutions/ characteristics)	Examples/case studies	References
Underlying philosophy	Exploitative, extractive, linear production model, dominion over nature	Reciprocal, regenerative, circular processes, harmony with nature	"The Crisis Beneath Our Feet," "From Green Revolution to Regenerative Renaissance"	Veltmeyer (2023), Yadav and Yadav (2024), Famiglietti and Ferguson (2021), Eagmark (2022)
Soil management	Soil degradation, loss of organic matter, reliance on synthetic inputs (fertilizers, pesticides)	Soil health restoration, carbon sequestration, enhanced biodiversity, reduced chemical inputs, improved water retention	Microbiomes, mycorrhizal fungi, conservation tillage, cover cropping, organic amendments, biochar, vermiculture, deep-rooting perennials	Seitz et al. (2024), Liang et al. (2025), Azhar et al. (2024)
Water management	Over-extraction of groundwater, pollution from runoff, inefficient irrigation, and lack of ecological integration	Water conservation, improved water quality, decentralized storage, integrated watershed management, respectful water governance	Agroecological hydrology, rainwater harvesting (RWH), managed aquifer recharge (MAR), contour farming, vegetative buffers, agroforestry systems	Abd-Elaty et al. (2024), Otieno (2024), Seif et al. (2024)
Energy systems	High reliance on fossil fuels for machinery, transportation, and input production	Transition to renewable energy, integration of solar energy with agriculture, and energy efficiency	Agrivoltaics (combining crop cultivation with PV installations), PM-KUSUM scheme (solar pumps in India), bifacial solar panels, AI modeling for photosynthesis	Widmer et al. (2024), Asa'a et al. (2024), Pal A 93–100, Mutum et al. (2024), Nomura et al. (2023), Varghese et al. (2023), Pu et al. (2022), Jouttijärvi et al. (2022)

(continued)

Table 2 (continued)

Aspect of comparison	Conventional agriculture (challenges/characteristics)	Regenerative agriculture (pathways/solutions/characteristics)	Examples/case studies	References
Biodiversity	Monoculture, habitat destruction, decline in pollinators and beneficial organisms, genetic erosion	Enhanced biodiversity, habitat creation, genetic diversification, and ecological resilience	Pollinator pathways, habitat corridors, green overpasses, crop rotation, landrace cultivation, varietal mixtures, intercropping, Barahnaja system	Chamoli (2024), Bourke et al. (2021), Moreira et al. (2024), Ghosh et al. (2004), Sharma et al. (2024), Sahu et al. (2025), Maitra et al. (2021)
Waste and resources	Linear "take-make-dispose" approach, nutrient loss, accumulation of agricultural waste	Circular economy principles, nutrient cycling, waste valorization, rèsource efficiency	Zero-waste farming systems, composting, anaerobic digestion, black soldier fly larvae, fungal valorization of lignocellulosic waste, algal bioremediation	Koyunoğlu (2024), Chaurasia et al. (2024), Tawfik et al. (2022), Bhatia and Sindhu (2024), Lin et al. (2018)
Technology use	Industrial-scale, chemical-intensive, often top-down, focus on maximizing yield	Precision agriculture for resource optimization, data-driven insights, AI for agroecology, ethical considerations, and farmer-led innovation	AI, IoT, remote sensing, digital twins, blockchain for transparent supply chains, AI-driven phenotyping platforms, digital commons for farmers	Verdouw (2022), Kisliuk et al. (2023), Banerjee et al. (2023), Lee et al. (2025), Lawes et al. (2022), Escribà-Gelonch et al. (2024), Ali et al. (2025)
Economic model	Externalization of environmental costs, focus on short-term profits, market volatility, dependence on subsidies	True cost accounting, sustainable finance, diversified income streams, local food systems, community wealth building	Valuing ecosystem services, green bonds, impact investing, community-supported agriculture (CSAs), food cooperatives, farmers' markets	Tay et al. (2024), Desalegn et al. (2024), Savarese et al. (2020), Xu et al. (2025a), Onyszkiewicz and Sylla (2965)

(continued)

Table 2 (continued)

Aspect of comparison	Conventional agriculture (challenges/characteristics)	Regenerative agriculture (pathways/solutions/characteristics)	Examples/case studies	References
Policy and governance	Sectoral approaches, top-down mandates, lack of farmer participation, misaligned incentives	Integrated policy frameworks, participatory governance, international agreements, and decentralized decision-making	Aligning with SDGs, CBD Post-2020 global biodiversity framework, Paris agreement, food sovereignty policies, multi-stakeholder platforms	Friedman et al. (2022), CCICED. in Green Consensus and High Quality Development: CCICED Annual Policy (2020)
Leadership and culture	Command-and-control, hierarchical, reactive problem-solving, anthropocentric views	Living systems leadership, transdisciplinary collaboration, ethical imagination, narrative agency, ecological attunement, humility	Learning from ecosystems (trees, ants), fostering mutuality, regeneration, systemic change, co-creating policies, storytelling	Asa'a et al. (2024), Kirina et al. (2022), Wyckhuys et al. (2024), Kumar et al. (2025a), Cabello et al. (2021), Dönmez et al. (2024), Frankel-Goldwater et al. (2023), Prasad et al. (2025), Coggan et al. (2021), Cartagena et al. (2025)

organs, influencing microbial community structure and soil organic carbon (SOC) stabilization (Zhou et al. 2025).

However, nitrogen fertilization has been shown to suppress AMF abundance, thereby altering carbon allocation patterns and microbial turnover rates—a phenomenon with profound implications for agroecosystem resilience (Enebe and Babalola 2021). The interplay between microbial networks and carbon cycling is further shaped by substrate stoichiometry and spatial heterogeneity. ECM systems, characterized by recalcitrant litter and nutrient imbalances, foster fungal-dominated networks that constrain microbial growth efficiency (Enebe and Babalola 2021). In contrast, AM systems support more diverse bacterial networks that promote synergistic interactions and efficient resource utilization (Zhou et al. 2025). These findings underscore the need to integrate microbial ecology into the design of climate-smart soil management strategies, recognizing that microbial traits and interactions are pivotal to long-term carbon stabilization and ecosystem resilience.

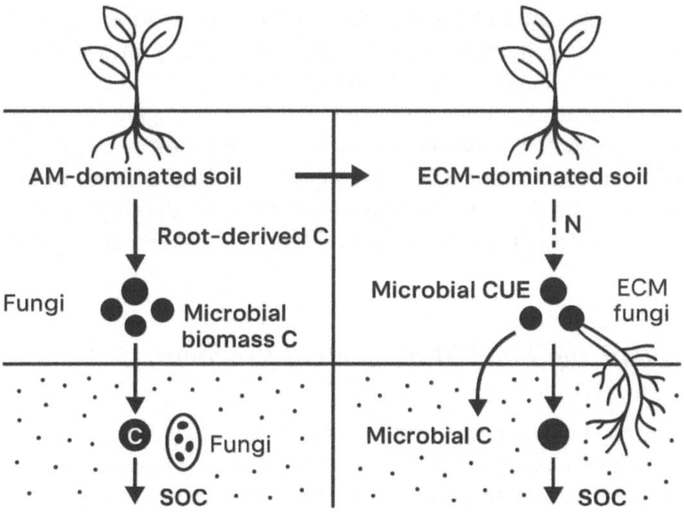

Fig. 2 Microbial carbon cycling pathways in arbuscular mycorrhizal (AM) versus ectomycorrhizal (ECM) systems. The diagram illustrates differences in carbon allocation, microbial carbon use efficiency (CUE), and soil organic carbon (SOC) stabilization between AM- and ECM-dominated soils. In AM systems, root-derived carbon supports diverse bacterial networks and efficient microbial biomass formation, enhancing SOC accumulation. ECM systems, characterized by recalcitrant litter and nutrient imbalances, exhibit lower CUE due to higher metabolic costs and fungal-bacterial competition. Nitrogen inputs further modulate these dynamics, often suppressing AMF abundance and altering carbon flow in the rhizosphere

3.2 Soil Health as the Foundation of Resilience

Soil health is increasingly recognized as the cornerstone of sustainable agriculture and climate resilience. Defined as the capacity of soil to function as a living ecosystem that sustains plants, animals, and humans, soil health encompasses physical, chemical, and biological attributes. Degradation of these attributes—through erosion, compaction, salinization, and loss of organic matter—threatens food security and ecosystem services globally (Whalen and Gul 2025).

Xu et al. (2025a, b) developed a global soil health dataset integrating biological indicators such as bacterial and fungal richness, nematode and earthworm abundance, alongside conventional physical and chemical metrics (Xu et al. 2025b). Their analysis revealed a consistent positive correlation between soil health scores and crop yield as well as nitrogen use efficiency (NUE), suggesting that enhancing soil health can simultaneously improve productivity and reduce environmental footprints (Whalen and Gul 2025). However, the benefits of soil health interventions are modulated by climatic conditions and agricultural management intensity. In regions with low input use, such as parts of Africa and central Europe, soil health accounted for over 30% of yield variability, highlighting its strategic importance in low-resource settings. Sustainable practices such as conservation tillage, cover cropping, and organic amendments have demonstrated efficacy in restoring soil structure, increasing SOC, and promoting biodiversity (Ţopa et al. 2025). These practices align with ecological principles and offer pathways to regenerate degraded soils while enhancing adaptive capacity. Importantly, soil health is not a static attribute but a dynamic outcome of interactions among plant, microbial, and human activities. Monitoring soil health through integrated indicators and aligning management with local ecological contexts is essential for building resilient agroecosystems.

3.3 Decolonizing Land Management Practices

The discourse on soil health and sustainability must also grapple with the epistemic dimensions of land management. Historically, scientific land management frameworks have marginalized Indigenous and local knowledge systems, perpetuating colonial legacies in agricultural institutions. Decolonizing land management entails recognizing and integrating diverse epistemologies that honor relational understandings of land, stewardship, and ecological reciprocity. O'Brien and Mudaliar (2022) examined land management plans across institutions of higher education and found a pervasive dominance of Western scientific knowledge, with minimal inclusion of Indigenous perspectives (O'Brien and Mudaliar 2022). This exclusion undermines the potential for resilient socio-ecological systems and calls for a reimagining of land governance that centers Indigenous sovereignty and ecological ethics. Indigenous land practices, often rooted in deep ecological knowledge and spiritual relationships with the land, offer valuable insights into sustainable soil stewardship.

In the Indian context, efforts to decolonize management theory have highlighted the need to move beyond Eurocentric paradigms and embrace pluriversal approaches that reflect local ontologies and epistemologies (Ojha and Venkateswaran 2025). Integrating Indigenous Indian knowledge systems into soil and land management can enrich agroecological design and foster cultural resilience. This includes recognizing traditional soil classification systems, crop-livestock integration practices, and community-led conservation models. Decolonization also involves structural shifts in policy and education. Land-grant universities and agricultural research institutions must critically examine their historical roles in land dispossession and knowledge exclusion. Rematriating land and resources to Indigenous communities, supporting participatory research, and embedding Indigenous scholars in decision-making processes are vital steps toward equitable and regenerative land management (O'Brien and Mudaliar 2022).

4 Water Wisdom

4.1 Agroecological Hydrology: Beyond Irrigation

Water in agroecosystems is not merely a resource—it is a dynamic medium that shapes ecological interactions, nutrient flows, and resilience to climate variability. Agroecological hydrology reframes water management by emphasizing ecological principles, landscape-level integration, and socio-cultural dimensions. Unlike conventional irrigation-centric approaches, agroecological hydrology seeks to harmonize water use with natural cycles, biodiversity, and community needs.

Recent studies underscore the importance of hydrological connectivity and soil–water-plant interactions in agroecological systems. For instance, integrated watershed management approaches that combine contour farming, vegetative buffers, and decentralized water storage have demonstrated improved infiltration, reduced runoff, and enhanced groundwater recharge in semi-arid regions (Seif et al. 2024). Moreover, agroforestry systems with deep-rooted perennials have been shown to stabilize water tables and buffer seasonal variability, offering a climate-resilient alternative to monoculture irrigation schemes. The concept of "hydro-ecological ethics" is gaining traction, advocating for water governance models that respect ecological thresholds and community rights. This includes recognizing the role of wetlands, riparian zones, and floodplains as multifunctional landscapes that support biodiversity, water purification, and cultural heritage. Agroecological hydrology thus calls for a paradigm shift—from extractive water use to regenerative water stewardship.

4.2 Rainwater Harvesting and Aquifer Ethics

Rainwater harvesting (RWH) is a cornerstone of agroecological water management, offering decentralized, low-energy solutions for water security. While RWH systems have ancient roots, their modern resurgence is driven by climate adaptation imperatives and the need to reduce dependence on overexploited aquifers. Studies have shown that well-designed RWH systems can significantly supplement irrigation needs, recharge aquifers, and mitigate urban flooding (Yildirim et al. 2022; Alam et al. 2022).

However, the ethics of aquifer recharge demand careful consideration. Managed aquifer recharge (MAR) using harvested rainwater must account for hydrogeological suitability, water quality, and community consent. A GIS-based multi-criteria decision analysis conducted in Saudi Arabia identified optimal MAR sites by integrating aquifer characteristics, land use, and water availability, demonstrating the potential for strategic aquifer storage in arid regions (Seif et al. 2024). In India, community-led RWH initiatives have shown mixed outcomes. While check dams and percolation tanks have improved water availability in some regions, inequitable access and lack of maintenance have undermined long-term benefits. A farm-scale social survey in

Gujarat revealed that nearly 40% of farmers reported no benefits from check dams, with proximity to the structure being a key determinant of access. Moreover, 72.8% of farmers reported no maintenance activity, raising concerns about sustainability and ownership (Alam et al. 2022).

Aquifer ethics also intersect with questions of intergenerational equity and ecological integrity. Over-reliance on groundwater extraction has led to declining water tables, salinization, and ecosystem degradation. Ethical aquifer governance must therefore integrate scientific assessments with participatory decision-making, ensuring that recharge interventions do not compromise long-term aquifer health or marginalize vulnerable communities.

4.3 Climate-Smart Water Governance

Water governance in the Anthropocene demands agility, inclusivity, and ecological foresight. Climate-smart water governance integrates technological innovation, policy coherence, and community engagement to navigate the complexities of water scarcity, variability, and conflict. It moves beyond sectoral silos to embrace systems thinking and adaptive management. Smart water management systems, leveraging IoT, AI, and digital twins, are transforming urban and agricultural water governance (Lidder et al. 2025; Pace et al. 2025; Banerjee et al. 2023; Lee et al. 2025; Escribà-Gelonch et al. 2024; Ali et al. 2025). Case studies from Singapore and Chinese cities demonstrate how real-time monitoring and predictive analytics can reduce non-revenue water, optimize distribution, and enhance resilience to climate-induced disruptions (Dai et al. 1932). However, these systems also face challenges related to data interoperability, cybersecurity, and equity of access.

At the policy level, climate-smart governance requires harmonization across scales and sectors. A bibliometric analysis of global water adaptation policies revealed significant gaps in risk assessment and governance regimes, particularly in developing countries. The study emphasized the need for integrated strategies that combine top-down policy frameworks with bottom-up community initiatives, addressing socio-economic, institutional, and biophysical complexities (Majahana et al. 2025). Importantly, climate-smart governance must be rooted in justice and ecological ethics. This includes recognizing water as a commons, safeguarding Indigenous water rights, and embedding resilience in infrastructure design. Participatory platforms, transdisciplinary research, and inclusive policy dialogues are essential to co-create water futures that are both adaptive and equitable (Fig. 3).

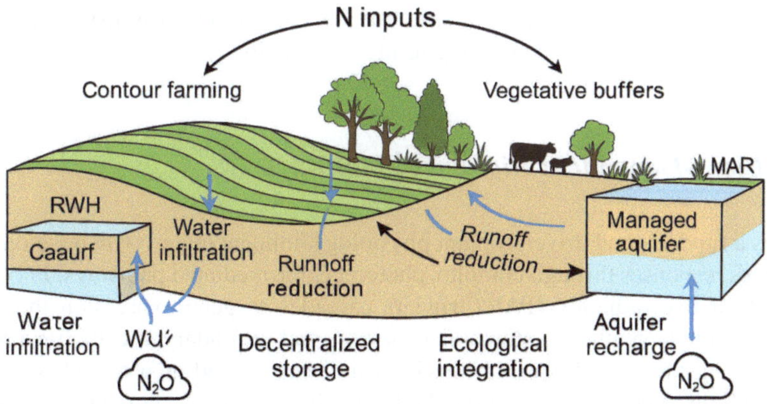

Fig. 3 Agroecological hydrology: linking landscape design to water resilience. Reinforces the water management section by showing how regenerative practices restore hydrological balance and improve water governance

5 The Sunlit Equation

5.1 Renewable Energy and Solar Farming

Solar energy stands at the forefront of renewable energy transitions, offering a decentralized, low-carbon alternative to fossil fuels. In agricultural contexts, solar farming not only powers irrigation and processing systems but also redefines land use through agrivoltaics—integrated systems that combine crop cultivation with photovoltaic (PV) installations (Widmer et al. 2024; Asa'a et al. 2024). Agrivoltaic systems have demonstrated dual benefits: enhancing land productivity and improving microclimatic conditions for crops. A meta-analysis of agrivoltaic trials across Europe and Asia revealed yield improvements in shade-tolerant crops and significant reductions in evapotranspiration under PV arrays (Asa'a et al. 2024).

Moreover, solar energy adoption in agriculture is increasingly supported by policy incentives and technological innovation. In India, the PM-KUSUM scheme has catalyzed the deployment of solar pumps, reducing diesel dependence and improving irrigation access for smallholders. Studies show that solar pump adoption correlates with increased cropping intensity and reduced groundwater depletion when coupled with water governance reforms (Mutum et al. 2024). However, challenges remain in terms of grid integration, land tenure conflicts, and lifecycle sustainability of PV modules.

Emerging research also explores the use of bifacial solar panels and dynamic tracking systems to optimize energy capture in agricultural landscapes. These technologies, while capital-intensive, offer higher efficiency and adaptability to diverse agroecological zones. The integration of solar energy into farming systems thus represents not just a technological shift but a socio-ecological transformation that demands inclusive planning and long-term stewardship.

5.2 Light-Dependent Biological Pathways

Light is a fundamental driver of plant physiology, influencing growth, development, and stress responses through complex photoreceptor-mediated pathways. Photosynthetically active radiation (PAR) triggers cascades of gene expression that regulate chlorophyll synthesis, stomatal behavior, and secondary metabolite production. Among these, anthocyanin biosynthesis has emerged as a model for understanding light-dependent regulation. Anthocyanins, flavonoid pigments with antioxidant properties, are synthesized in response to light intensity and quality, mediated by transcription factors such as MYB, bHLH, and WD40 complexes (Ma et al. 2021).

Recent studies have elucidated the role of UV-B light in activating the UVR8 photoreceptor, which interacts with COP1 and HY5 to upregulate anthocyanin biosynthesis genes. This pathway not only enhances photoprotection but also contributes to abiotic stress tolerance, including salinity and drought. In Arabidopsis and grapevine, light-induced anthocyanin accumulation has been linked to improved resilience and commercial quality of fruits (Ma et al. 2021).

Beyond anthocyanins, light regulates a suite of metabolic pathways including carotenoid synthesis, circadian rhythms, and photomorphogenesis. The integration of light signals with hormonal pathways—such as auxin and gibberellin—further modulates plant architecture and resource allocation. Understanding these light-dependent biological pathways is crucial for optimizing crop performance under variable light environments, especially in controlled systems like vertical farms and greenhouses.

5.3 Photosynthetic Innovation and AI Modeling

Photosynthesis, the cornerstone of terrestrial life, is undergoing a renaissance through synthetic biology and computational modeling (Kaur et al. 2023a, 2025a; Abdi et al. 2024; Dhar et al. 2024; Parashar et al. 2023; Debnath et al. 2023; Verma et al. 2022). Traditional photosynthetic pathways, while efficient in natural ecosystems, exhibit limitations in agricultural productivity due to photorespiration and suboptimal light utilization. Innovations such as the introduction of synthetic carbon fixation cycles and engineered Rubisco variants aim to overcome these bottlenecks. For example, the CETCH cycle—a synthetic pathway developed to fix CO_2 more efficiently than the natural Calvin cycle—has shown promise in in vitro systems (Luo et al. 2022; Osmanoglu et al. 2021).

AI modeling is revolutionizing the study of photosynthesis by enabling high-resolution simulations of light capture, energy transfer, and carbon assimilation.

Machine learning algorithms trained on hyperspectral and chlorophyll fluorescence data can predict photosynthetic efficiency under varying environmental conditions. These models are being integrated into digital twins of crop systems, allowing real-time optimization of light regimes in precision agriculture (Ghorbani et al. 2025; Pace et al. 2025; Escribà-Gelonch et al. 2024; Ali et al. 2025).

Moreover, AI-driven phenotyping platforms are accelerating the discovery of photosynthetically superior genotypes. By correlating genomic data with physiological traits, researchers can identify alleles associated with enhanced light use efficiency and stress tolerance. This convergence of photosynthetic innovation and AI modeling holds transformative potential for climate-resilient agriculture, enabling crops to thrive under fluctuating light and temperature regimes (Pu et al. 2022; Yin et al. 2024; Li et al. n.d.; Roeber et al. 2021; Ma et al. 2021).

6 Living Biodiversity, Thriving Farms

6.1 Pollinator Pathways and Habitat Corridors

Pollinators are foundational to agricultural productivity and ecosystem resilience, yet their populations are in decline globally due to habitat fragmentation, pesticide exposure, and climate change. The integration of pollinator pathways and habitat corridors into agricultural landscapes offers a promising strategy to reverse this trend and enhance biodiversity. These corridors—linear strips of vegetation connecting fragmented habitats—facilitate the movement of pollinators and improve plant reproductive success.

Recent studies underscore the nuanced role of green corridors in urban and peri-urban agroecosystems. For instance, Breitbart et al. (2023) found that proximity to green corridors increased reproductive effort in *Asclepias syriaca* (common milkweed), although pollinator diversity was paradoxically lower within the corridor itself, suggesting that corridor design and surrounding land use significantly influence outcomes (Breitbart et al. 2023). In a broader landscape context, Pardee et al. (2023) demonstrated that pollinator foraging specialization increased with habitat heterogeneity, while generality decreased with semi-natural habitat cover. These findings suggest that maintaining native forbs and enhancing landscape-level heterogeneity can support pollinator networks across urbanizing ecosystems (Pardee et al. 2023). Moreover, Wang et al. (2021) used computational fluid dynamics to model floral scent dispersion across highway corridors, revealing that green overpasses planted with native flowers can preserve scent trails and facilitate pollinator movement, thereby improving habitat connectivity and resilience (Wang et al. 2021).

These studies collectively highlight that pollinator pathways are not merely ecological constructs but socio-technical interventions requiring thoughtful design, species selection, and community engagement. Integrating such corridors into agroecological planning can bolster pollination services, enhance crop yields, and contribute to biodiversity conservation.

6.2 Crop Rotation and Genetic Diversification

Crop rotation and genetic diversification are time-tested agroecological strategies that underpin soil health, pest management, and climate resilience. Diverse rotations disrupt pest and disease cycles, enhance nutrient cycling, and foster microbial diversity. Malik et al. (2025) emphasized that crop rotation improves resource use efficiency and breaks pathogen transmission cycles, contributing to long-term soil and farm sustainability (Malik et al. 2025). The inclusion of legumes in rotations, for example, enhances nitrogen fixation and reduces dependence on synthetic fertilizers. It is also reported that mungbean integration into maize-wheat rotations increased nitrogen uptake by 34%, phosphorus by 46%, and potassium by 36%, demonstrating the synergistic effects of legume-cereal systems (Malik et al. 2025). Genetic diversification within crops—through landrace cultivation, varietal mixtures, and intercropping—further strengthens agroecosystem resilience. Hufnagel et al. (2020) reviewed diverse approaches to crop diversification, noting that intercropping maize with climbing beans not only diversified rotations but also improved silage quality and reduced input dependency (Hufnagel et al. 2020).

Crop diversification also enhances yield stability under climate variability. Diversified rotations sustains cereal yields despite changing climatic conditions, while country-level crop diversity studies correlate with greater harvest stability across years (Perino et al. 2022; Alletto et al. 2022). However, barriers to diversification persist, including market limitations, policy disincentives, and knowledge gaps. Mortensen and Smith (2020) argued that monocultures are entrenched by economic feedback loops and lack of research on diversified systems, calling for "virtuous feedbacks" to support agroecological transitions (Mortensen and Smith 2020).

To overcome these barriers, integrated frameworks that combine crop rotation, genetic diversification, and participatory research are essential. Such frameworks can align ecological benefits with farmer livelihoods, policy incentives, and market development.

6.3 Indigenous Knowledge and Agrobiodiversity

Indigenous and local knowledge systems are vital repositories of agrobiodiversity and ecological wisdom. These systems, shaped by centuries of co-evolution with local landscapes, offer context-specific insights into crop selection, soil management, and climate adaptation. Swiderska et al. (2022) emphasized that Indigenous food systems sustain much of the world's cultivated and wild biodiversity through agroecological practices rooted in cultural and spiritual values (Swiderska et al. 2022).

In India, traditional cropping systems such as Barahnaja in Uttarakhand exemplify the integration of multiple crops—millets, legumes, oilseeds—within a single field, enhancing resilience and nutritional diversity. Such systems are not only ecologically robust but also culturally embedded, offering models for regenerative agriculture (Chamoli 2024; Ghosh et al. 2004; Gururani et al. 2021). Moreover, Indigenous knowledge contributes to seed conservation and varietal adaptation. Mattalia et al. (2025) found that Indigenous farmers in Mexico and India maintain crop diversity through informal seed networks and cultural rituals, despite pressures from industrial agriculture and climate change (Mattalia et al. 2025). Bibliometric analyses reveal a growing interest in agrobiodiversity and traditional knowledge, yet transnational collaboration remains limited. Liu et al. (2022) called for interdisciplinary research and stakeholder cooperation to enhance policy frameworks and safeguard traditional knowledge systems (Liu et al. 2022). Importantly, Indigenous knowledge is dynamic and adaptive. Sherpa (2022) applied Protection Motivation Theory to assess Indigenous perceptions of climate change in Nepal, finding that traditional agricultural practices are evolving in response to changing climatic conditions and remain effective for adaptation (Sherpa 2022). Decolonizing research and policy frameworks is crucial for honoring and integrating Indigenous epistemologies. This includes recognizing land rights, supporting community-led conservation, and embedding Indigenous scholars in decision-making processes. As Swiderska et al. (2022) argue, decolonial and interdisciplinary approaches can revitalize Indigenous food systems and contribute to global sustainability transformations (Swiderska et al. 2022).

7 Circularity in Practice

7.1 Zero Waste Farming Systems

Zero waste farming systems represent a paradigm shift from linear resource consumption to regenerative, closed-loop agricultural models. These systems aim to eliminate waste by reintegrating all outputs—organic residues, wastewater, and by-products—into productive cycles. The concept aligns with circular agriculture, which emphasizes nutrient recycling, ecological balance, and minimal external inputs. Nenciu et al. (2022) proposed a synergistic zero-waste food production system integrating aquaculture and horticulture, demonstrating enhanced resource efficiency and reduced operational costs. Their model combined a closed-loop fish pond, straw bale vegetable cultivation, and composting, resulting in a 12% improvement in energy efficiency, 18% increase in food production, and 25% reduction in expenses compared to conventional systems (Nenciu et al. 2022). Circular agriculture also leverages traditional practices such as composting and vermicomposting to transform organic waste into soil amendments. Stefi and Vorgias (2025) highlighted that composting agricultural residues—such as crop stalks, fruit peels, and livestock manure—enhances soil fertility and reduces reliance on synthetic fertilizers. Innovations like in-vessel composting and biochar production further improve nutrient retention and carbon sequestration (Stefi and Vorgias 2025). Moreover, zero-waste systems are increasingly supported by policy frameworks and urban planning initiatives. Kumar et al. (2025b) emphasized the role of decentralized waste valorization technologies—such as anaerobic digestion, pyrolysis, and biodigesters—in building climate-positive circular cities. These technologies not only reduce landfill dependency but also generate renewable energy and biofertilizers from organic waste streams (Kumar et al. 2025b).

Despite their promise, zero-waste farming systems face challenges related to scalability, infrastructure, and behavioral change. Gavrilaş et al. (2025) noted that while biowaste valorization offers economic and ecological benefits, implementation is hindered by high costs, lack of farmer awareness, and limited access to sustainable technologies. Their review called for increased investment in education, stakeholder engagement, and policy support to mainstream circular farming practices (Gavrilaş et al. 2025).

7.2 *Nutrient Cycling and Biowaste Valorization*

Nutrient cycling is the cornerstone of agroecological resilience, enabling the reuse of essential elements—nitrogen, phosphorus, and potassium—within farming systems (Singh et al. 2025). Biowaste valorization complements this process by transforming organic residues into value-added products such as biofertilizers, biogas, and biopolymers. Maher and von Blanckenburg (2023) explored the circular nutrient economy of terrestrial ecosystems, emphasizing that efficient recycling strategies—such as microbial decomposition and root resorption—can minimize nutrient limitations and enhance plant productivity. Their model demonstrated that thicker regolith and erosion-driven mineral rejuvenation support long-term nutrient availability, while excessive chemical weathering may deplete essential elements like phosphorus (Maher and von Blanckenburg 2023). Microbial systems play a pivotal role in biowaste valorization. Paul et al. (2023) reviewed microbial cell factories that convert agricultural waste into biofuels, pigments, enzymes, and fertilizers. For instance, red yeast (*Rhodosporidium* sp.) cultivated on market waste produced carotenoids with anti-cancer properties, while bacterial consortia enhanced the efficacy of organic fertilizers derived from horticultural residues (Paul et al. 2023). Suárez Valdés et al. (2023) identified 25 circular technologies for urban biowaste valorization, including hydrothermal processing, anaerobic digestion, and biorefineries. These technologies enable the conversion of municipal organic waste and wastewater sludge into bioplastics, biofertilizers, and nutraceuticals, supporting urban bioeconomy transitions (Suárez Valdés et al. 2024). In agricultural contexts, biowaste valorization enhances soil health and reduces greenhouse gas emissions. Stefi and Vorgias (2025) noted that anaerobic digestion of crop residues and livestock manure produces biogas and nutrient-rich digestate, which can replace synthetic fertilizers. Similarly, biochar derived from pyrolyzed agricultural waste improves soil structure, water retention, and carbon sequestration, particularly in arid regions (Stefi and Vorgias 2025). However, effective nutrient cycling requires integrated management of organic and inorganic inputs. Imbalanced nutrient flows—due to excessive fertilization or poor waste management—can lead to eutrophication, soil degradation, and biodiversity loss (Kaur et al. 2023a, b, 2025b; Mudgal et al. 2025; Seetharaman et al. 2025; Balu et al. 2024; Fotina et al. 2024; Sharma et al. 2021; Devi et al. 2021). Therefore, monitoring nutrient budgets and aligning them with ecological thresholds is essential for sustainable intensification.

7.3 The Role of Fungi and Algae in Circular Bioeconomy

Fungi and algae are emerging as key agents in circular bioeconomy, offering versatile applications in waste recycling, bioproduct synthesis, and ecosystem restoration. Their unique metabolic capabilities enable the transformation of low-value residues into high-value commodities.

Microalgae, in particular, have gained prominence for their ability to assimilate nutrients from wastewater and produce biofuels, bioplastics, and biofertilizers. It is also highlighted that microalgae can fix CO_2, recover nitrogen and phosphorus from effluents, and generate biomass rich in lipids and proteins. Integrated biorefineries utilizing microalgae have demonstrated potential for biodiesel production, methane generation, and carbon credit accumulation (Moreira et al. 2023). Ezhumalai et al. (2024) emphasized that microalgal biorefineries can reduce production costs by utilizing wastewater nutrients and generating multiple co-products. Their review proposed a holistic approach to circular bioeconomy, integrating biofuel production, biosorption of heavy metals, and feed supplementation within a single platform (Ezhumalai et al. 2024). Fungi also contribute to circularity through biodegradation, composting, and biopolymer synthesis. Gavrilaş et al. (2025) noted that fungal enzymes decompose lignocellulosic residues, enabling the production of organic fertilizers and soil conditioners. Moreover, fungal biomass can be used to create biodegradable plastics, offering alternatives to petroleum-based materials (Gavrilaş et al. 2025).

Microalgae-derived bioproducts and biomaterials, including pigments, biochar, and polyhydroxyalkanoates (PHAs) have also been studied extensiely. Microalgae cultivation on non-arable land using wastewater streams supports carbon–neutral production and enhances circularity in food, feed, and material systems (Thevarajah et al. 2024).

Despite their promise, fungi and algae-based systems face challenges related to scalability, strain optimization, and regulatory frameworks. Fabris et al. (2020) called for advancements in synthetic biology, phenomics, and IoT-enabled cultivation to enhance the productivity and economic viability of algal biofactories (Fabris et al. 2020).

8 Technology and Tradition—An Interwoven Future

8.1 Precision Agriculture, AI and Remote Sensing

Precision agriculture (PA) represents a transformative approach to farming, leveraging digital technologies to optimize resource use, enhance productivity, and reduce environmental impact. At its core, PA integrates geospatial tools, sensor networks, and data analytics to enable site-specific management of crops, soils, and inputs. The rapid evolution of remote sensing platforms—from satellite imagery to unmanned aerial vehicles (UAVs)—has significantly expanded the scope and granularity of agricultural monitoring.

Sishodia et al. (2020) provided a comprehensive review of remote sensing applications in PA, highlighting the use of vegetation indices, thermal imaging, and hyperspectral data for crop health assessment, irrigation scheduling, and yield prediction (Sishodia et al. 2020). These technologies allow farmers to detect stress conditions early, apply inputs precisely, and monitor field variability in real time. The integration of AI and machine learning (ML) further enhances the capabilities of PA. AI algorithms can process vast datasets from sensors, satellites, and weather stations to generate predictive models for disease outbreaks, nutrient deficiencies, and optimal harvest timing. Bhatti et al. (2023) emphasized the role of AI in enabling smart precision agriculture, noting its potential to improve decision-making, reduce input costs, and enhance sustainability (Bhatti et al. 2023). IoT-enabled smart sensors are central to this transformation. These devices measure soil moisture, pH, temperature, and plant stress, transmitting data to cloud-based platforms for analysis and intervention. Mansoor et al. (2025) reviewed trends in IoT integration, highlighting its role in real-time monitoring, automated irrigation, and pest control systems (Mansoor et al. 2025).

Despite these advances, challenges persist. Data interoperability, connectivity in remote areas, and high initial costs limit adoption, particularly among smallholder farmers. Addressing these barriers requires inclusive design, capacity building, and policy support to ensure equitable access to digital agriculture.

8.2 Digital Commons for Farmer-Led Innovation

The concept of digital commons refers to shared digital resources and platforms that enable collaborative innovation, knowledge exchange, and participatory governance. In agriculture, digital commons empower farmers to co-create solutions, share best practices, and access open-source tools tailored to local contexts. Barman and Singh (2025) conducted a bibliometric analysis of farmer-led innovation, revealing a growing body of research that recognizes farmers as active agents of change rather than passive recipients of technology. Their study emphasized the importance of grassroots knowledge, policy support, and digital infrastructure in fostering entrepreneurship and sustainability (Barman and Singh 2025).

Digital platforms such as farmer field schools, community knowledge hubs, and open-access repositories facilitate peer-to-peer learning and innovation diffusion. These platforms support the documentation of traditional practices, adaptation of technologies, and collective problem-solving. Sen et al. (2025) reviewed interactive digital extension services, identifying opportunities for cost-effective knowledge delivery, network collaboration, and enhanced productivity. However, they also noted barriers such as poor connectivity, lack of technical know-how, and limited two-way interaction (Sen et al. 2025). Gumbi et al. (2023) highlighted the potential of digital agriculture ecosystems to support smallholder farmers, but cautioned that many remain beyond the reach of existing solutions due to affordability, digital literacy, and infrastructure gaps. Their review called for business model innovation, inclusive design, and targeted capacity building to bridge the digital divide (Gumbi et al. 2023).

Digital commons also intersect with data sovereignty and ethical governance. Ensuring that farmers retain control over their data and benefit from its use is essential for trust and equity. Participatory data governance models, community-led platforms, and transparent algorithms can help safeguard rights and foster inclusive innovation.

8.3 Crafting Agroecological AI Ethics

As AI becomes increasingly embedded in agricultural systems, ethical considerations must guide its development and deployment. Agroecological AI ethics emphasizes fairness, transparency, accountability, and ecological integrity in the design and use of intelligent technologies. Kapoor (2025) explored ethical frameworks for AI in the agri-food industry, highlighting concerns around algorithmic bias, data privacy, and socio-economic impacts. She argued that ethical AI must address the needs of smallholder farmers, protect biodiversity, and promote equitable access to resources (Kapoor 2025). Dara et al. (2022) proposed a six-principle framework for ethical AI in agriculture: fairness, transparency, accountability, sustainability, privacy, and robustness. Their recommendations included participatory design, bias mitigation, and inclusive policy development to ensure responsible innovation (Dara et al. 2022).

The principle of fairness requires that AI systems do not reinforce existing inequalities or marginalize vulnerable groups. This includes addressing biases in training data, ensuring equitable access to digital tools, and supporting diverse farming systems. Transparency involves making algorithms understandable and decisions explainable, enabling farmers to trust and engage with AI systems. Accountability ensures that stakeholders are responsible for the outcomes of AI applications, particularly when errors or harms occur. Sustainability emphasizes that AI should support ecological resilience, resource efficiency, and long-term viability. Privacy safeguards farmers' data and ensures informed consent, while robustness ensures that AI systems perform reliably under diverse conditions.

These ethical principles must be embedded in the design, regulation, and governance of agroecological AI. This includes interdisciplinary collaboration, stakeholder engagement, and continuous evaluation to align technology with ecological and social values.

9 Economies that Nourish

9.1 Rethinking Value: True Cost Accounting

Conventional economic models often fail to account for the hidden costs and benefits embedded in agricultural systems—externalities that affect ecosystems, human health, and social equity (Balu et al. 2024; Harika et al. 2025). True Cost Accounting (TCA) emerges as a transformative framework that seeks to internalize these externalities by evaluating agri-food systems through the lens of four capitals: natural, human, social, and produced. This multidimensional approach enables policymakers and stakeholders to make informed decisions that reflect the real costs and benefits of food production and consumption.

Bellon et al. (2024) demonstrated the application of TCA using Life Cycle Assessment (LCA) methods to compare palm, rapeseed, and coconut oils. Their study revealed that while environmental impacts of production were relatively modest, the nutritional and health benefits varied significantly across oils, underscoring the importance of integrating health metrics into sustainability evaluations (Bellon et al. 2024). The Food and Agriculture Organization (FAO) estimates that global externalities from agri-food systems amount to at least USD 10 trillion annually, including costs related to greenhouse gas emissions, biodiversity loss, and public health burdens. TCA provides a pathway to visualize and monetize these impacts, enabling the design of policies that promote regenerative practices and discourage harmful ones. Moreover, TCA aligns with systems thinking and holistic evaluation, incorporating both Environmental LCA (E-LCA) and Social LCA (S-LCA) to assess impacts across the entire value chain. While E-LCA has been widely used to quantify natural capital impacts, S-LCA remains underutilized despite its potential to capture human and social dimensions such as labor rights, community development, and cultural integrity (Bellon et al. 2024). Michalk's study argue that embedding TCA within corporate accounting systems can enhance transparency and accountability, allowing businesses to align their operations with sustainability goals and stakeholder expectations (Michalk and Kunz 2025).

9.2 *Sustainable Finance for Agro-innovation*

Sustainable finance plays a pivotal role in catalyzing agro-innovation by directing capital toward environmentally and socially responsible agricultural practices. It encompasses a range of instruments—including green bonds, impact investing, and blended finance—that support transitions to low-carbon, resilient food systems (Mudgal et al. 2025; Fotina et al. 2024).

Yao et al. (2025) examined the impact of science and technology finance on agricultural green development in China, finding that targeted financial mechanisms significantly enhanced rural innovation and environmental performance. Their econometric analysis revealed that rural human capital acted as a key mediator, while agricultural industrial agglomeration had a masking effect, suggesting the need for nuanced financial strategies (Yao et al. 2025). Green finance also fosters regional spillovers, promoting sustainable development beyond the immediate beneficiaries. Huang (2024) highlighted that green finance initiatives in one province can stimulate innovation and environmental improvements in neighboring regions, reinforcing the systemic nature of agroecological transitions (Raman et al. 2025). Innovative instruments such as social impact bonds (SIBs) and green fintech platforms are gaining traction. These tools enable investors to support projects that deliver measurable environmental and social outcomes, while also generating financial returns. Raman et al. (2025) emphasized the role of green fintech in democratizing access to sustainable finance, particularly in developing countries where institutional barriers often limit adoption (Raman et al. 2025).

Springer's *Sustainable Finance series* further explores the integration of digital technologies, blockchain, and AI in enhancing transparency and accountability in green investments. These innovations facilitate real-time tracking of ESG metrics, enabling investors and regulators to assess the impact and integrity of agro-innovation projects. However, challenges persist. High transaction costs, fragmented regulatory frameworks, and limited capacity in the Global South hinder the scalability of sustainable finance. Addressing these barriers requires coordinated efforts across sectors, including policy harmonization, capacity building, and the development of de-risking mechanisms to attract private capital.

9.3 *Local Food Systems and Community Sovereignty*

Local food systems are increasingly recognized as engines of resilience, equity, and ecological stewardship. They prioritize short supply chains, community engagement, and culturally appropriate food production, aligning closely with the principles of food sovereignty.

Rivera et al. (2024) conducted a systematic review of food sovereignty initiatives, identifying five key domains: agroecological management, localization, social justice, traditional knowledge, and self-determination. Their analysis revealed that initiatives often pursue overlapping goals such as food security, environmental stewardship, and crisis preparedness, supported by 29 identified drivers including networks and holistic approaches (Rivera et al. 2024). Community-based food sovereignty assessments (FSAs) offer participatory tools for evaluating local food systems. Steckley reviewed FSAs across diverse contexts, highlighting their role in capturing socio-ecological dynamics, cultural values, and health outcomes. These assessments emphasize local control, traditional knowledge, and multi-scalar analysis, providing a counter-narrative to top-down food security metrics (Steckley 2025). Glaros and Newell (2025) proposed an indicator framework based on the Climate, Biodiversity, Health, and Justice (CBHJ) nexus, enabling integrated planning and policy evaluation for local food systems. Their scoping review revealed gaps in biodiversity and justice metrics, calling for expanded data collection and interdisciplinary collaboration (Glaros and Newell 2025). Moreover, local food systems contribute to economic localization, reducing dependence on volatile global markets and enhancing community sovereignty. Lacy emphasized that empowered communities, citizen science, and democratic governance are essential for redirecting unsustainable agendas toward just and livable societies. Local food systems serve as platforms for civic engagement, knowledge co-production, and transformative action (Lacy 2023).

However, scaling local food systems requires supportive infrastructure, policy alignment, and market development. This includes investment in regional processing facilities, cooperative models, and public procurement programs that prioritize local sourcing. Integrating local food systems into urban planning and climate adaptation strategies can further enhance their impact and sustainability.

10 Policy Landscapes and Global Agreements

10.1 SDGs, Biodiversity Frameworks, and Climate Goals

Global policy frameworks such as the Sustainable Development Goals (SDGs), the Kunming-Montreal Global Biodiversity Framework (GBF), and the Paris Agreement represent interconnected efforts to address planetary challenges through integrated governance. These frameworks are not isolated mandates but synergistic instruments that shape agricultural and environmental policy at multiple scales. The SDGs, adopted in 2015, encompass 17 goals and 169 targets aimed at balancing economic development, social inclusion, and environmental sustainability. Biodiversity is embedded across several SDGs, including SDG 2 (Zero Hunger), SDG 13 (Climate Action), SDG 14 (Life Below Water), and SDG 15 (Life on Land), reflecting its foundational role in food systems, ecosystem services, and climate resilience (CBD 2030).

The Kunming-Montreal GBF, adopted in 2022, sets ambitious targets for biodiversity conservation, including restoring 30% of degraded ecosystems and protecting

30% of terrestrial and marine areas by 2030. It explicitly links biodiversity goals with climate adaptation and mitigation, recognizing that ecosystem integrity underpins carbon sequestration and resilience to climate extremes (UNNFCC 2024). Similarly, the Paris Agreement emphasizes limiting global temperature rise to below 1.5 °C, with nature-based solutions and ecosystem restoration increasingly recognized as critical pathways. The 2023 UAE Framework for Global Climate Resilience integrates biodiversity into climate adaptation strategies, reinforcing the convergence of these global agendas (UNNFCC 2024). Hernandez et al. (2020) proposed an Innovation Systems Roadmap to align renewable energy development with both climate goals and SDGs. Their framework identifies six research themes—ranging from environmental sustainability to social equity—and maps their linkages to all 17 SDGs, offering a holistic approach to policy coherence (Hernandez et al. 2020).

However, operationalizing these synergies remains a challenge. A bibliometric analysis by Xin et al. (2024) revealed fragmented research efforts and limited integration across SDG indicators, particularly in environmental dimensions. The study calls for multi-party cooperation, innovative technologies, and localized indicator systems to enhance SDG implementation (Xin et al. 2024).

10.2 Harmonizing Regulation Across Borders

Cross-border regulatory harmonization is essential for addressing transboundary environmental issues, facilitating trade in sustainable goods, and ensuring consistency in biodiversity and climate governance. Yet, regulatory fragmentation persists due to divergent national priorities, legal systems, and institutional capacities. Schmidt explored regulatory integration across borders through public–private cooperation, highlighting the emergence of transnational governance networks that stabilize normative expectations without formal hierarchies. These networks—such as ISO standards and Olympic environmental norms—demonstrate how decentralized authority can foster regulatory coherence (Meerssche 2021).

In the context of digital agriculture and data governance, Chin and Zhao (2022) examined international trade agreements governing cross-border data flows, noting the lack of a coherent global framework. They advocate for a thin, narrowly scoped WTO agreement that balances privacy, national security, and free data movement, enabling interoperability across jurisdictions (Chin and Zhao 2022). The OECD has also mapped commonalities in regulatory approaches to cross-border data transfers, identifying unilateral mechanisms, plurilateral arrangements, and trade agreements as key instruments. Their analysis supports international dialogue on predictable and transparent data governance, which is critical for precision agriculture and climate-smart technologies (Casalini et al. 2021).

In the EU context, Twigg-Flesner (2012) proposed a cross-border-only regulation for consumer transactions, arguing that such an approach would benefit e-commerce and reduce legal conflicts without disrupting domestic law. This model could inform agricultural trade policies that respect local sovereignty while enabling global cooperation (Twigg-Flesner 2012). Despite these efforts, regulatory integration faces structural and political barriers. Van Den Meerssche (2021) cautioned that transnational regulatory networks may lack democratic accountability and reinforce power asymmetries. He calls for normative frameworks that balance efficiency with legitimacy, particularly in environmental and agricultural governance (Meerssche 2021).

10.3 Power, Access, and Decentralized Decision-Making

Decentralized decision-making is increasingly recognized as a strategy for enhancing resilience, equity, and participatory governance in agricultural and environmental systems. It reallocates authority from centralized entities to distributed networks, enabling context-specific solutions and community empowerment. Namany and group reviewed resilience strategies in the Energy-Water-Food (EWF) nexus, emphasizing the role of decentralization in reducing single points of failure and enabling adaptive responses. Their analysis of 84 studies found that multi-criteria decision-making frameworks and composite indicators are effective in managing risks and uncertainties across nexus systems (Haji et al. 2024). Tavares et al. (2024) conducted a case study on decentralized managers in a large industrial company, revealing that middle managers significantly influence strategic decisions and adapt policies to local needs. Their findings underscore the importance of decentralization in complex organizational structures, including agricultural cooperatives and regional planning bodies (Tavares et al. 2024). In the realm of scientific research, decentralized science (DeSci) platforms are democratizing access to resources and reducing centralized control. Lustenberger et al. (2025) analyzed trends in Decentralized Autonomous Organizations (DAOs), highlighting their potential to foster equitable governance, transparency, and innovation in agri-tech and environmental monitoring (Lustenberger et al. 2025).

However, decentralization must be accompanied by robust accountability mechanisms. Morelli and coworkers emphasized that decision-making processes—whether centralized or decentralized—require transparency, feedback loops, and inclusive participation to avoid bias and ensure legitimacy (Morelli et al. 2022). The challenge lies in balancing local autonomy with global coordination. As highlighted by the UNFCCC and CBD, integrated governance frameworks must enable decentralized implementation while maintaining coherence with international commitments. This includes aligning national biodiversity strategies with the GBF and embedding climate adaptation into local land-use planning (UNNFCC 2024).

11 Future-Proof Farming—Scenarios and Solutions

11.1 Climate Modeling and Adaptive Risk Management

Climate modeling has evolved into a cornerstone of future-proof farming, offering predictive insights into weather variability, crop performance, and long-term sustainability. Advanced frameworks now integrate General Circulation Models (GCMs), Earth System Models (ESMs), and AI-enhanced simulations to capture the nonlinear dynamics of climate-agriculture interactions. Chahed (2025) emphasized that refining sub-grid parameterizations and incorporating socio-economic dimensions into climate models enhances their relevance for agricultural planning and risk mitigation (Chahed 2025). The integration of machine learning (ML) and artificial intelligence (AI) has significantly improved forecast accuracy, particularly in capturing cloud dynamics, hydrological cycles, and crop-climate feedbacks. Schneider et al. (2017) and Baño-Medina et al. (2024) demonstrated that AI-driven models outperform traditional statistical approaches in short-term forecasting and scenario generation (Baño-Medina et al. 2024; Schneider et al. 2017). However, Reichstein et al. (2019) cautioned that empirical models must be grounded in physical laws to maintain interpretability and reliability (Reichstein et al. 2019).

Adaptive risk management complements climate modeling by translating forecasts into actionable strategies. Khan et al. (2025) conducted a global bibliometric review of climate risk management, identifying key themes such as flood risk, vulnerability assessment, and uncertainty modeling. Their findings underscore the shift from reactive to anticipatory governance, where adaptation measures are embedded in long-term planning and policy frameworks (Khan et al. 2025). Moreover, insurance mechanisms and financial instruments are increasingly used to buffer climate risks. Schäfer's group proposed climate risk insurance as a tool to navigate adaptation frontiers, offering financial protection and incentivizing risk reduction. Their framework integrates hazard exposure, vulnerability mapping, and contingency planning, aligning with the Sendai Framework and Paris Agreement goals (Schäfer et al. 2019).

The convergence of climate modeling and adaptive risk management thus enables farmers, policymakers, and researchers to co-create resilient farming systems that anticipate and respond to climatic uncertainties.

11.2 Urban Agriculture and Vertical Farming Futures

Urban agriculture and vertical farming are redefining the spatial and technological boundaries of food production. As urbanization accelerates and arable land diminishes, these systems offer decentralized, resource-efficient solutions for food security and ecological integration. Zhang and Zha described urban agriculture as a multifunctional resource encompassing ecological, recreational, and high-tech modalities, with vertical farming emerging as a key innovation in building-integrated agriculture (Zhang 2022). Vertical farming systems utilize hydroponics, aeroponics, and controlled environment agriculture (CEA) to produce crops year-round with minimal land and water use. Sowmya et al. (2024) highlighted that vertical farming addresses urban food demand, biodiversity loss, and climate resilience by enabling high-density, soilless cultivation in urban settings (Sowmya et al. 2024).

Technological integration is central to vertical farming. Siregar et al. (2022) reviewed the role of AI, IoT, and deep learning in optimizing vertical farming operations, including intelligent irrigation, disease monitoring, and yield prediction. Their study emphasized that smart vertical farming represents a new era of precision agriculture, transforming traditional practices into data-driven systems (Siregar et al. 2022). However, vertical farming faces challenges related to energy consumption, scalability, and consumer acceptance. Zhou et al. (2022) explored stakeholder perceptions in Shanghai, revealing that sustainability narratives and health benefits influence adoption, while concerns about artificiality and cost persist (Zhou et al. 2022). Integrating renewable energy sources and circular resource flows can mitigate these concerns and enhance system viability. Urban agriculture also contributes to climate adaptation and social equity. Rooftop gardens and community farms reduce urban heat, improve air quality, and foster community resilience. These systems align with nature-based solutions and support the localization of food systems, particularly during crises such as pandemics and supply chain disruptions.

In sum, urban agriculture and vertical farming represent scalable, adaptive solutions for future-proofing food systems in densely populated and climate-vulnerable regions.

11.3 Transdisciplinary Collaboration for Sustainable Agriculture

Transdisciplinary collaboration is increasingly recognized as essential for addressing the complex, interconnected challenges of sustainable agriculture. It transcends disciplinary boundaries, integrating knowledge from plant science, engineering, social sciences, and local communities to co-create solutions that are ecologically sound and socially just. Soriano Sánchez et al. (2024) explored collaborative, situated, and critical methodologies in agroecology, highlighting how local knowledge systems contribute to regenerative practices, seed sovereignty, and biodiversity conservation. Their review identified five methodological clusters, including popular pedagogies, sustainability assessments, and participatory marketing, which support intercultural dialogue and epistemic justice (Sánchez et al. 2024). Watabe et al. (2025) emphasized the role of transdisciplinary communication in sustainable food systems, noting that integrating perspectives from agriculture, nutrition, ecology, and policy enhances resilience and innovation. Their editorial synthesized findings from diverse studies, including consumer preferences, microbiological safety, and cultivated meat, illustrating the multifaceted nature of food system transformation (Watabe et al. 2025). Manjula and Rengalakshmi (2021) analyzed transdisciplinary engagement in agricultural research, identifying key drivers such as conceptual clarity, shared goals, and openness to cross-fertilization. Their comparative study of climate forecasting and biofertilization projects revealed that participatory design and stakeholder inclusion are critical for adoption and scalability (Manjula and Rengalakshmi 2021). These studies collectively affirm that transdisciplinary collaboration fosters innovation, equity, and resilience in agricultural systems. It enables the co-production of knowledge, democratizes decision-making, and aligns research with societal needs.

12 Conclusion—Regenerating More Than Land

The transition to sustainable agriculture is not solely a technical or ecological endeavor—it is a profound cultural and cognitive shift. At its core lies the transformation of mindsets, the internal frameworks through which individuals and societies interpret their relationship with land, food, and community. This shift requires moving from extractive paradigms to regenerative ones, where agriculture is seen not as a means of domination over nature but as a co-creative process with living systems. Narvaez and coworkers argue that moral imagination—the capacity to envision ethical futures and act upon them—is essential for cultivating regenerative mindsets. Moral imagination involves empathy, foresight, and the ability to perceive oneself as part of a larger ecological and social whole (Narvaez and Mrkva 2014). It enables individuals to navigate complexity and ambiguity, fostering values such as reciprocity, stewardship, and intergenerational responsibility.

Movements such as agroecology, food sovereignty, and regenerative agriculture exemplify the power of collective moral imagination. They challenge dominant narratives of industrial efficiency and offer alternative visions rooted in equity, biodiversity, and cultural resilience. These movements are not monolithic but pluralistic, drawing from Indigenous knowledge, feminist ethics, and ecological science to reframe agriculture as a site of healing and transformation.

Moreover, the cultivation of regenerative mindsets requires educational and institutional support. Programs that integrate ecological literacy, systems thinking, and ethical reasoning can empower future leaders to navigate the moral and practical dimensions of sustainability. Storytelling is a powerful tool for shaping agricultural transitions, offering a medium through which values, experiences, and visions are communicated and co-created. It bridges the gap between data and meaning, enabling communities to articulate their histories, aspirations, and relationships with the land. Yueh and Zheng (2019) developed a validated scale to measure the effectiveness of storytelling in agricultural marketing, identifying four dimensions: narrative processing, affect, brand attitude, and purchase intention. Their findings suggest that storytelling enhances consumer engagement and fosters emotional connections with agricultural products, particularly in contexts lacking strong branding (Yueh and Zheng 2019).

Beyond marketing, storytelling plays a critical role in extension communication and participatory research. Digital storytelling in Malawi enabled smallholder farmers and fishers to share their experiences with agricultural technologies, fostering empowerment and mutual learning. Their image-based methodology emphasized beneficiary perspectives and facilitated meaningful dialogue between stakeholders (Yueh and Zheng 2019). Storytelling also supports knowledge co-production and epistemic justice. Giagnocavo et al. (2022) explored agroecological transitions in Spain, highlighting how niche initiatives reconnect farmers with nature through experimentation and shared narratives. Their study emphasized the role of Agricultural Knowledge and Innovation Systems (AKIS) in facilitating transitions and fostering human-nature connectedness (Giagnocavo et al. 2022). In mentoring and education, storytelling cultivates creative mindsets and metacognitive awareness. Jia et al. (2025) found that metacognitive disfluency experiences—triggered by storytelling tasks—enhanced originality in divergent thinking, particularly among individuals with growth-oriented creative mindsets. Their study underscores the dynamic interplay between narrative, cognition, and creativity in agricultural innovation (Jia et al. 2025). Ultimately, storytelling is not merely a communication tool—it is a cultural practice that shapes identity, agency, and collective action. It enables communities to envision regenerative futures, honor ancestral wisdom, and navigate the uncertainties of transition. To regenerate more than land, we must embrace Living Systems Leadership—a holistic approach that draws inspiration from nature's patterns, processes, and principles. This leadership paradigm recognizes that agricultural systems are embedded within complex socio-ecological networks and that transformation requires adaptive, collaborative, and ethical engagement.

References

Abd-Elaty I, Kuriqi A, Ahmed A, Ramadan EM (2024) Enhanced groundwater availability through rainwater harvesting and managed aquifer recharge in arid regions. Appl Water Sci 14:121. https://doi.org/10.1007/s13201-024-02166-7

Abdi G et al (2024) In: Singh V (ed) Advances in genomics: methods and applications. Springer Nature Singapore, pp 181–197

Abhilash PC et al (2022) Introducing 'anthropocene science': a new international journal for addressing human impact on the resilience of planet earth. Anthropocene Sci 1:1–4. https://doi.org/10.1007/s44177-021-00001-1

Addorisio R, Spadoni R, Maesano G (2025) Adoption of innovative technologies for sustainable agriculture: a scoping review of the system. Domain 17:4224

Ahern D (2025) The new anticipatory governance culture for innovation: regulatory foresight, regulatory experimentation and regulatory learning. Eur Bus Organ Law Rev 26:241–283. https://doi.org/10.1007/s40804-025-00348-7

Ahmad S et al (2022) Organic amendments and conservation tillage improve cotton productivity and soil health indices under arid climate. Sci Rep 12:14072. https://doi.org/10.1038/s41598-022-18157-0

Akchaya K et al (2025) Boosting resource use efficiency, soil fertility, food security, ecosystem services, and climate resilience with legume intercropping: a review 9. https://doi.org/10.3389/fsufs.2025.1527256

Alam MF, McClain ME, Sikka A, Daniel D, Pande S (2022) Benefits, equity, and sustainability of community rainwater harvesting structures: an assessment based on farm scale social survey 10. https://doi.org/10.3389/fenvs.2022.1043896

Ali S et al (2024) Groundwater quality assessment using water quality index and principal component analysis in the Achnera block, Agra district, Uttar Pradesh, Northern India. Sci Rep 14:5381. https://doi.org/10.1038/s41598-024-56056-8

Ali ZA et al (2025) Digital twins: cornerstone to circular economy and sustainability goals. Environ Dev Sustain. https://doi.org/10.1007/s10668-025-06221-4

Alletto L, Celette F, Drexler D, Plaza-Bonilla D, Reckling M (2022) Editorial: crop diversification, a key pillar for the agroecological transition 4. https://doi.org/10.3389/fagro.2022.950822

Ambus JV, Awe GO, Faccio Carvalho PCD, Reichert JM (2023) Integrated crop-livestock systems in lowlands with rice cultivation improve root environment and maintain soil structure and functioning. Soil Tillage Res 227:105592. https://doi.org/10.1016/j.still.2022.105592

Ankomah G et al (2024) Integrating cover crops and organic amendments to mitigate the limitations of tillage on soil health and cotton productivity 116:2615–2629. https://doi.org/10.1002/agj2.21649

Asa'a S et al (2024) A multidisciplinary view on agrivoltaics: future of energy and agriculture. Renew Sustain Energy Rev 200:114515. https://doi.org/10.1016/j.rser.2024.114515

Atta HS, Omar MA-S, Tawfik AM (2022) Water quality index for assessment of drinking groundwater purpose case study: area surrounding Ismailia Canal, Egypt. J Eng Appl Sci 69:83. https://doi.org/10.1186/s44147-022-00138-9

Azhar MF, Ali E, Aziz A (2024) In: Kumar S, Meena RS, Sheoran P, Jhariya MK (eds) Regenerative agriculture for sustainable food systems. Springer Nature Singapore, pp 423–451

Balu UR et al (2024) Linking EDC-laden food consumption and modern lifestyle habits with preeclampsia: a non-animal approach to identifying early diagnostic biomarkers through biochemical alterations. Food Chem Toxicol 194:115073. https://doi.org/10.1016/j.fct.2024.115073

Banerjee U, Karmakar R, Basu D, Sen A, Saha T (2023) Use of IoT in agriculture: a systematic review. In: National conference on control instrumentation system conference. Springer Nature Singapore, Singapore, pp 781–790

Baño-Medina J, Iturbide M, Fernández J, Gutiérrez JM (2024) Transferability and explainability of deep learning emulators for regional climate model projections: perspectives for future applications %J artificial intelligence for the earth systems. Artif Intell Earth Syst 3:e230099. https://doi.org/10.1175/AIES-D-23-0099.1

Barman B, Singh R (2025) Farmer-led innovations and their role in entrepreneurship: a bibliometric exploration. Discov Agricult 3:90. https://doi.org/10.1007/s44279-025-00257-w

Bellon MR, Benard N, Vizcaino M, Merrigan K, Wharton C (2024) True cost accounting using life cycle assessment methods and data: a case study comparing palm, rapeseed, and coconut oils for sustainability and nutrition. Sustainability 16:10366. https://doi.org/10.3390/su162310366

Bhagat R et al (2024) The integrated farming system is an environmentally friendly and cost-effective approach to the sustainability of agri-food systems in the modern era of the changing climate: a comprehensive review 13:e534. https://doi.org/10.1002/fes3.534

Bhanye J (2025) Flood-tech frontiers: smart but just? a systematic review of AI-driven urban flood adaptation and associated governance challenges. Discov Glob Soc 3:59. https://doi.org/10.1007/s44282-025-00190-9

Bhatia T, Sindhu SS (2024) Sustainable management of organic agricultural wastes: contributions in nutrients availability, pollution mitigation and crop production. Discov Agricult 2:130. https://doi.org/10.1007/s44279-024-00147-7

Bhattacharya S (2024) In: Jatav HS et al (eds) Agroforestry to combat global challenges: current prospects and future challenges. Springer Nature Singapore, pp 479–502

Bhattacharyya R et al (2016) Soil conservation issues in India 8:565
Bhatti UA, Masud M, Bazai SU, Tang H (2023) Editorial: investigating AI-based smart precision agriculture techniques 14. https://doi.org/10.3389/fpls.2023.1237783
Bhunia GS, Keshavarzi A, Shit PK, Omran E-SE, Bagherzadeh A (2018) Evaluation of groundwater quality and its suitability for drinking and irrigation using GIS and geostatistics techniques in semiarid region of Neyshabur, Iran. Appl Water Sci 8:168. https://doi.org/10.1007/s13201-018-0795-6
Bian Y et al (2022) PIFs- and COP1-HY5-mediated temperature signaling in higher plants. Stress Biol 2:35. https://doi.org/10.1007/s44154-022-00059-w
Bibri SE, Huang J, Jagatheesaperumal SK, Krogstie J (2024) The synergistic interplay of artificial intelligence and digital twin in environmentally planning sustainable smart cities: a comprehensive systematic review. Environ Sci Ecotechnol 20:100433. https://doi.org/10.1016/j.ese.2024.100433
Birla D, Yadav SL, Gajanand, Patel RA, Sanodiya P (2024) In: Kumar S et al (eds) Agroforestry solutions for climate change and environmental restoration. Springer Nature Singapore, pp 437–462
Bjornlund V, Bjornlund H (2025) Reviewing the green revolution strategy in view of lessons from Mexico and India–Africa. Int J Water Resour Dev 41:229–273. https://doi.org/10.1080/07900627.2024.2363954
Bonilla Valverde JP, Blank C, Roidt M, Schneider L, Stefan C (2016) Application of a GIS multi-criteria decision analysis for the identification of intrinsic suitable sites in costa rica for the application of managed aquifer recharge (MAR) through spreading methods 8:391
Borsari B (2021) Food production and Earth's limits to growth in the anthropocene. J Agric Food Syst Commun Dev 11:211–213. https://doi.org/10.5304/jafscd.2021.111.007
Bourke PM et al (2021) Breeding beyond monoculture: putting the "intercrop" into crops. Front Plant Sci 12:734167. https://doi.org/10.3389/fpls.2021.734167
Breitbart S, Tomchyshyn A, Wagner HH, Johnson MTJ (2023) Urbanization and a green corridor influence reproductive success and pollinators of common milkweed. Urban Ecosyst 26:31–43. https://doi.org/10.1007/s11252-022-01278-9
Cabello V, Romero D, Musicki A, Guimarães Pereira Â, Peñate B (2021) Co-creating narratives for WEF nexus governance: a quantitative story-telling case study in the Canary Islands. Sustain Sci 16:1363–1374. https://doi.org/10.1007/s11625-021-00933-y
Calheiros CSC, Stefanakis AI (2021) Green roofs towards circular and resilient cities. Circ Econ Sustain 1:395–411. https://doi.org/10.1007/s43615-021-00033-0
Carrara JE et al (2021) Differences in microbial community response to nitrogen fertilization result in unique enzyme shifts between arbuscular and ectomycorrhizal-dominated soils. Glob Change Biol 27:2049–2060. https://doi.org/10.1111/gcb.15523
Cartagena LB, Hegger D, Tittonell P, Runhaar H (2025) How transformative is agroecological knowledge co-creation? insights from a systematic literature review. Agroecol Sustain Food Syst 49:124–150. https://doi.org/10.1080/21683565.2024.2405885
Casalini F, López González J, Nemoto T (2021) Mapping commonalities in regulatory approaches to cross-border data transfers. OECD Trade Policy Papers, OECD Publishing, Paris. https://doi.org/10.1787/ca9f974e-en
CBD (2018) All information for this press brief drawn from the policy brief: biodiversity and the 2030 agenda for sustainable development. www.cbd.int/development/doc/biodiversity2030-agenda-policy-brief-en.pdf 4 (Secretariat of the Convention on Biological Diversity, 413, Saint Jacques Street, suite 800 Montreal, Quebec, H2Y 1N9 Canada)
CCICED (2022) In: Green consensus and high quality development: CCICED annual policy report 2020 63–115. Springer Singapore
Chahed J (2025) Advanced climate modeling frameworks: state-of-the-art techniques, uncertainties, and the principle of responsibility. Model Earth Syst Environ 11:331. https://doi.org/10.1007/s40808-025-02525-6
Chamoli N, Prabha D (2024) 1517–1531

Charatsari C, Lioutas ED, De Rosa M, Vecchio Y (2022) Technological innovation and agrifood systems resilience: the potential and perils of three different strategies 6. https://doi.org/10.3389/fsufs.2022.872706

Chauhan N, von Wehrden H (2025) A critical analysis of the policy potential for sustainable agriculture in India. Discov Sustain 6:233. https://doi.org/10.1007/s43621-025-01032-z

Chaurasia PK et al (2024) Fungal pretreatment methods for organic wastes: advances and challenges in biomass valorization. RSC Sustain 3:1234–1266. https://doi.org/10.1039/d4su00582a

Chen L et al (2024) Green building practices to integrate renewable energy in the construction sector: a review. Environ Chem Lett 22:751–784. https://doi.org/10.1007/s10311-023-01675-2

Chin Y-C, Zhao J (2022) Governing cross-border data flows: international trade agreements and their limits. Laws 11:63. https://doi.org/10.3390/laws11040063

Chmelíková L, Schmid H, Anke S, Hülsbergen K-J (2021) Nitrogen-use efficiency of organic and conventional arable and dairy farming systems in Germany. Nutr Cycl Agroecosyst 119:337–354. https://doi.org/10.1007/s10705-021-10126-9

Coggan A, Carwardine J, Fielke S, Whitten S (2021) Co-creating knowledge in environmental policy development. An analysis of knowledge co-creation in the review of the significant residual impact guidelines for environmental offsets in Queensland, Australia. Environ Chall 4:100138. https://doi.org/10.1016/j.envc.2021.100138

Cougnon M, Durand J-L, Julier B, Barre P, Litrico I (2022) Using perennial plant varieties for use as living mulch for winter cereals. A review. Agronomy Sustain Dev 42:110. https://doi.org/10.1007/s13593-022-00844-x

Covington MF, Maloof JN, Straume M, Kay SA, Harmer SL (2008) Global transcriptome analysis reveals circadian regulation of key pathways in plant growth and development. Genome Biol 9:R130. https://doi.org/10.1186/gb-2008-9-8-r130

Dai Y, Huang Z, Khan N, Labbo MS (2025) Smart water management: governance innovation, technological integration, and policy pathways toward economic and ecological sustainability 1932

Dai W et al (2025) Impact of winter cover crop and soil amendments on total and aggregate-associated nitrogen in a no-till corn field. NPJ Sustain Agric 3:39. https://doi.org/10.1038/s44264-025-00077-x

Dang R et al (2024) Soil microbial carbon use efficiency and the constraints. Ann Microbiol 74:37. https://doi.org/10.1186/s13213-024-01780-9

Dara R, Hazrati Fard SM, Kaur J (2022) Recommendations for ethical and responsible use of artificial intelligence in digital agriculture 5.https://doi.org/10.3389/frai.2022.884192

Das SK, Mallick S (2024) Repossession through Community Participation: a Study of Vrihi Community Seed Bank in Odisha. 54:508–525. https://doi.org/10.1177/00490857241257110

Debnath S et al (2023) The enhanced affinity of WRKY reinforces drought tolerance in solanum Lycopersicum L.: an innovative bioinformatics study. Plants 12:762

Desalegn G, Tangl A, Fekete-Farkas M, Gudisa G, Boros A (2024) Linking policies and regulations to sustainable finance for the promotion of urban agriculture: evidence from micro and small businesses. Heliyon 10:e31938. https://doi.org/10.1016/j.heliyon.2024.e31938

Devi A, Mudgal G, Khan ZA (2021) In: Kesari KK, Jha NK (eds) Free radical biology and environmental toxicity. Springer International Publishing, pp 115–130

Dhar SK et al (2024) Novel Bacillus and Prestia isolates from Dwarf century plant enhance crop yield and salinity tolerance. Sci Rep 14:14645. https://doi.org/10.1038/s41598-024-65632-x

Dijkgraaf KH, Goddek S, Keesman KJ (2019) Modeling innovative aquaponics farming in Kenya. Aquacult Int 27:1395–1422. https://doi.org/10.1007/s10499-019-00397-z

Dönmez D, Isak MA, İzgü T, Şimşek Ö (2024) Green horizons: navigating the future of agriculture through sustainable practices 16:3505

Douglas MM et al (2019) Conceptualizing hydro-socio-ecological relationships to enable more integrated and inclusive water allocation planning. One Earth 1:361–373. https://doi.org/10.1016/j.oneear.2019.10.021

Eagmark (2022). https://eagmark.net/

Eliazer Nelson ARL, Ravichandran K, Antony U (2019) The impact of the green revolution on indigenous crops of India. J Ethnic Foods 6:8. https://doi.org/10.1186/s42779-019-0011-9
Enebe MC, Babalola OO (2021) Soil fertilization affects the abundance and distribution of carbon and nitrogen cycling genes in the maize rhizosphere. AMB Express 11:24. https://doi.org/10.1186/s13568-021-01182-z
Escribà-Gelonch M et al (2024) Digital twins in agriculture: orchestration and applications. J Agric Food Chem 72:10737–10752. https://doi.org/10.1021/acs.jafc.4c01934
Estévez S, González-García S, Feijoo G, Moreira MT (2022) How decentralized treatment can contribute to the symbiosis between environmental protection and resource recovery. Sci Total Environ 812:151485. https://doi.org/10.1016/j.scitotenv.2021.151485
Ezhumalai G, Arun M, Manavalan A, Rajkumar R, Heese K (2024) A holistic approach to circular bioeconomy through the sustainable utilization of microalgal biomass for biofuel and other value-added products. Microb Ecol 87:61. https://doi.org/10.1007/s00248-024-02376-1
Fabris M et al (2020) Emerging technologies in algal biotechnology: toward the establishment of a sustainable, algae-based bioeconomy 11. https://doi.org/10.3389/fpls.2020.00279
Famiglietti JS, Ferguson G (2021) The hidden crisis beneath our feet 372:344–345. https://doi.org/10.1126/science.abh2867
Farias GD et al (2020) Integrated crop-livestock system with system fertilization approach improves food production and resource-use efficiency in agricultural lands. Agron Sustain Dev 40:39. https://doi.org/10.1007/s13593-020-00643-2
Fotina NV et al (2024) Enhancement of wheat growth by plant growth-stimulating bacteria during phytopathogenic inhibition. Biocatal Agric Biotechnol 60:103294. https://doi.org/10.1016/j.bcab.2024.103294
Frankel-Goldwater L, Wojtynia N, Dueñas-Ocampo S (2023) Healthy people, soils, and ecosystems: uncovering primary drivers in the adoption of regenerative agriculture by US farmers and ranchers 7. https://doi.org/10.3389/fsufs.2023.1070518
Friedman K et al (2022) The CBD post-2020 biodiversity framework: People's place within the rest of nature 4:1475–1484. https://doi.org/10.1002/pan3.10403
Frimpong F et al (2023) Water-smart farming: review of strategies, technologies, and practices for sustainable agricultural water management in a changing climate in West Africa 7. https://doi.org/10.3389/fsufs.2023.1110179
Galiwango H, Miiro R, Turyahabwe N, Egeru A (2022) Does the climate-smart village extension model enhance farmer empowerment? Outlook Agric 51:281–292. https://doi.org/10.1177/00307270221113600
Gamage A et al (2024) Advancing sustainability: the impact of emerging technologies in agriculture. Curr Plant Biol 40:100420. https://doi.org/10.1016/j.cpb.2024.100420
Gangappa SN, Botto JF (2016) The multifaceted roles of HY5 in plant growth and development. Mol Plant 9:1353–1365. https://doi.org/10.1016/j.molp.2016.07.002
Garbisu C, Unamunzaga O, Alkorta I (2025) Generating regenerative agriculture 9. https://doi.org/10.3389/fsufs.2025.1545811
Garrido-Baserba M et al (2024) Using water and wastewater decentralization to enhance the resilience and sustainability of cities. Nat Water 2:953–974. https://doi.org/10.1038/s44221-024-00303-9
Gavazov K et al (2022) Plant-microbial linkages underpin carbon sequestration in contrasting mountain tundra vegetation types. Soil Biol Biochem 165:108530. https://doi.org/10.1016/j.soilbio.2021.108530
Gavrilaş S, Raţ M, Munteanu F-D (2025) Biowaste valorisation and its possible perspectives within sustainable food chain development. Processes 13:2085
Geels FW, McMeekin A, Mylan J, Southerton D (2015) A critical appraisal of sustainable consumption and production research: the reformist, revolutionary and reconfiguration positions. Glob Environ Chang 34:1–12. https://doi.org/10.1016/j.gloenvcha.2015.04.013

Ghiglieno I et al (2025) Development of an integrated indicator for assessing management impacts on soil quality: a case study in organic viticulture. Environ Sustain Indic 27:100732. https://doi.org/10.1016/j.indic.2025.100732

Ghorbani J, Aghdasi S, Nazem M, McCartney JS, Kodikara J (2025) Augmented intelligence framework for real-time ground assessment under significant uncertainty. Eng Comput. https://doi.org/10.1007/s00366-025-02108-0

Ghosh P, Dhyani, Baranaaja PP (2004) The traditional mixed cropping system of the Central Himalaya, vol 33, pp 261–266, https://doi.org/10.5367/0000000042664756

Giagnocavo C et al (2022) Reconnecting farmers with nature through agroecological transitions: interacting niches and experimentation and the role of agricultural knowledge and innovation systems. Agriculture 12:137. https://doi.org/10.3390/agriculture12020137

Giller KE, Hijbeek R, Andersson JA, Sumberg J (2021) Regenerative agriculture: an agronomic perspective 50:13–25. https://doi.org/10.1177/0030727021998063

Glaros A, Newell R (2025) The state of local food systems and integrated planning and policy research: an application of the climate, biodiversity, health, and justice nexus. Agric Hum Values 15:718. https://doi.org/10.3390/agriculture15070718

Gollin D, Hansen CW, Wingender AM (2021) Two blades of grass: the impact of the green revolution 129:2344–2384. https://doi.org/10.1086/714444

Gordon HSJ et al (2023) Integrating indigenous traditional ecological knowledge of land into land management through indigenous-academic partnerships. Land Use Policy 125:106469. https://doi.org/10.1016/j.landusepol.2022.106469

Govindasamy P et al (2023) Nitrogen use efficiency—a key to enhance crop productivity under a changing climate 14. https://doi.org/10.3389/fpls.2023.1121073

Grard BJP et al (2017) Rooftop farming on urban waste provides many ecosystem services. Agron Sustain Dev 38:2. https://doi.org/10.1007/s13593-017-0474-2

De Greef JA, Fredericq H (1983) In: Shropshire W, Mohr H (eds) Photomorphogenesis. Springer Berlin Heidelberg, pp 401–427

Gumbi N, Gumbi L, Twinomurinzi H (2023) Towards sustainable digital agriculture for smallholder farmers: a systematic literature review 15:12530. https://doi.org/10.3390/su151612530

Gururani K et al (2021) Mainstreaming Barahnaja cultivation for food and nutritional security in the Himalayan region. Biodivers Conserv 30:551–574. https://doi.org/10.1007/s10531-021-02123-9

Haji M, Namany S, Al-Ansari T (2024) Strengthening resilience: decentralized decision-making and multi-criteria analysis in the energy-water-food nexus systems. Front Sustain 5. https://doi.org/10.3389/frsus.2024.1367931

Harika R et al (2025) Sweetener synapse: exploring non-nutritive sweeteners' role in zebrafish glucose homeostasis. Environ Technol Innov 39:104322. https://doi.org/10.1016/j.eti.2025.104322

Hassan R, Di Martino M, Daher B (2025) Quantifying sustainability and resilience in food systems: a systematic analysis for evaluating the convergence of current methodologies and metrics 9. https://doi.org/10.3389/fsufs.2025.1479691

Hernandez RR, Jordaan SM, Kaldunski B, Kumar N (2020) Aligning climate change and sustainable development goals with an innovation systems roadmap for renewable power. Front Sustain 1. https://doi.org/10.3389/frsus.2020.583090

Hill KX et al (2024) Rematriation and climate justice: intersections of indigenous health and place. J Clim Change Health 18:100314. https://doi.org/10.1016/j.joclim.2024.100314

Hiscock KM et al (2024) Developing managed aquifer recharge (MAR) to augment irrigation water resources in the sand and gravel (Crag) aquifer of coastal Suffolk, UK. J Environ Manage 351:119639. https://doi.org/10.1016/j.jenvman.2023.119639

Huang T et al (2024) Cultivar mixtures increase crop yields and temporal yield stability globally. A meta analysis. Agron Sustain Dev 44:28. https://doi.org/10.1007/s13593-024-00964-6

Hufnagel J, Reckling M, Ewert F (2020) Diverse approaches to crop diversification in agricultural research. A review. Agron Sustain Dev 40:14. https://doi.org/10.1007/s13593-020-00617-4

Hussein H, El Maghraby MMMS, Abu Salem HS (2024) Application of water quality index and statistical-hydrochemical techniques in groundwater assessment of the Quaternary aquifer, southwest Nile Delta of Egypt. Appl Water Sci 14:143. https://doi.org/10.1007/s13201-024-02189-0

Issac J, Newell R, Das R (2025) A systems thinking approach to examine local food systems planning through a climate-biodiversity-health lens: a Comox Valley case study. Environ Sci Policy 163:103972. https://doi.org/10.1016/j.envsci.2024.103972

Jackson-Smith D, Veisi H (2023) A typology to guide design and assessment of participatory farming research projects. Socio-Ecol Pract Res 5:159–174. https://doi.org/10.1007/s42532-023-00149-7

Jagustović R et al (2019) Contribution of systems thinking and complex adaptive system attributes to sustainable food production: example from a climate-smart village. Agric Syst 171:65–75. https://doi.org/10.1016/j.agsy.2018.12.008

Jia X, Li P, Li W (2025) The role of creative mindsets in the relationship between metacognitive experience and divergent thinking: a metacognitive perspective. Intelligence 13:27. https://doi.org/10.3390/jintelligence13030027

John DA, Babu GR (2021) Lessons from the aftermaths of green revolution on food system and health. Front Sustain Food Syst 5. https://doi.org/10.3389/fsufs.2021.644559

Jouttijärvi S, Lobaccaro G, Kamppinen A, Miettunen K (2022) Benefits of bifacial solar cells combined with low voltage power grids at high latitudes. Renew Sustain Energy Rev 161:112354. https://doi.org/10.1016/j.rser.2022.112354

Jug D et al (2019) Effect of conservation tillage on crop productivity and nitrogen use efficiency. Soil Tillage Res 194:104327. https://doi.org/10.1016/j.still.2019.104327

Kabir M et al (2023) Climate change due to increasing concentration of carbon dioxide and its impacts on environment in 21st century; a mini review. J King Saud Univ Sci 35:102693. https://doi.org/10.1016/j.jksus.2023.102693

Kallenbach CM, Wallenstein MD, Schipanksi ME, Grandy AS (2019) Managing agroecosystems for soil microbial carbon use efficiency: ecological unknowns, potential outcomes, and a path forward 10. https://doi.org/10.3389/fmicb.2019.01146

Kapoor P (2025) In: Dutta PK, Hamad A, Haghi AK, Prabhakar PK (eds) Food and industry 5.0: transforming the food system for a sustainable future. Springer Nature Switzerland, pp 305–315

Karunarathne A, Gunnell D, Konradsen F, Eddleston M (2020) How many premature deaths from pesticide suicide have occurred since the agricultural Green Revolution? Clin Toxicol 58:227–232. https://doi.org/10.1080/15563650.2019.1662433

Kataria S, Jajoo A, Guruprasad KN (2014) Impact of increasing Ultraviolet-B (UV-B) radiation on photosynthetic processes. J Photochem Photobiol, B 137:55–66. https://doi.org/10.1016/j.jphotobiol.2014.02.004

Kaur J et al (2023a) GC-MS validated phytochemical up-leveling with in vitro-raised Sansevieria trifasciata [Prain]: the mother in law's tongue gets more antibacterial. Curr Plant Biol 35–36:100308. https://doi.org/10.1016/j.cpb.2023.100308

Kaur J et al (2023b) Reactive black-5, Congo red and methyl orange: chemical degradation of azo-dyes by agrobacterium. Water 15:1664

Kaur J, Mudgal G, Dhar SK (2025) In: Latef AAHA, Zayed EM, Omar AA (eds) Sustainable remediation for pollution and climate resilience. Springer Nature Singapore, pp 695–724

Kaur S, Thakur H, Singh A, Ramasamy V, Mudgal G (2025) In: Prakash C, Kesari KK, Negi A (eds) Sustainable development goals towards environmental toxicity and green chemistry: environment and sustainability. Springer Nature Switzerland, pp 167–200

Khan KD, Guo X, Khan T, Saeed M (2025) Risk management strategies and climate change adoption policy: a global literature review using bibliometric analysis. Nat Hazards 121:10601–10622. https://doi.org/10.1007/s11069-025-07188-1

Kimmerer R (2011) In: Egan D, Hjerpe EE, Abrams J (eds) Human dimensions of ecological restoration: integrating science, nature, and culture. Island Press/Center for Resource Economics, pp 257–276

Kirina T, Groot A, Shilomboleni H, Ludwig F, Demissie T (2022) Scaling climate smart agriculture in East Africa: experiences and lessons 12:820
Kisliuk B et al (2023) AI in current and future agriculture: an introductory overview. KI—Künstliche Intelligenz 37:117–132. https://doi.org/10.1007/s13218-023-00826-5
Koyunoğlu C (2024) Biofuel production utilizing black soldier fly (Hermetia illucens): a sustainable approach for organic waste management. Int J Thermofluids 23:100754. https://doi.org/10.1016/j.ijft.2024.100754
Kubis A, Bar-Even A (2019) Synthetic biology approaches for improving photosynthesis. J Exp Bot 70:1425–1433. https://doi.org/10.1093/jxb/erz029
Kumar A, Antoniella G, Blasi E, Chiti T (2025a) Recent advances in regenerative sustainable agricultural strategies for managing soil carbon and mitigating climate change consequences. CATENA 258:109208. https://doi.org/10.1016/j.catena.2025.109208
Kumar R et al (2025b) Pioneering zero-waste technologies utilization and its framework on sustainable management: international, national and state level. Discov Appl Sci 7:224. https://doi.org/10.1007/s42452-025-06693-z
Kumar V et al (2025) In: Babu S et al (eds) Ecological solutions to agricultural land degradation. Springer Nature Singapore, pp 29–61
Kumari K, Mirzakhani Nafchi A, Mirzaee S, Abdalla A (2025) AI-driven future farming: achieving climate-smart and sustainable agriculture. AgriEngineering 7:89
Kusz D (2014) Modernization of agriculture vs sustainable agriculture. Sci Pap Ser Manage Econ Eng Agricult Rural Dev 14:171–178
Lacy W (2023) Local food systems, citizen and public science, empowered communities, and democracy: hopes deserving to live. Agric Hum Values 40:1–17. https://doi.org/10.1007/s10460-022-10398-z
Lal R (2023) Farming systems to return land for nature: it's all about soil health and re-carbonization of the terrestrial biosphere. Farming Syst 1:100002. https://doi.org/10.1016/j.farsys.2023.100002
Lalani B, Gray S, Mitra-Ganguli T (2023) Systems thinking in an era of climate change: does cognitive neuroscience hold the key to improving environmental decision making? a perspective on climate-smart agriculture 17. https://doi.org/10.3389/fnint.2023.1145744
Lam V, Rousselot Y (2024) Anthropocene, planetary boundaries and tipping points: interdisciplinarity and values in Earth system science. Eur J Philos Sci 14:18. https://doi.org/10.1007/s13194-024-00579-4
Himshikha et al (2024) In: Kumar S, Meena RS, Sheoran P, Jhariya MK (eds) Regenerative agriculture for sustainable food systems. Springer Nature Singapore, pp 135–174
Lawes R, Mata G, Richetti J, Fletcher A, Herrmann C (2022) Using remote sensing, process-based crop models, and machine learning to evaluate crop rotations across 20 million hectares in Western Australia. Agron Sustain Dev 42:120. https://doi.org/10.1007/s13593-022-00851-y
Lebrini Y, Ayerdi Gotor A (2024) Crops disease detection, from leaves to field: what we can expect from artificial intelligence. Agronomy 14:2719
Lee T, Ramp D, Bless A (2025) Unlocking digital twin planning for grazing industries with farmer centred design. Agric Hum Values. https://doi.org/10.1007/s10460-025-10752-x
Li J, Liu C, Zhang Z, Zhang N, Xu W (n/a) Decoding the multifunctionality of B-Box proteins: bridging light, stress and developmental networks in diverse plant species. https://doi.org/10.1111/pce.70061
Li Z, Qi D (2025) The response mechanism of soil microbial carbon use efficiency to land-use change: a review 17:7023
Liang X, Yu S, Ju Y, Wang Y, Yin D (2025) Integrated management practices foster soil health, productivity, and agroecosystem resilience. Agronomy 15:1816. https://doi.org/10.3390/agronomy15081816
Lidder P, Cattaneo A, Chaya M (2025) Innovation and technology for achieving resilient and inclusive rural transformation. Glob Food Sec 44:100827. https://doi.org/10.1016/j.gfs.2025.100827

Lin L, Xu F, Ge X, Li Y (2018) Improving the sustainability of organic waste management practices in the food-energy-water nexus: a comparative review of anaerobic digestion and composting. Renew Sustain Energy Rev 89:151–167. https://doi.org/10.1016/j.rser.2018.03.025
Liu Y, Ren X, Lu F (2022) Research status and trends of agrobiodiversity and traditional knowledge based on bibliometric analysis (1992–Mid-2022). Diversity 14:950
Lorenz K, Lal R (2023) In: Organic agriculture and climate change. Springer International Publishing, pp 39–127
Lucas S (2025) Against a pluriversal approach in global bioethics. J Bioethical Inq 22:463–465. https://doi.org/10.1007/s11673-025-10434-7
Luo S et al (2022) A cell-free self-replenishing CO2-fixing system. Nat Catal 5:154–162. https://doi.org/10.1038/s41929-022-00746-x
Lustenberger M, Spychiger F, Küng L, Bassi E, Wollenschläger S (2025) DAO research trends: reflections and learnings from the first European DAO workshop (DAWO). Appl Sci 15:3491. https://doi.org/10.3390/app15073491
Ma Y, Ma X, Gao X, Wu W, Zhou B (2021) Light induced regulation pathway of anthocyanin biosynthesis in plants 22:11116
MacPherson J, Rosman A, Helming K, Burkhard B (2025) A participatory impact assessment of digital agriculture: A Bayesian network-based case study in Germany. Agric Syst 224:104222. https://doi.org/10.1016/j.agsy.2024.104222
Maher K, von Blanckenburg F (2023) The circular nutrient economy of terrestrial ecosystems and the consequences for rock weathering 10. https://doi.org/10.3389/fenvs.2022.1066959
Maitra S et al (2021) Intercropping—a low input agricultural strategy for food and environmental security 11:343
Majahana JM, Kalumba AM, Zhou L, Mazinyo SP, Afuye GA (2025) Global trend analysis of climate change adaptation policy and governance in the water sector. Discov Sustain 6:241. https://doi.org/10.1007/s43621-025-01111-1
Malik AI, Bell R, Zang H, Boitt G, Whalley WR (2025) Exploring the plant and soil mechanisms by which crop rotations benefit farming systems. Plant Soil 507:1–9. https://doi.org/10.1007/s11104-024-06994-z
Manjula M, Rengalakshmi R (2021) Making research collaborations: learning from processes of transdisciplinary engagement in agricultural research 26:25–39. https://doi.org/10.1177/09722661211007589
Manono BO, Gichana Z (2025) Agriculture-livestock-forestry nexus: pathways to enhanced incomes, soil health, food security and climate change mitigation in Sub-Saharan Africa 6:74
Mansoor S et al (2025) Integration of smart sensors and IOT in precision agriculture: trends, challenges and future prospectives 6. https://doi.org/10.3389/fpls.2025.1587869
Marambe B, Jayawardena SSBDG, Weerakoon WMW, Wijewardena H (2020) In: Marambe B, Weerahewa J, Dandeniya WS (eds) Agricultural research for sustainable food systems in Sri Lanka: Volume 1: a historical perspective. Springer Singapore, pp 215–248
Marin A, Rousselot S, Jeavons E, Joannon A, Chable V (2025) Contrasting performances of modern and landrace cultivars in wheat mixtures and wheat-legume intercrops in organic and low input conditions. Euphytica 221:20. https://doi.org/10.1007/s10681-024-03456-6
Masha M, Bojago E, Ngare I (2024) Determinants of adoption of urban agriculture (UA) as climate-smart agriculture (CSA) practices and its impact on food security: evidence from Wolaita Sodo city, South Ethiopia. Discov Sustain 5:168. https://doi.org/10.1007/s43621-024-00365-5
Mattalia G et al (2025) Crop diversity trends captured by Indigenous and local knowledge: introduction to the symposium. Agric Hum Values. https://doi.org/10.1007/s10460-025-10751-y
McAlvay AC et al (2022) Cereal species mixtures: an ancient practice with potential for climate resilience. A review. Agron Sustain Dev 42:100. https://doi.org/10.1007/s13593-022-00832-1
Michalk A, Kunz J, Gaugler T (2025) In true cost accounting 23
Mishra V (2025) Rethinking urban future: towards sustainable cities and communities. Anthropocene Sci. https://doi.org/10.1007/s44177-025-00088-w

Mohamed Shaffril HA et al (2024) Diversification of agriculture practices as a response to climate change impacts among farmers in low-income countries: a systematic literature review. Clim Serv 35:100508. https://doi.org/10.1016/j.cliser.2024.100508

Moreira JB et al (2023) Role of microalgae in circular bioeconomy: from waste treatment to biofuel production. Clean Technol Environ Policy 25:427–437. https://doi.org/10.1007/s10098-021-02149-1

Moreira B et al (2024) Intercropping systems: an opportunity for environment conservation within nut production 14:1149

Morelli M, Casagrande M, Forte G (2022) Decision making: a theoretical review. Integr Psychol Behav Sci 56:609–629. https://doi.org/10.1007/s12124-021-09669-x

Mortensen DA, Smith RG (2020) Confronting barriers to cropping system diversification 4. https://doi.org/10.3389/fsufs.2020.564197

Mudgal G et al (2025) In: Latef AAHA, Zayed EM (2025) In: Latef AAHA, Zayed EM, Omar AA (eds) Sustainable remediation for pollution and climate resilience. Springer Nature Singapore, pp 139–206

Mudigere Sannegowda UB, Garkoti SC (2022) Traditional community-led seed system for maintaining crop vigour, diversity and socio-cultural network in view of the changing climate: a case study from western Himalaya, India. Clim Action 1:19. https://doi.org/10.1007/s44168-022-00020-7

Musvoto C et al (2015) Applying a transdisciplinary process to define a research agenda in a smallholder irrigated farming system in South Africa. Agric Syst 137:39–50. https://doi.org/10.1016/j.agsy.2015.03.008

Mutum L et al (2024) In: Workshop on envisioning future scaling of clean energy business models through the lens of nexus approach. International Water Management Institute, New Delhi, p 29

Mzungu I, Adekola O, Terdoo F, Anjembe CB (2024) In: Aransiola SA, Atta HI, Maddela NR (eds) Soil microbiome in green technology sustainability. Springer Nature Switzerland, pp 273–288

Narvaez D, Mrkva K (2014) In: Moran S, Cropley D, Kaufman JC (eds) The ethics of creativity. Palgrave Macmillan UK, pp 25–45

Naveed M et al (2024) In: Khamparia A, Pandey B, Pandey DK, Gupta D (eds) Microbial data intelligence and computational techniques for sustainable computing. Springer Nature Singapore, pp 235–258

Nenciu F et al (2022) Zero-waste food production system supporting the synergic interaction between aquaculture and horticulture 14:13396

Ngwenya ZD, Mohammed M, Dakora FD (2024) Monocropping and intercropping of maize with six food legumes at Malkerns in Eswatini: their effects on plant growth, grain yield and N2 fixation, measured using the 15N natural abundance and ureide techniques. Symbiosis 92:257–269. https://doi.org/10.1007/s13199-024-00971-x

Noel L-A et al (2023) Pluriversal futures for design education. She Ji: J des Econ Innov 9:179–196. https://doi.org/10.1016/j.sheji.2023.04.002

Nomura K et al (2023) Hybrid AI model for estimating the canopy photosynthesis of eggplants. Photosynth Res 155:77–92. https://doi.org/10.1007/s11120-022-00974-z

O'Brien L, Mudaliar P (2022) Decolonizing land management in institutions of higher education. J Land Use Sci 17:195–210. https://doi.org/10.1080/1747423X.2021.2011969

Ojha AK, Venkateswaran RT (2025) Indigenous Indian management: decolonizing management knowledge in India—promoting pluriversalism through indigenous knowledge paradigms. Decision 52:149–153. https://doi.org/10.1007/s40622-025-00441-3

Okolo CC, Gebresamuel G, Retta AN, Zenebe A, Haile M (2019) Advances in quantifying soil organic carbon under different land uses in Ethiopia: a review and synthesis. Bull Natl Res Centre 43:99. https://doi.org/10.1186/s42269-019-0120-z

Omer E et al (2024) Farming practice variability and its implications for soil health in agriculture: a review 14:2114

Omrany H, Mehdipour A, Oteng D, Al-Obaidi KM (2025) The uptake of urban digital twins in the built environment: a pathway to resilient and sustainable cities. Comput Urban Sci 5:20. https://doi.org/10.1007/s43762-025-00177-x

Onyszkiewicz M, Sylla M (2025) Current status of the community-supported agriculture model in Poland—exploring key areas of sustainable operations 17:2965

Osmanoglu Ö et al (2021) Topological analysis of the carbon-concentrating CETCH cycle and a photorespiratory bypass reveals boosted CO(2)-sequestration by plants. Front Bioeng Biotechnol 9:708417. https://doi.org/10.3389/fbioe.2021.708417

Otieno M (2024) Sustainable agroecological practices in Sub-Saharan Africa in the face of climate change. Springer Nature Switzerland, pp 15–19

Pace R et al (2025) Artificial intelligence in soil microbiome analysis: a potential application in predicting and enhancing soil health—a review. Discov Appl Sci 7:85. https://doi.org/10.1007/s42452-024-06381-4

Pal A 93–100. Springer Nature Singapore

Panda T et al (2023) Wild and cultivated plant collections from marketplaces of Odisha, India and their role in livelihood support. https://ethnobotanyjournal.org/index.php/era/article/view/5001

Pandian K et al (2024) Synergistic conservation approaches for nurturing soil, food security and human health towards sustainable development goals. J Hazard Mater Adv 16:100479. https://doi.org/10.1016/j.hazadv.2024.100479

Parashar M et al (2023) Two novel plant-growth-promoting Lelliottia amnigena isolates from euphorbia prostrata Aiton enhance the overall productivity of wheat and tomato. Plants 12:3081

Pardee GL et al (2023) Local and landscape factors influence plant-pollinator networks and bee foraging behavior across an urban corridor 12:362

Patel SK, Sharma A, Singh GS (2020) Traditional agricultural practices in India: an approach for environmental sustainability and food security. Energy, Ecol Environ 5:253–271. https://doi.org/10.1007/s40974-020-00158-2

Paul D, Bohacz J, Bhatia SK (2023) Editorial: biowaste valorization utilizing microbial systems 14. https://doi.org/10.3389/fmicb.2023.1213598

Paut R, Garreau L, Ollivier G, Sabatier R, Tchamitchian M (2024) A global dataset of experimental intercropping and agroforestry studies in horticulture. Sci Data 11:5. https://doi.org/10.1038/s41597-023-02831-7

Peddyreddy SR (2023) Industrial agriculture: complication to solution PhD thesis, Department of Sustainable Agriculture program at Selinus University Faculty of Life & Earth Science

Perino A et al (2022) Biodiversity post-2020: closing the gap between global targets and national-level implementation 15:e12848. https://doi.org/10.1111/conl.12848

Pienaah CKA, Baruah B, Kansanga M, Luginaah I (2024) The impact of community-led conservation models on women's nature-based livelihood outcomes in semi-arid Northern Ghana. Discov Environ 2:45. https://doi.org/10.1007/s44274-024-00073-x

Pingali PL (2012) Green revolution: impacts, limits, and the path ahead. Proc Natl Acad Sci 109:12302–12308. https://doi.org/10.1073/pnas.0912953109

Płaskonka B, Zych M, Mazurkiewicz M, Skłodowski M, Roguz K (2024) Pollinator-mediated connectivity in fragmented urban green spaces—tracking pollen grain movements in the city center. Acta Oecol 123:103985. https://doi.org/10.1016/j.actao.2024.103985

Plunkett BJ et al (2019) Apple B-box factors regulate light-responsive anthocyanin biosynthesis genes. Sci Rep 9:17762. https://doi.org/10.1038/s41598-019-54166-2

Prasad CS, Dutta D, Chakraborty A (2025) Academic institutions as innovation intermediaries; co-creating knowledge and building the Gujarat agroecology learning alliance (GALA). Environ Innov Soc Transit 57:101033. https://doi.org/10.1016/j.eist.2025.101033

Prasad M, Kumar S, Kumar A, Khan N, Shukla SK (2025) In: Acharya CL, Chaudhary RS, Gurumurthy P, Subba Rao A (eds) Blending Indian farmers' traditional knowledge in agriculture with modern scientific technologies: a way forward. Springer Nature Singapore, pp 353–374

Pu L, Li Y, Gao P, Zhang H, Hu J (2022) A photosynthetic rate prediction model using improved RBF neural network. Sci Rep 12:9563. https://doi.org/10.1038/s41598-022-12932-9

Qoseem IO et al (2024) Unlocking the potentials of digital twins for optimal healthcare delivery in Africa. Oxford Open Digit Health 2. https://doi.org/10.1093/oodh/oqae039

Raman R, Ray S, Das D, Nedungadi P (2025) Innovations and barriers in sustainable and green finance for advancing sustainable development goals. Front Environ Sci 12. https://doi.org/10.3389/fenvs.2024.1513204

Reichstein M et al (2019) Deep learning and process understanding for data-driven earth system science. Nature 566:195–204. https://doi.org/10.1038/s41586-019-0912-1

Rivera I, Díaz de León D, Pérez-Salazar MR (2024) Drivers of the food system based on food sovereignty domains: an integrative systematic literature review 8. https://doi.org/10.3389/fsufs.2024.1450321

Roeber VM, Bajaj I, Rohde M, Schmülling T, Cortleven A (2021) Light acts as a stressor and influences abiotic and biotic stress responses in plants 44:645–664. https://doi.org/10.1111/pce.13948

Sahu SK et al (2025) In: 2025 IEEE international students' conference on electrical, electronics and computer science (SCEECS), pp 1–6

Saikia N et al (2024) In: Babu S et al (ed) Agricultural diversification for sustainable food production. Springer Nature Singapore, pp 215–239

Sánchez MAS, Moreno-Calles AI, Hernández López JDJ, Casas A (2024) Collaborative, situated, and critical methodologies in transdisciplinary agroecologies for life sustainability. Discov Sustain 5:256. https://doi.org/10.1007/s43621-024-00479-w

Savarese M, Chamberlain K, Graffigna G (2020) Co-creating value in sustainable and alternative food networks: the case of community supported agriculture in New Zealand 12:1252

Schäfer L, Warner K, Kreft S (2019) In: Mechler R et al. (eds) Loss and damage from climate change: concepts, methods and policy options. Springer International Publishing, pp 317–341

Schneider T, Lan S, Stuart A, Teixeira J (2017) Earth system modeling 2.0: a blueprint for models that learn from observations and targeted high-resolution simulations. Geophys Res Lett 44:12:396–312, 417. https://doi.org/10.1002/2017GL076101

Seetharaman B et al (2025) In: Prakash C, Kesari KK, Negi A (eds) Sustainable development goals towards environmental toxicity and green chemistry: environment and sustainability. Springer Nature Switzerland, pp 291–313

Seif AK et al (2024) Identifying managed aquifer recharge and rain water harvesting sites and structures for storing non-conventional water using GIS-based multi-criteria decision analysis approach. Appl Water Sci 14:181. https://doi.org/10.1007/s13201-024-02246-8

Seitz VA et al (2024) Cover crop root exudates impact soil microbiome functional trajectories in agricultural soils. Microbiome 12:183. https://doi.org/10.1186/s40168-024-01886-x

Sen LTH et al (2025) The opportunities and barriers in developing interactive digital extension services for smallholder farmers as a pathway to sustainable agriculture: a systematic review. Sustainability 17:3007. https://doi.org/10.3390/su17073007

Sengupta N, Ghosh K (2024) Assessment of groundwater suitability using water quality index and health risk analysis in upper catchment area of Kangsabati river, India. Total Environ Adv 11:200114. https://doi.org/10.1016/j.teadva.2024.200114

Sharma A, Chand K, Singh GB, Mudgal G (2021) In: Kesari KK, Jha NK (eds) Free radical biology and environmental toxicity. Springer International Publishing, pp 305–325

Sharma P, Thilakarathna I, Fennell A (2024) Hyperspectral imaging and artificial intelligence enhance remote phenotyping of grapevine rootstock influence on whole vine photosynthesis 15. https://doi.org/10.3389/fpls.2024.1409821

Shemshad M, Synowiec A, Kopyra M, Benedek Z (2025) The community-driven ecosystem resilience and equity framework: a novel approach for social resilience in ecosystem services 17:3452

Sherpa TO (2023) Indigenous people's perception of indigenous agricultural knowledge for climate change adaptation in Khumbu, Nepal 4. https://doi.org/10.3389/fclim.2022.1067630

Singh GB, Dey S, Mudgal G (2025) In: Kaur S, Dhiman S, Tripathi M (eds) Microbial metabolomics: recent developments, challenges and future opportunities. Springer Nature Singapore, pp 419–439
Siregar RRA, Seminar KB, Wahjuni S, Santosa E (2022) Vertical farming perspectives in support of precision agriculture using artificial intelligence: a review. Computers 11:135. https://doi.org/10.3390/computers11090135
Sishodia RP, Ray RL, Singh SK (2020) Applications of remote sensing in precision agriculture: a review. Remote Sens 12:3136. https://doi.org/10.3390/rs12193136
Smoliak BV, Gelobter M, Haley JT (2022) Mapping potential surface contributions to reflected solar radiation. Environ Res Commun 4:065003. https://doi.org/10.1088/2515-7620/ac7a25
Snapp SS, DeDecker J, Davis AS (2019) Farmer Participatory Research Advances Sustainable Agriculture: Lessons from Michigan and Malawi 111:2681–2691. https://doi.org/10.2134/agronj2018.12.0769
Sowmya C, Anand M, Indu Rani C, Amuthaselvi G, Janaki P (2024) Recent developments and inventive approaches in vertical farming. Front Sustain Food Syst. https://doi.org/10.3389/fsufs.2024.1400787
Steckley M (2025) Community-based food sovereignty assessments (FSAs): a review. Food Secur 17:257–273. https://doi.org/10.1007/s12571-024-01500-w
Stefi AL, Vorgias KE (2025). Springer Berlin Heidelberg, pp 1–36
Suárez Valdés MÁ, Soriano Disla JM, Gambuzzi E, Castejón Martínez G (2024) Innovative circular biowaste valorisation—state of the art and guidance for cities and regions 16:8963
Sun G, Hallema D, Asbjornsen H (2017) Ecohydrological processes and ecosystem services in the anthropocene: a review. Ecol Process 6:35. https://doi.org/10.1186/s13717-017-0104-6
Swiderska K et al (2022) Indigenous peoples' food systems and biocultural heritage: addressing indigenous priorities using decolonial and interdisciplinary research approaches 14:11311
Tahvili Z et al (2021) Locating suitable sites for rainwater harvesting (RWH) in the central arid region of Iran. Sustain Water Resour Manage 7:10. https://doi.org/10.1007/s40899-021-00491-2
Tavares MC, Vale J, Costa A (2024) The role of decentralised managers in decision-making in a large industrial company. Adm Sci 14:202. https://doi.org/10.3390/admsci14090202
Tawfik A, Eraky M, Alhajeri NS, Osman AI, Rooney DW (2022) Cultivation of microalgae on liquid anaerobic digestate for depollution, biofuels and cosmetics: a review. Environ Chem Lett 20:3631–3656. https://doi.org/10.1007/s10311-022-01481-2
Tay M-J, Ng T-H, Lim Y-S (2024) Fostering sustainable agriculture: an exploration of localised food systems through community supported agriculture. Environ Sustain Ind 22:100385. https://doi.org/10.1016/j.indic.2024.100385
Thevarajah B et al (2024) In: Garg VK, Kataria N (eds) Bioeconomy for sustainability. Springer Nature Singapore, pp 125–162
Tittonell P et al (2022) Regenerative agriculture—agroecology without politics? 6. https://doi.org/10.3389/fsufs.2022.844261
Ţopa D-C, Căpşună S, Calistru A-E, Ailincăi C (2025) Sustainable practices for enhancing soil health and crop quality in modern agriculture: a review 15:998
Tripathi SK et al (2024) In: Tripathi SK, Brearley FQ (eds) Shifting cultivation systems: from indigenous practices to soil fertility in North–East India. Springer Nature Switzerland, pp 29–48
Trujillo Bilbao F (2024) Green revolution (not) for all: wheat and irrigation for the history of technology. Tapuya: Latin Am Sci Technol Soc 7:2313876. https://doi.org/10.1080/25729861.2024.2313876
Twigg-Flesner C (2012) A cross-border-only regulation for consumer transactions in the EU. Springer, New York, NY
UNNFCC (2024) Deutsche Gesellschaft für Internationale Zusammenarbeit (GIZ) GmbH
Van Den Meerssche D (2021) Regulatory integration across borders: public-private cooperation in transnational regulation. Eur J Int Law 31:1561–1567. https://doi.org/10.1093/ejil/chaa102%
Varghese R et al (2023) Machine learning in photosynthesis: prospects on sustainable crop development. Plant Sci 335:111795. https://doi.org/10.1016/j.plantsci.2023.111795

Vasconcelos MW et al (2020) In: Hasanuzzaman M, Araújo S, Gill SS (eds) The plant family fabaceae: biology and physiological responses to environmental stresses. Springer Singapore, pp 3–25
Veltmeyer H (2023) In: Bowles P, Andrews N (eds) Extractive bargains: natural resources and the state-society nexus. Springer International Publishing, pp 31–50
Verdouw C (2022) In: Zhang Q (ed) Encyclopedia of smart agriculture technologies. Springer International Publishing, pp 1–9
Verma S, Negi NP, Pareek S, Mudgal G, Kumar D (2022) Auxin response factors in plant adaptation to drought and salinity stress 174:e13714. https://doi.org/10.1111/ppl.13714
Waghaye AM et al (2023) Identification of suitable zones and sites for rainwater harvesting using GIS and multicriteria decision analysis. Environ Monit Assess 195:279. https://doi.org/10.1007/s10661-022-10801-6
Walters SA, Stoelzle Midden K (2018) Sustainability of urban agriculture: vegetable production on green roofs 8:168
Wang Y et al (2021) Planning considerations of green corridors for the improvement of biodiversity resilience in suburban areas. J Infrastruct Preserv Resilience 2:6. https://doi.org/10.1186/s43065-021-00023-4
Ward CR, Chadwick DR, Hill PW (2023) Potential to improve nitrogen use efficiency (NUE) by use of perennial mobile green manures. Nutr Cycl Agroecosyst 125:43–62. https://doi.org/10.1007/s10705-022-10253-x
Wassie SB (2020) Natural resource degradation tendencies in Ethiopia: a review. Environ Syst Res 9:33. https://doi.org/10.1186/s40068-020-00194-1
Watabe S, Hu Y, Maeda H, Yuan C (2025) Editorial: transdisciplinary communication for sustainable food systems. Front Sustain Food Syst 9. https://doi.org/10.3389/fsufs.2025.1563312
Whalen JK, Gul S (2025) Managing soil health for resilient crop production. Nat Food 6:535–536. https://doi.org/10.1038/s43016-025-01186-z
Widmer J, Christ B, Grenz J, Norgrove L (2024) Agrivoltaics, a promising new tool for electricity and food production: a systematic review. Renew Sustain Energy Rev 192:114277. https://doi.org/10.1016/j.rser.2023.114277
Wopereis TM, Dijkstra C, Wierda JJ, Rongen FC, Poelman MP (2024) Systems thinking for local food environments: a participatory approach identifying leverage points and actions for healthy and sustainable transformations. Health Res Policy Syst 22:101. https://doi.org/10.1186/s12961-024-01199-3
Wyckhuys KAG et al (2024) Restoring functional integrity of the global production ecosystem through biological control. J Environ Manage 370:122446. https://doi.org/10.1016/j.jenvman.2024.122446
Xin S et al (2024) Bibliometric analysis of research hotspots and frontiers in progress towards the sustainable development goals. Sustainability 16:2005. https://doi.org/10.3390/su16052005
Xu L, Liu X, Gatto A, Vasa L, Zhao X (2025a) Valuation of ecosystem services from forests in Chinese rural areas based on forest resource investment. Hum Soc Sci Commun 12:583. https://doi.org/10.1057/s41599-025-04674-6
Xu J et al (2025b) Soil health contributes to variations in crop production and nitrogen use efficiency. Nat Food 6:597–609. https://doi.org/10.1038/s43016-025-01155-6
Yadav V, Yadav N (2024) Beyond sustainability, toward resilience, and regeneration: an integrative framework for archetypes of regenerative innovation. Glob J Flex Syst Manag 25:849–879. https://doi.org/10.1007/s40171-024-00418-8
Yao S, Wu G, Zhang S (2025) How does science and technology finance affect the agricultural green development: an interpretation from the perspective of rural human capital and agricultural industrial agglomeration. Front Sustain Food Syst 9. https://doi.org/10.3389/fsufs.2025.1540645
Yildirim G, Alim MA, Rahman A (2022) Review of rainwater harvesting research by a bibliometric analysis 14:3200

Yin H, Wang L, Wu Y, Xi Z (2024) Roles of non-visible light and temperature in the regulation of anthocyanin synthesis in fruits and vegetables 5:1968–1983. https://doi.org/10.1002/fft2.426

You L et al (2023) Global mean nitrogen recovery efficiency in croplands can be enhanced by optimal nutrient, crop and soil management practices. Nat Commun 14:5747. https://doi.org/10.1038/s41467-023-41504-2

Yu X et al (2024) In: Fu B, Stafford-Smith M (eds) Dryland social-ecological systems in changing environments. Springer Nature Singapore, pp 69–107

Yu W et al (2025a) Soil Microbial carbon use efficiency in natural terrestrial ecosystems 14:348

Yu J et al (2025b) Soil microbial carbon use efficiency differs between mycorrhizal trees: insights from substrate stoichiometry and microbial networks. ISME Commun 5:ycae173. https://doi.org/10.1093/ismeco/ycae173

Yu K et al (n/a) Enhancing food production through modern agricultural technology. https://doi.org/10.1111/pce.15299

Yueh HP, Zheng YL (2019) Effectiveness of storytelling in agricultural marketing: scale development and model evaluation. Front Psychol 10. https://doi.org/10.3389/fpsyg.2019.00452

Zainol Abidin MA, Mahyuddin MN, Mohd Zainuri MAA (2021) Solar photovoltaic architecture and agronomic management in agrivoltaic system: a review 13:7846

Žák J, Kraus M, Machová P, Plachý J (2020) Smart green bridge—wildlife crossing bridges of new generation. IOP Conf Ser Mater Sci Eng 728:012010. https://doi.org/10.1088/1757-899X/728/1/012010

Zhai Y, Peng H, Neff MM, Pappu HR (2020) Emerging molecular links between plant photomorphogenesis and virus resistance 11https://doi.org/10.3389/fpls.2020.00920

Zhang J, Zha L (2022) In: Zhang Q (ed) Encyclopedia of smart agriculture technologies. Springer International Publishing, pp 1–16

Zhou H, Specht K, Kirby CK (2022) Consumers' and Stakeholders' acceptance of indoor agritecture in Shanghai (China). Sustainability 14:2771. https://doi.org/10.3390/su14052771

Zhou J et al (2025) Linking microbial community dynamics to rhizosphere carbon flow depend on arbuscular mycorrhizae and nitrogen fertilization. Biol Fertil Soils 61:791–804. https://doi.org/10.1007/s00374-025-01897-2

Zupancic N (2023) Systematic literature review: inter-reletedness of innovation, resilience and sustainability—major, emerging themes and future research directions. Circular Econ Sustain 3:1157–1185. https://doi.org/10.1007/s43615-022-00187-5

Sustainable Developments Towards 3D Printing Replacing Agricultural Wastes

Garima, Jaspreet Kaur, Sanjoy K. Dhar, Panjanathan Radha, Abhijeet Singh, and Gaurav Mudgal

Abstract The urgent pursuit of sustainable development in manufacturing has sparked innovative efforts to repurpose agricultural waste—long considered an environmental burden—into valuable resources, particularly through 3D printing technologies. As global industries grapple with climate change, material scarcity, and escalating pollution, conventional production models are increasingly unsustainable. Agricultural byproducts like rice husks, wheat straw, coconut shells, and other lignocellulosic residues present a promising avenue for eco-conscious material substitution in additive manufacturing. By leveraging these organic remnants, 3D printing offers a dual advantage: reducing environmental degradation while promoting circular economy principles. It is interesting for how the fusion of biowaste and precision manufacturing is opening new doors for decentralized material sourcing, breathing life into rural economies, and sparking innovation at the grassroots level. With recent breakthroughs, there's a noticeable shift toward embracing hybrid filaments, where biodegradable polymers like PLA are blended with finely ground agricultural residues. These materials aren't just strong and durable; they

Garima · G. Mudgal
University Institute of Biotechnology, Chandigarh University, Mohali, Punjab, India
e-mail: bhardwajgarima1508@gmail.com

G. Mudgal
e-mail: gauravmdgl@gmail.com

J. Kaur (✉)
Kanya Maha Vidyalaya, Jalandhar, Punjab, India
e-mail: kaurjaspreet2308@gmail.com

S. K. Dhar
Banophool Adibashi Green Heart College, Mirpur, Dhaka, Bangladesh
e-mail: sanjoydharctg@gmail.com

P. Radha
Bioprocess Principles Laboratory, Department of Biotechnology, Faculty of Engineering and Technology, SRM Institute of Science and Technology, Chennai, Tamil Nadu, India
e-mail: radhap@srmist.edu.in

A. Singh
School of Bio Sciences and Technology (SBST), VIT University, Vellore, Tamil Nadu, India
e-mail: abhijeet040202@gmail.com

K. K. Kesari et al. (eds.), *Smart Sustainable Materials Advancing Re-manufacturing and Technology*, World Sustainability Series, https://doi.org/10.1007/978-3-032-17907-4_3

also bring visual appeal and a lighter environmental footprint, making them ideal for sustainable, locally-driven fabrication. Open-source platforms and localized fabrication centers have further democratized experimentation, allowing innovators to reimagine waste as utility, artistry, and infrastructure. Looking forward, the integration of these sustainable feedstocks into mainstream production workflows is poised to disrupt supply chains and recalibrate how materials are valued, shifting focus from extractive to regenerative practices. However, challenges such as material consistency, regulatory frameworks, and long-term durability remain and will require robust interdisciplinary collaboration. This chapter explores the underlying environmental imperatives, technological readiness, and case studies showcasing successful implementation. It also addresses the ethical and infrastructural hurdles that must be overcome to scale this innovation meaningfully. Ultimately, it envisions a future where agrarian economies evolve into decentralized biofabrication hubs—transforming agricultural detritus into social, economic, and ecological assets through the fusion of sustainability and digital craftsmanship.

Keywords Biomanufacturing · Feedstock · Rheology · Valorization · Decentralization · Composites · Lifespan · Fabrication · Inclusivity · Nanotechnology

1 The Need for Sustainable Innovation in Agricultural Waste Management Through 3D Printing

The global manufacturing sector is at a crossroads. Traditional production models—largely linear and extractive—have led to alarming levels of resource depletion, greenhouse gas emissions, and waste accumulation. Municipal solid waste generation is projected to rise from 2.1 billion tonnes in 2023 to 3.8 billion tonnes by 2050, with associated costs potentially doubling to USD 640.3 billion annually (Bernardes et al. 2024). Manufacturing alone accounts for nearly 20% of global GHG emissions and consumes almost half of global energy resources (Grozier 2025). This unsustainable trajectory has prompted a paradigm shift toward circular economy models, where waste is reimagined as a resource (Singh et al. 2025). In this context, agricultural waste—a massive and underutilized byproduct of global food systems—offers a compelling opportunity for sustainable innovation, especially when paired with additive manufacturing technologies like 3D printing.

Agriculture, while essential for food security and economic stability, is also one of the largest contributors to biomass waste (Seetharaman et al. 2025; Dhar et al. 2024). The FAO estimates that 13% of food—equivalent to 931 million tonnes—is lost in the supply chain, while 1.05 billion tonnes are wasted at the consumer and retail levels (FAO n.d.). Moreover, the *Driven to Waste* report by WWF (WWF-UK, Driven to Waste 2021) reveals that 2.5 billion tonnes of food go uneaten globally each year, with 40% of all food grown never consumed.

Agricultural waste includes crop residues such as rice husks, wheat straw, corn stalks, animal manure, fruit peels, shells, and agro-industrial byproducts. According to Nath and coworkers (Nath et al. 2023), Asia alone produces over 5 billion tonnes of agricultural waste annually, with India contributing significantly due to its vast agrarian economy. These residues are often burned, dumped, or landfilled, leading to air pollution and smog (e.g., stubble burning in North India), soil degradation and loss of fertility, water contamination from pesticide runoff, and methane emissions from organic decomposition (Sharma et al. 2021; Vinayak et al. 2021). Despite being largely non-toxic, the volume and mismanagement of agrowaste pose serious environmental and public health risks (Nath et al. 2023; Kaur et al. 2025a).

The maladies associated with agricultural waste mismanagement are multifaceted. Agricultural waste contributes to 8–10% of global GHG emissions, particularly methane, a gas 28–36 times more potent than CO_2 over 100 years (FAO/IPCC). The open dumping and burning of moist agricultural waste not only degrade the environment but also invite disease-carrying pests like flies and mosquitoes, heightening the risk of vector-borne illnesses (Yadav et al. 2025). As this waste breaks down, it releases leachates and pesticide-laden runoff that seep into groundwater and surface water sources, posing serious threats to ecosystems and human health (Balu et al. 2024; Singh et al. 2024). What's more, valuable biomass is often squandered—lost to decay instead of being transformed into energy, materials, or nutrients. These pressing challenges highlight the need for innovative, scalable, and sustainable approaches to agrowaste management (Mudgal et al. 2025).

At the heart of these solutions lies the concept of waste valorization—reimagining waste as a resource and converting it into useful products like energy, chemicals, or materials. Technologies such as anaerobic digestion and pyrolysis are gaining ground, turning agrowaste into biogas, bio-oil, and biochar (Ufitikirezi et al. 2024). Agricultural residues are now being repurposed into high-value products like nanocellulose, biodegradable plastics, and biochar, with applications spanning packaging, cosmetics, and medicine (Liu et al. 2025; Jayaprakash and Kesari 2025; Kaur et al. 2023a; Parashar et al. 2023). Even citrus peels, seed extracts, and fruit pulps, once discarded, are being tapped for their rich bioactive compounds and antioxidants, making them valuable in pharmaceutical and nutraceutical sectors (Yadav et al. 2025). Biosorbents derived from coconut shells, corn cobs, and sugarcane bagasse are also proving effective in removing heavy metals and pollutants from wastewater (Yadav et al. 2025). These valorization pathways not only reduce environmental harm but also unlock new economic opportunities for rural communities and agro-industries (Harika et al. 2025).

In parallel, 3D printing—also known as additive manufacturing—is emerging as a game-changing technology in the agrowaste valorization landscape. Unlike conventional manufacturing, 3D printing enables layer-by-layer fabrication using tailored feedstocks, allowing for decentralized production, minimal material waste, and design flexibility suited to local needs. Recent studies have shown that agrowaste powders like rice husk, wheat straw, and sugarcane bagasse can be blended with biodegradable binders such as PLA or methyl cellulose to create printable pastes and filaments (Karsoun et al. 2024). These materials have already been used to fabricate furniture parts, packaging solutions, agricultural tools, and biodegradable containers and seed trays. The idea is even extending to food-grade applications—where agri-food processing waste is being transformed into edible cutlery, sustainable packaging, and nutrient-rich snacks (Yoha and Moses 2023a, 2023b).

This chapter delves into the dynamic intersection of sustainability, material science, and digital fabrication, with a focus on how 3D printing can reshape agricultural waste management. It covers material characterization—examining the physical, chemical, and rheological traits of agrowaste-based feedstocks suitable for 3D printing; technological readiness—assessing current techniques like FDM and LDM and their compatibility with bio-based materials; case studies—spotlighting successful implementations from India, Egypt, Europe, and Latin America; environmental and economic analysis—evaluating life cycle impacts, carbon footprint reduction, and cost-effectiveness; and policy and ethical considerations—addressing regulatory gaps, safety standards, and cultural acceptance of biofabricated products.

In response to the global sustainability crisis and the mounting burden of agricultural waste, this chapter presents 3D printing not merely as a tool but as a philosophy for circular growth. By integrating agrowaste valorization with additive manufacturing, we envision a future where rural communities become biofabrication hubs, waste streams are revalorized into economic assets, and production is localized, inclusive, and regenerative. The subsequent sections will delve into material science and feedstock development, technological innovations and printability challenges, global case studies and prototypes, environmental and economic impact assessments, policy frameworks and ethical dimensions, and a future roadmap for scaling and integration. Through this exploration, the chapter aims to contribute to the discourse on sustainable manufacturing and offer actionable insights for researchers, policymakers, and entrepreneurs.

2 Agricultural Waste as a Resource: Composition and Potential

2.1 Understanding Agricultural Waste as a Material Stream

Agricultural waste, often dismissed as a nuisance or discarded without second thought, is increasingly being recognized as a rich and diverse material stream with significant potential for sustainable applications. These residues—ranging from crop stalks, husks, shells, and peels to agro-industrial byproducts—are primarily lignocellulosic, composed of cellulose, hemicellulose, lignin, and minor constituents like proteins, waxes, and minerals (Mudgal and Kesari 2025; Thakur and Kesari 2025). Their abundance and renewability make them attractive candidates for material substitution in various sectors, including construction, packaging, and notably, additive manufacturing (Das and Kesari 2025). India alone generates over 400 million tonnes of agricultural waste annually, with paddy straw accounting for nearly 130 million tonnes, much of which is either burned or landfilled, contributing to severe air pollution and soil degradation (Biswas 2022). The valorization of these residues is not merely a matter of waste management—it is a strategic pivot toward circularity. With the right processing, agricultural waste can be turned into powders, pastes, or composite fillers that work beautifully in 3D printing—offering a biodegradable, low-carbon alternative to petroleum-based materials. But this transformation isn't just about mixing ingredients; it depends on a deep understanding of the waste's physicochemical traits. Factors like particle size, moisture content, thermal stability, and flow behavior all play a critical role in determining how well the material prints and how it performs in its final form.

2.2 *Types of Agricultural Residues Suitable for 3D Printing*

A wide range of agricultural residues have shown promising potential for use in additive manufacturing (Table 1). Materials like rice husk, wheat straw, sugarcane bagasse, coconut shells, corn cobs, and date palm fibers are among the most extensively studied. Typically, these residues are ground into fine powders and blended with biodegradable binders such as methyl cellulose or flour to create printable pastes or filaments. Figure 1 visually captures this transformation—tracing the journey of agrowaste like rice husks and wheat straw from raw form through grinding, sieving, binder integration, and extrusion modeling, all the way to biodegradable product design.

One compelling example comes from Karsoun et al. (2024), who demonstrated that wheat straw and sugarcane fillers, when processed to particle sizes below 0.053 mm and combined with methyl cellulose, achieved elastic moduli comparable to wood-based reference samples, highlighting their suitability for structural applications. Rice husk has drawn particular interest due to its high silica content and favorable thermal behavior. Morales et al. developed composite filaments using rice husk and recycled polypropylene, which helped reduce warping and enhance mechanical stability during fused filament fabrication (FFF) (Morales et al. 2021). Beyond structural applications, agrowaste continues to prove its versatility—coconut shell-derived activated carbon, for instance, has been successfully explored for water filtration systems (Krishna et al. 2025).

2.3 *Material Science Fundamentals and Printability Considerations*

The ability to successfully 3D print materials derived from agricultural waste depends on a delicate balance of mechanical strength, thermal behavior, and flow properties. Lignocellulosic residues naturally offer stiffness and biodegradability, making

Table 1 Comparative Attributes of Selected Agricultural Feedstocks for 3D Printing. The table highlights compositional traits, thermal and rheological properties, regional availability, and end-use potential. The table was prepared using a source in published literature (Liu et al. 2025; Karsoun et al. 2024; Yoha and Moses 2023a; Darekar et al. 2024; Soriano-Cuadrado et al. 2024; Silva et al. 2024; Pemas et al. 2024)

Feedstock	Cellulose Content (%)	Lignin Content (%)	Thermal Stability Range (°C)	Avg. Particle Size (μm)	Moisture Sensitivity	Rheological Behavior	Printability Score†	Regional Abundance	Typical Applications
Rice Husk	35–40	20–25	280–350	100–150	Low	Good gel elasticity	High	India, Southeast Asia	Filaments, packaging, biochar
Wheat Straw	30–35	15–20	260–330	80–120	Moderate	Moderate paste flow	Medium	North India, Europe	Furniture panels, seed trays, and insulation
Coconut Shell	25–30	30–35	300–360	150–200	Low	High-stiffness slurry	Medium	South India, Philippines	Activated carbon, molded components
Corn Cob	32–37	18–22	250–320	90–140	Moderate	Moderate elasticity	Medium–High	Americas, South Asia	Biomedical scaffolds, packaging
Sugarcane Bagasse	40–45	20–25	270–340	120–160	High	Viscous dispersion	Medium	Brazil, India	Construction panels, compostable containers

(continued)

Table 1 (continued)

Feedstock	Cellulose Content (%)	Lignin Content (%)	Thermal Stability Range (°C)	Avg. Particle Size (μm)	Moisture Sensitivity	Rheological Behavior	Printability Score†	Regional Abundance	Typical Applications
Tomato Stem Waste	28–32	~18	240–310	~100	Low	Stable filament melt	High	Mediterranean, Southern Europe	Bioplastic blends, antioxidant packaging
Coffee Husk	22–28	~20	230–310	100–130	Moderate	Variable rheology	Medium	Central America, India	Capsules, bio-enriched packaging

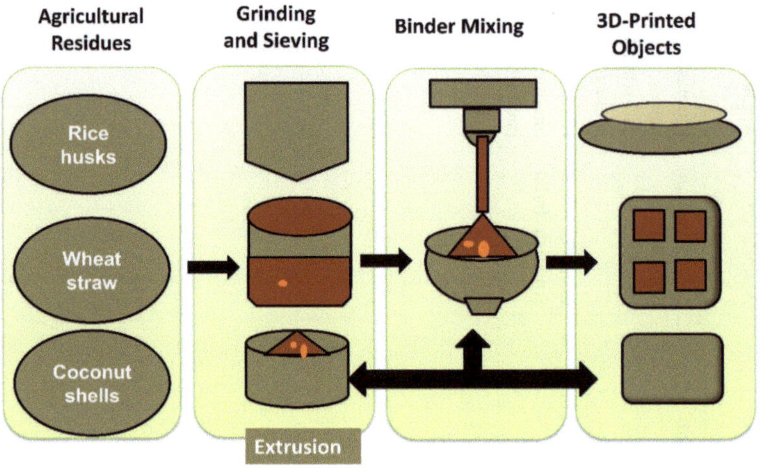

Fig. 1 Transformative Process of Agricultural Residues into 3D Printable Feedstocks

them attractive feedstocks—but their inconsistent composition can make it tricky to maintain smooth flow and strong layer bonding during printing (Rao and Kesari 2025). Rheological studies using Bohlin rheometers have shed light on this variability, showing that the phase angle and elastic modulus of agrowaste pastes shift significantly based on particle size and the type of binder used—factors that directly influence shape retention and extrusion stability (Karsoun et al. 2024). Thermal stability is equally important. For instance, thermogravimetric analysis of rice husk composites revealed early degradation linked to lignin content, underscoring the need for precise temperature control during the printing process (Morales et al. 2021). Moreover, the compatibility between agrowaste fillers and polymer matrices influences interfacial bonding and mechanical strength. Surface treatments, such as alkali or silane modification, have been proposed to enhance adhesion and reduce porosity in printed parts (Darekar et al. 2024; Darekar and Goyal 2025).

2.4 Valorization Pathways and Emerging Applications

The valorization of agricultural residues through 3D printing is not limited to structural components. Recent innovations have explored their use in food packaging, biodegradable containers, and even edible products. For example, the BARBARA project in the EU developed bioplastics from lemon rinds, almond husks, and corn by-products for automotive and construction applications, demonstrating high thermal resistance and mechanical strength (Commission n.d.; Machado et al. 2024). In India, rice husk and wheat straw have been used to fabricate seed trays, irrigation components, and low-cost housing panels, aligning with rural development goals and circular economy principles (Biswas 2022). The potential to localize production and reduce dependency on imported materials makes agrowaste-based 3D printing a strategic tool for sustainable development.

Agricultural waste, once relegated to the margins of industrial relevance, is now emerging as a cornerstone of sustainable material innovation. Its integration into 3D printing workflows offers a pathway to reduce environmental impact, create decentralized production models, and empower rural economies (Singh et al. 2025; Fotina et al. 2024). However, realizing this potential requires a nuanced understanding of material science, processing techniques, and application-specific requirements.

A visual representation of the valorization pipeline where agricultural waste such as rice husks, wheat straw, and coconut shells undergo processing stages, including grinding, sieving, binder blending, and extrusion modeling to become functional inputs for sustainable 3D printing. The flow illustrates how biomass transitions from raw residues to biodegradable products, emphasizing circular material logic and eco-innovation.

3 Evolution of 3D Printing with Natural Feedstocks

3.1 Historical Emergence of Biocomposites in Additive Manufacturing

The evolution of 3D printing with natural feedstocks is deeply intertwined with the broader movement toward sustainable materials science. Additive manufacturing (AM) technologies like stereolithography (SLA) and fused deposition modeling (FDM) were originally designed to work with synthetic polymers such as ABS and PETG, materials known for their strength and thermal resilience. But as concerns grew over the environmental toll of petroleum-based plastics, including their non-biodegradability, high carbon emissions, and toxic disposal—researchers began seeking greener alternatives. A major shift came in the early 2000s with the rise of polylactic acid (PLA), a biodegradable filament made from corn starch and sugarcane. Its ease of use and compostability quickly made it a favorite in desktop 3D printing (Kasahun Tsegaye et al. 2024). Still, pure bioplastics like PLA often fell short in industrial settings, especially where mechanical strength was critical. This led to the emergence of biocomposites—materials that combine a polymer matrix with natural fibers or fillers. Researchers started blending agricultural residues such as rice husks, wheat straw, and coconut shells into PLA, creating hybrid filaments that not only improved tensile strength and reduced warping but also enhanced biodegradability (Andanje et al. 2023; Sabana Ara et al. 2023).

3.2 Key Breakthroughs in Material Development

A major leap forward in this field came with the development of PLA composites reinforced with biofillers, materials that outperformed neat PLA in terms of mechanical strength. Researchers found that adding components like lignin, cellulose nanocrystals, and graphene nanoplatelets could significantly boost tensile strength, Young's modulus, and thermal stability, making these composites viable for demanding, load-bearing applications in sectors like automotive and aerospace (Vijayan et al. 2024). Another exciting milestone was the successful use of biosourced resins in photopolymer-based 3D printing, opening up new possibilities for sustainable fabrication in precision applications. Researchers at the University of Birmingham developed a recyclable resin from lipoic acid, enabling high-resolution printing and closed-loop recycling—a major advancement in sustainable stereolithography (Birmingham 2025). The exploration of all-cellulose 3D printing also gained momentum, with cellulose slurries and nanocolloids being used to fabricate biodegradable structures without synthetic binders. These innovations opened new possibilities for medical implants, packaging, and textiles, aligning with circular economy principles (Singh et al. 2025; Albelo et al. 2024; Singh and Kesari 2025).

3.3 Research Trajectory and Global Trends

The trajectory of research in biocomposites for additive manufacturing has accelerated dramatically over the past decade. A scientometric review revealed a 50.92% annual growth rate in publications related to polymer composite AM between 2013 and 2023, with India emerging as a leading contributor (Ashok et al. 2025). The focus has shifted from prototyping to functional end-use products, with increasing interest in multi-material printing, functionally graded materials, and AI-driven process optimization. Recent studies have emphasized the importance of fiber orientation, interfacial bonding, and void reduction in enhancing the performance of printed biocomposites.

Advanced techniques like curved layer deposition and voxel-level printing have brought a new level of precision to how materials are structured, allowing researchers to fine-tune anisotropic properties and enhance overall structural integrity (Yang et al. 2025). At the same time, the inclusion of natural fibers and minerals—such as basalt, hemp, and clay—into thermoplastic matrices has broadened the horizon of biocomposites well beyond traditional agricultural residues. These additions bring unique benefits like thermal insulation, flame resistance, and sound absorption, making them ideal for use in construction and everyday consumer products. The journey of 3D printing with natural feedstocks reflects a larger movement toward sustainable innovation in materials science. What began with early trials using PLA has evolved into a sophisticated landscape of biocomposite architectures. This progress has not only tackled the mechanical limitations of bio-based materials but also laid the foundation for decentralized, low-carbon manufacturing systems with real-world applications and impact.

4 Current Material Trends and Technological Readiness

4.1 Rise of Hybrid Filaments and Bio-Based Composites

Innovations like curved layer deposition and voxel-level printing have brought a new level of precision to additive manufacturing, allowing researchers to fine-tune material architecture and improve directional strength and overall structural integrity (Yang et al. 2025). At the same time, the integration of natural fibers and minerals, such as basalt, hemp, and clay, into thermoplastic matrices has expanded the possibilities of biocomposites well beyond agricultural residues. These materials offer distinct advantages, including thermal insulation, flame resistance, and sound absorption, making them ideal for use in construction and consumer products. The evolution of 3D printing with natural feedstocks reflects a broader shift toward sustainable innovation in materials science. From early experiments with PLA to the development of advanced biocomposite structures, the field has matured into a dynamic research domain with tangible real-world impact. These breakthroughs haven't just solved mechanical limitations—they've paved the way for decentralized, low-carbon manufacturing systems that align with both environmental goals and local production needs.

These fillers not only improve mechanical strength and thermal stability but also reduce the overall carbon footprint of the composite material (Soriano-Cuadrado et al. 2024; Vengadesan et al. 2025). Recent studies have demonstrated that incorporating up to 20% rice husk into PLA can enhance the elastic modulus beyond that of virgin PLA, despite heterogeneity in filler particle sizes. This suggests that agricultural residues, when properly processed, do not significantly compromise printability or structural integrity (Soriano-Cuadrado et al. 2024). Moreover, hybridization with biocarbon—produced via pyrolysis of agrowaste, has been shown to further improve tensile and flexural strength, water resistance, and thermal stability. For instance, PLA composites reinforced with 10% rice husk biocarbon achieved tensile strengths of 61.2 MPa and flexural strengths of 135.8 MPa, surpassing neat PLA by over 100% (Vengadesan et al. 2025).

4.2 Processing Techniques and Feedstock Preparation

The success of agrowaste-based 3D printing hinges on meticulous preparation and processing techniques (Fig. 2). Agricultural residues must be ground into fine powders—typically below 250 μm—and sieved to ensure uniformity. These powders are then blended with biodegradable binders such as methyl cellulose, flour, or starch to form printable pastes or filaments. Liquid Deposition Modeling (LDM) has proven particularly effective for such formulations, allowing extrusion of viscous pastes with high filler content (Karsoun et al. 2024). Rheological studies using Bohlin rheometers have revealed that the elastic modulus and phase angle of agrowaste pastes are comparable to wood-based references, indicating good shape retention and extrusion stability. For example, wheat straw and sugarcane fillers with particle sizes below 0.053 mm exhibited elastic moduli in the range of 20–40 kPa and phase angles below 45°, confirming their suitability for LDM printing (Karsoun et al. 2024). Additionally, chemical treatments such as alkali or silane modification have been employed to enhance interfacial bonding between fillers and polymer matrices, reducing porosity and improving mechanical performance (Karsoun et al. 2024).

4.3 Printability Challenges and Material Compatibility

While the results so far are encouraging, printing with agrowaste composites still comes with its share of challenges. One of the biggest hurdles is the natural variability in filler particles, their size, shape, and composition can differ widely. This heterogeneity often disrupts flow consistency, causes nozzle blockages, and compromises layer bonding during the printing process, making it difficult to achieve reliable and high-quality outputs. Scanning Electron Microscopy (SEM) analyses have shown significant variability in filler distribution and size, even when sieved through uniform mesh sizes (Soriano-Cuadrado et al. 2024). This affects not only the mechanical properties but also the aesthetic quality of printed parts. Another challenge is the compatibility between hydrophilic agrowaste fillers and hydrophobic polymer matrices. Poor interfacial bonding can result in stress concentration points and early material failure. While bonding agents and surface treatments can mitigate these issues, they may also alter other material properties, such as flexibility and biodegradability (Vengadesan et al. 2025). Moreover, high filler loading—beyond 30%—often leads to brittleness, reduced crystallinity, and diminished mechanical strength, necessitating careful optimization of composite formulations.

4.4 Applications and Performance Benchmarks

Agrowaste-based filaments and pastes have been successfully used to fabricate a wide range of products, including furniture components, packaging materials, seed trays, and even biomedical devices (Kaur et al. 2025b; UKHI 2025). For instance, soybean hulls incorporated into PLA/PBAT blends improved flexural strength by 35% and maintained impact resistance, making them suitable for rigid packaging and biomedical applications (Shafahat 2025). Similarly, coffee husk and mineral

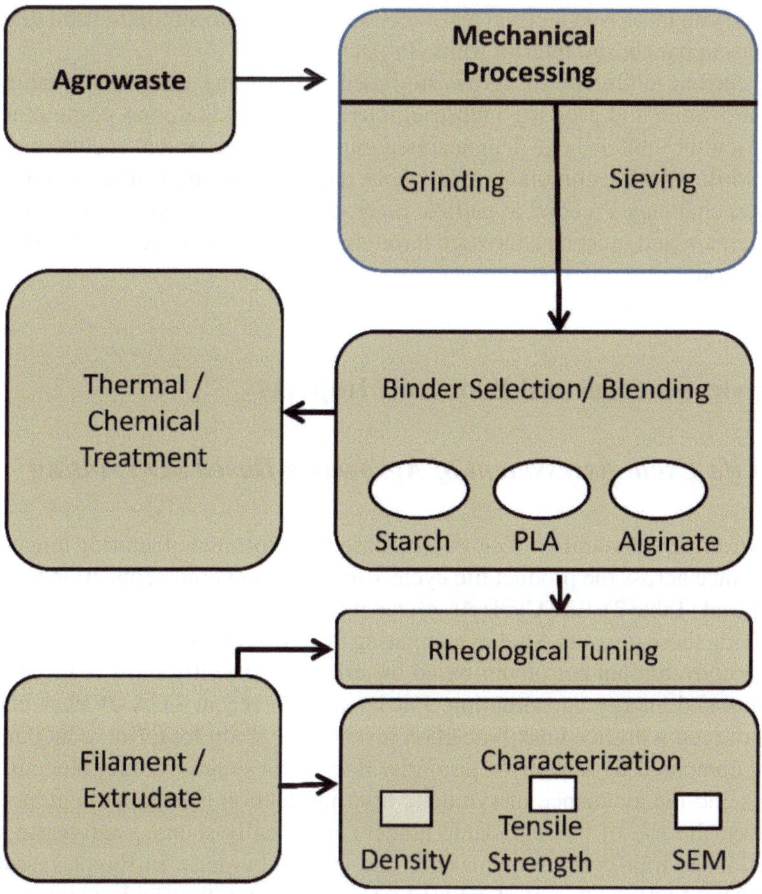

Fig. 2 Integrated Feedstock Preparation Workflow for Sustainable 3D Printing. A comprehensive schematic illustrating the transformation of agricultural residues into printable bio-composites. The transformation of agricultural waste into printable materials involves a series of carefully orchestrated steps. It starts with mechanical refinement—grinding and sieving the raw biomass to achieve uniform particle size. This is followed by biochemical and thermal conditioning of the lignocellulosic components to enhance their compatibility with printing processes. Strategic blending with biodegradable binders like PLA or starch helps tailor the material's flow properties for extrusion-based 3D printing. To ensure reliability and performance, quality checkpoints such as SEM imaging for surface texture and tensile testing for mechanical strength are built into the workflow, reinforcing the integrity of eco-friendly fabrication

fillers have been used to produce biodegradable capsules with enhanced degradation rates and soil-enriching properties (Silva et al. 2024).

The BARBARA project in Europe exemplifies the large-scale valorization of agri-food waste into bioplastics for the automotive and construction sectors. By extracting bio-additives from lemon, broccoli, and almond peels, researchers developed hybrid filaments with improved thermal resistance and mechanical strength, validating their use in fascia panels and interior trims (Project 2024).

The current landscape of agrowaste-based 3D printing materials is marked by rapid innovation and growing industrial interest. Hybrid filaments combining PLA with agricultural fillers have demonstrated impressive mechanical and thermal properties, while LDM techniques offer viable pathways for high-filler formulations. However, challenges related to particle heterogeneity, interfacial bonding, and scalability remain and must be addressed through interdisciplinary research and process optimization.

5 Environmental and Economic Impacts

5.1 Life Cycle Assessment of Agrowaste-Based 3D Printing

The environmental viability of agrowaste-based additive manufacturing hinges on its performance across the product life cycle—from raw material acquisition to end-of-life disposal (Table 2). Life Cycle Assessment (LCA) has emerged as a critical tool for quantifying these impacts. Studies comparing PLA-based composites with agrowaste fillers to conventional petroleum-based materials consistently show reduced carbon footprints and energy consumption. For instance, a recent LCA of PLA composites reinforced with rice husk biocarbon revealed a carbon footprint reduction of up to 40% compared to neat PLA, primarily due to the sequestration potential of the biomass and the avoidance of synthetic filler production (Vengadesan et al. 2025). Moreover, the use of biodegradable binders and locally sourced agrowaste significantly lowers transportation emissions and landfill burdens. In Egypt, researchers explored liquid deposition modeling (LDM) using wheat straw and sugarcane fillers, and found that the printed paste showed elastic modulus and phase angle values on par with wood-based materials, highlighting both its mechanical strength and environmental compatibility (Karsoun et al. 2024). These results reinforce the idea that agrowaste valorization isn't just technically feasible, it can also play a meaningful role in climate mitigation while advancing sustainable manufacturing practices.

5.2 Carbon Footprint and Emission Metrics

Quantitative studies on carbon emissions linked to agrowaste-based composites are showing encouraging results. One review highlights that geopolymer composites made from agro-industrial waste can cut CO_2 emissions by as much as 90% compared to traditional Portland cement materials (Alawi et al. 2023). This dramatic reduction stems from bypassing energy-intensive steps like calcination and clinker production, making these alternatives not just environmentally sound, but strategically aligned

Table 2 Comparative Environmental Impact Metrics of Agrowaste-Based vs Conventional 3D Printing Materials. The table is designed to visually reinforce sustainability comparisons between agrowaste-based composites and conventional materials

Metric	PLA + Agrowaste Composite	Virgin PLA	ABS (Petrochemical)	Reference Source
Carbon Footprint (kg CO_2e/kg)	0.8–1.2	~1.5	3.8–4.1	Vengadesan et al. (2025) Alexander et al. (2024)
Energy Consumption (MJ/kg)	45–55	~80	~110	Alawi et al. (2023)
Biodegradability (% after 180 days)	70–85	60–70	0	Silva et al. (2024) Soriano-Cuadrado et al. (2024)
Compostability	Yes (industrial and home composting)	Partially (industrial only)	No	Yoha and Moses (2023a)
Material Source	Agrowaste residues (renewable)	Corn/Sugarcane (renewable)	Petroleum derivatives (non-renewable)	Liu et al. (2025)
End-of-Life Outcome	Biodegrades or composts	Compostable under conditions	Persistent in landfills/oceans	Mohanta (2025)
Circular Economy Potential	High (local feedstock, closed loop)	Moderate (centralized production)	Low (linear value chain)	Barbara (n.d.)
Hazardous Byproducts (during process)	None	Minimal	VOCs and microplastics	Ufitikirezi et al. (2024)
Certifications Achievable	Compostability, Ecolabel, Bio-based	Biodegradable, Bio-based	Limited (RoHS only)	Prasad et al. (2024)

with low-carbon construction goals. Similarly, the use of rice husk ash and sugarcane bagasse in biocomposites has shown significant reductions in embodied energy and greenhouse gas emissions, especially when integrated into decentralized production models. In a comparative study of PLA-based 3D printing, accelerated UV testing contributed 333.87 kg CO_2e, highlighting the importance of optimizing post-processing steps to maintain environmental gains (Alexander et al. 2024). These metrics reinforce the need for holistic design strategies that consider not just material sourcing but also processing, usage, and disposal phases. These contextual points have been more comprehensively depicted in Table 2.

5.3 *Economic Feasibility and Cost Modeling*

From an economic standpoint, agrowaste-based additive manufacturing presents a compelling case for cost reduction and value creation. The raw materials—agricultural residues—are often freely available or low-cost, especially in agrarian economies. A study by Fraunhofer IPK emphasized that economic evaluation and feasibility analyses of additive manufacturing must account for material costs, processing efficiency, and scalability (Technology 2025). When agrowaste is used as filler, the cost of filament production can drop by up to 30–50%, depending on the binder and processing method. Furthermore, decentralized fabrication models—such as community-based fab labs—enable localized production, reducing logistics costs and empowering rural entrepreneurship. The Sustainable Agro-Waste Composite Market Report 2033 projects a CAGR of 12.8%, with market value expected to reach USD 9.1 billion by 2033, driven by demand for eco-friendly materials in construction, packaging, and consumer goods (Sharma n.d.).

5.4 *Circular Economy Integration and Resource Efficiency*

Agrowaste-based 3D printing aligns seamlessly with circular economy principles. By converting biomass residues into functional products, it closes material loops and reduces dependency on virgin resources. The *Waste and Biomass Valorization* journal highlights that bioplastics derived from agricultural waste—such as PLA, PHA, and starch-based polymers—offer biodegradability, resource efficiency, and reduced pollution, especially when integrated into circular production systems (Mohanta 2025). Additionally, the valorization of tomato stems, banana peels, and sugarcane bagasse into 3D-printable composites has demonstrated not only mechanical viability but also antioxidant activity and soil-enriching properties, suggesting multifunctional benefits beyond structural use (Pemas et al. 2024; Alawi et al. 2023). The environmental and economic impacts of agrowaste-based additive manufacturing are both profound and promising. Life cycle assessments confirm substantial reductions in carbon emissions and energy use, while cost modeling reveals significant savings and market potential. These benefits are amplified when integrated into circular economy frameworks, where waste becomes a resource and production becomes regenerative.

6 Decentralized Biofabrication and Rural Innovation

6.1 Rethinking Production: From Centralized Industry to Localized Fabrication

For decades, global production has relied on a centralized model, where raw materials are shipped to far-off industrial hubs for processing. But this approach is now facing growing criticism, as long supply chains drive up environmental costs, rural communities are left out of the economic loop, and mass production struggles to meet localized needs. In contrast, decentralized biofabrication, especially through additive manufacturing using agricultural waste—is emerging as a powerful alternative. By empowering local communities to transform biomass residues into useful products, this model promotes self-reliance, cuts down on transport-related emissions, and supports the principles of a circular economy. The idea of "making where you grow" is gaining momentum, particularly in agrarian regions where biomass is plentiful but often overlooked (Yoha and Moses 2023a; Karsoun et al. 2024).

6.2 Makerspaces and Fab Labs as Catalysts of Rural Innovation

Makerspaces and Fab Labs, collaborative, community-powered hubs equipped with digital fabrication tools, are redefining how innovation happens. These spaces empower individuals and local groups to prototype, build, and solve problems on their own terms, making decentralized manufacturing not just possible, but personal. These spaces democratize access to technology, allowing individuals to prototype, iterate, and produce locally relevant solutions. In rural contexts, they serve as incubators for agro-waste valorization, where farmers, artisans, and students collaborate to transform residues into functional products. The REFLOW project in Europe exemplifies how Fab Labs can catalyze circular urban ecosystems by reconfiguring material flows and empowering citizen-led production (Armstrong 2021). Similarly, in India, initiatives like UKHI and Craste are leveraging agricultural waste to produce biodegradable packaging and construction materials, creating green jobs and reducing stubble burning—a major contributor to air pollution in northern India (UKHI 2025; Insights 2023).

6.3 Empowering Communities Through Accessible Technology

Decentralized biofabrication is not merely a technological shift, it is a socio-economic movement. Empowering rural communities with affordable 3D printers and hands-on training in digital design can spark grassroots entrepreneurship and unlock new skill sets. In Egypt, for instance, researchers have used Liquid Deposition Modeling (LDM) with agrowaste pastes, specifically wheat straw and sugarcane fillers, to produce durable components whose mechanical properties rival those of wood-based materials (Karsoun et al. 2024). These kinds of innovations are especially meaningful in areas where industrial resources are scarce but biomass is abundant. What makes this approach even more powerful is the integration of open-source platforms and collaborative design networks, which enable rural makers to exchange ideas, solve problems together, and scale their solutions. This participatory model of innovation echoes the spirit of Fab Labs—spaces built on global knowledge-sharing and local problem-solving (Evaluating Fab Labs and Makerspaces n.d.).

6.4 Economic and Environmental Benefits of Localized Production

Localized biofabrication using agricultural waste is proving to be a win–win—both economically and environmentally. On the economic front, it helps reduce reliance on imported materials, cuts down production costs, and opens up new income opportunities for farmers and small-scale entrepreneurs. Environmentally, it tackles the damaging effects of open dumping and stubble burning, which are major contributors to air and soil pollution (Kaur et al. 2023b). By converting agricultural residues into useful products like bioplastics, packaging, and construction materials, this decentralized approach not only keeps waste out of landfills but also slashes the greenhouse gas emissions tied to conventional manufacturing. A compelling example is the BARBARA project in the EU, which showed that biopolymer-based car parts and building components made from lemon rinds and almond husks could reduce CO_2 emissions by 20% and production costs by 40% (Barbara n.d.).

6.5 Challenges and the Path Forward

Despite its enormous potential, decentralized biofabrication still faces a number of hurdles. Access to high-quality equipment remains limited in many regions, and the natural variability of agrowaste feedstocks can lead to inconsistent material performance. Policy support is also uneven, making it difficult to establish safety standards, scale operations, and gain market trust. Cultural acceptance and active community involvement are equally vital—without them, rural innovation hubs may struggle to take root. To bridge these gaps, we need targeted interventions: capacity-building programs to develop local expertise, public–private partnerships to share resources and risks, and subsidies that make equipment and training more accessible. Encouragingly, India's emerging biomanufacturing framework—aimed at building a $100 billion bioeconomy by 2024, could offer the momentum needed to scale decentralized fabrication models (Insights 2023; Kaur et al. 2022). At its core, decentralized biofabrication marks a fundamental shift in how we produce, consume, and value materials. By turning agricultural waste into opportunity and equipping communities with the tools to innovate locally, it paves the way for economies that are inclusive, regenerative, and resilient. Makerspaces and Fab Labs are the beating heart of this movement—spaces where global knowledge meets local ingenuity, and where sustainable innovation is born.

7 Case Studies—Global Implementation and Success Stories

7.1 Egypt: Liquid Deposition Modeling with Sugarcane and Wheat Straw

A standout example of agrowaste-driven additive manufacturing comes from Egypt, where researchers have successfully harnessed wheat straw and sugarcane residues in Liquid Deposition Modeling (LDM). Their method involved grinding the agricultural waste into fine powders, sieving it for uniformity, and blending it with biodegradable binders like methyl cellulose and flour. Rheological tests using Bohlin rheometers revealed that the resulting pastes had elastic moduli and phase angles comparable to wood-based materials—signaling excellent shape retention and printability. The final printed components proved both mechanically robust and biodegradable, affirming the potential of Egyptian biomass as a sustainable feedstock for eco-friendly fabrication (Karsoun et al. 2024).

7.2 Greece: Tomato Stem Waste in Fused Filament Fabrication

In Greece, researchers found a novel use for non-edible tomato stem waste by blending it into PLA-based filaments for Fused Filament Fabrication (FFF). This approach not only gave agricultural residues a second life but also advanced the development of sustainable, bio-enhanced materials for 3D printing. Composites with 5–10 wt.% tomato stem powder were developed and tested for mechanical and thermal properties. The 5% formulation maintained consistent tensile strength, while dynamic mechanical analysis revealed changes in storage modulus and damping factor. Notably, the composites exhibited antioxidant activity, which increased with higher filler content. This dual functionality—mechanical performance and bioactivity—underscores the potential of agrowaste in multifunctional applications (Pemas et al. 2024).

7.3 India: Rice Husk and Coffee Husk in Biodegradable Packaging

India has seen a surge in agrowaste valorization startups, with companies like Craste and UKHI pioneering the use of rice husk, wheat straw, and coffee husk in biodegradable packaging and construction materials. These enterprises have developed hybrid filaments and molded products that replace single-use plastics and reduce stubble burning. The materials are not only compostable but also customizable for various applications, including seed trays, food containers, and eco-friendly building panels. Their success has attracted attention from policymakers and investors, positioning India as a leader in agrowaste-based biofabrication (UKHI 2025; Insights 2023) (Fig. 3).

7.4 European Union: The BARBARA Project

The BARBARA project, funded by the EU, exemplifies large-scale agrowaste valorization for industrial applications. In a remarkable demonstration of circular innovation, researchers extracted polysaccharides from corn and bioactive compounds from lemon, broccoli, pomegranate, and almond peels to create bioplastics suitable for Fused Filament Fabrication (FFF). These food waste-derived materials were then used to fabricate automotive fascia panels and construction molds—applications that demand both high thermal resistance and mechanical strength. The success of this project not only validated the technical viability of biopolymers sourced from food waste, but also showcased the scalability of agrowaste-based additive manufacturing across Europe (Project 2024).

7.5 United States: Buckwheat Husk and Corn Cob Composites

In the United States, researchers have tapped into the potential of buckwheat husk and corn cob residues by incorporating them into PLA-based composites for Fused Deposition Modeling (FDM). These agrowaste-infused materials were rigorously

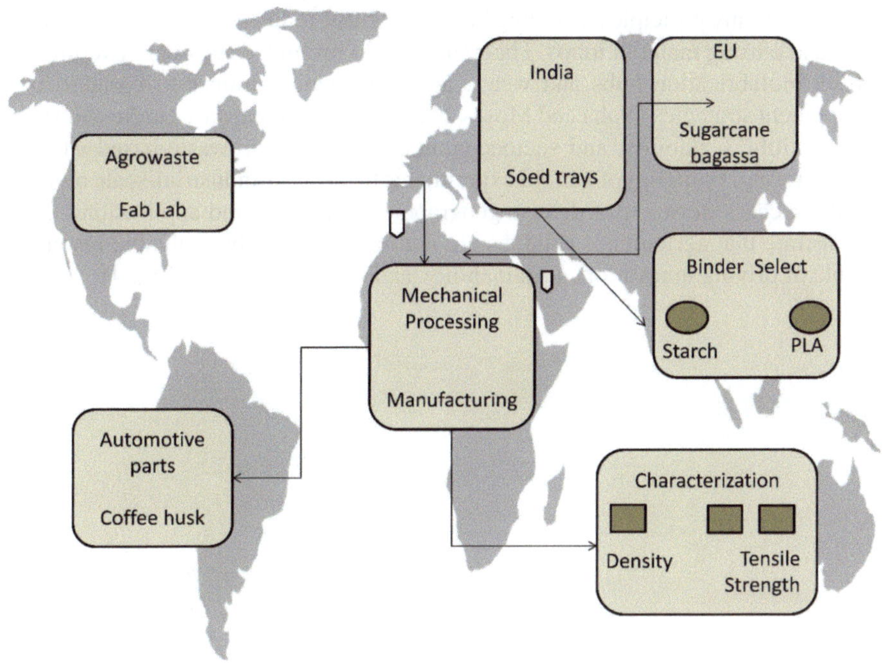

Fig. 3 Global Footprint of Agrowaste-Based 3D Printing Innovations. A geospatial infographic mapping real-world implementations of sustainable additive manufacturing using agricultural residues. Case highlights include Liquid Deposition Modeling (LDM) with wheat and sugarcane in Egypt, PLA composites with tomato stems in Greece, biodegradable packaging from rice and coffee husks in India, and automotive parts from lemon peels under the EU's BARBARA project. This figure underscores how diverse biomass sources across continents are being transformed into eco-functional products, reflecting localized ingenuity within a shared circular economy vision

tested for printability, mechanical strength, and biodegradability. Impressively, up to 20% of the PLA could be replaced with these natural fillers without compromising performance. Taking it a step further, hemicellulosic polymers extracted from corn cobs were successfully used to print biomedical scaffolds—highlighting the untapped promise of biopolymers in healthcare innovation (Rahman et al. 2023).

7.6 South America: Vine Shoots and Kenaf Fibers in Circular Economy Models

In South America, especially in Argentina and Brazil, researchers have been turning agricultural residues like vine shoots and kenaf fibers into PLA-based filaments for Fused Deposition Modeling (FDM). These bio-composites have been successfully used to create furniture components and packaging materials, showcasing both functionality and sustainability. What sets these projects apart is their commitment to circular economy principles: sourcing biomass locally, reducing transportation emissions, and closing material loops. Their success has prompted regional governments to back biofabrication hubs and weave agrowaste valorization into broader rural development strategies (Yoha and Moses 2023a). These global case studies illustrate the versatility, scalability, and socioeconomic impact of agrowaste-based additive manufacturing. From Egypt's LDM experiments to the EU's industrial-scale bioplastics, the success stories span diverse geographies, materials, and applications. They demonstrate that agricultural residues are not merely waste but valuable resources capable of driving innovation, sustainability, and inclusive growth (Fig. 4).

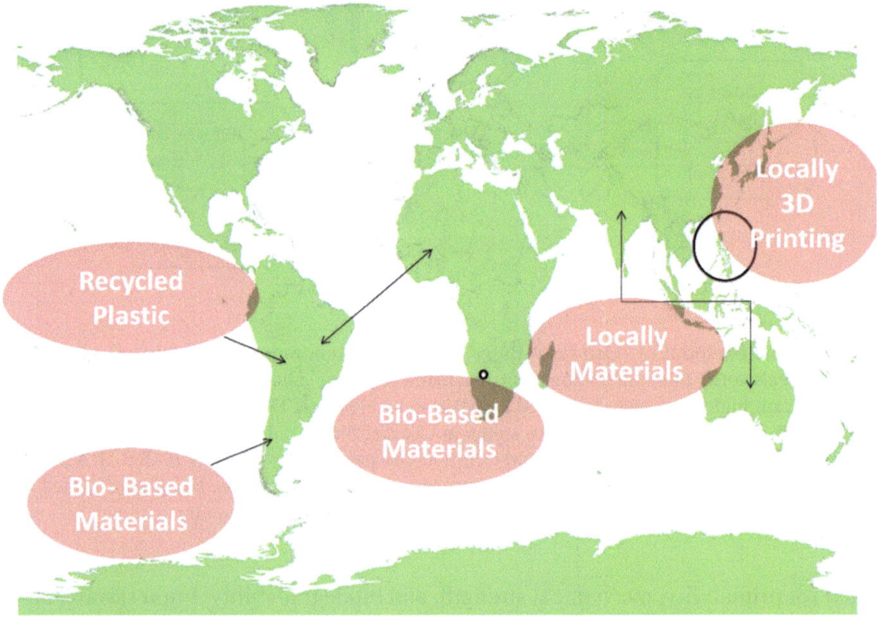

Fig. 4 Breakdown of regions where sustainable 3D printing is making significant waves. This representation was created using data from sources like Wohlers Report 2025 (Wohlers Report: https://www.astm.org/standards-and-solutions/market-insights-and-innovation/wohlers-report; https://wohlersassociates.com/; https://wohlersassociates.com/wr25-executive-summary/), Aranca (https://www.aranca.com/knowledge-library/articles; https://www.aranca.com/knowledge-library/articles/business-research/megatrends-in-3d-printing---sustainable-materials), Raise3D (https://www.raise3d.com/go-green-initiative/; https://www.raise3d.com/news/raise3d-announces-its-go-green-initiative/), World Economic Forum (https://www.weforum.org/stories/2023/06/how-3d-printing-helps-accelerate-sustainable-manufacturing/), Markforged Blog (https://markforged.com/resources/blog/3d-printing-and-the-environmental-impact-of-manufacturing), Nexa3D Blog (https://nexa3d.com/blog/sustainable-manufacturing/), Weerg (https://www.weerg.com/guides/3d-printing-sustainability) and Metal AM Magazine: https://www.metal-am.com/)

8 Challenges and Limitations

8.1 *Technical Constraints in Material Consistency and Printability*

One of the foremost challenges in agrowaste-based additive manufacturing lies in the inherent variability of agricultural residues. Unlike synthetic polymers, agrowaste feedstocks such as rice husk, wheat straw, or coconut shells exhibit significant heterogeneity in particle size, moisture content, lignin-cellulose ratios, and mineral composition. This inconsistency affects rheological behavior, extrusion stability, and layer adhesion during printing. SEM analyses of rice husk composites have revealed uneven filler distribution and porosity, leading to compromised mechanical strength and surface finish (Karsoun et al. 2024). High filler loading—beyond 30%—often results in brittleness and reduced crystallinity, necessitating precise formulation and processing control (Soriano-Cuadrado et al. 2024). Thermal degradation is another technical hurdle. Lignocellulosic residues tend to degrade at lower temperatures than conventional polymers, limiting their compatibility with high-temperature printing techniques such as SLS or FFF. Thermogravimetric studies have shown early onset of decomposition in wheat straw composites, requiring careful thermal management to prevent premature degradation during fabrication (Darekar et al. 2024).

8.2 *Regulatory Gaps and Policy Ambiguity*

Despite growing interest in agrowaste valorization, regulatory frameworks remain fragmented and underdeveloped. In the European Union, the lack of clear guidelines for achieving "end-of-waste" status has hindered the commercialization of bio-based materials derived from agricultural residues (Mohanta 2025). Without standardized certification pathways, many agrowaste-derived composites are still classified as waste, reducing investor and industry confidence. In India, policy efforts such as the National Biomanufacturing Strategy aim to promote innovation in sustainable materials, yet there remains no unified regulatory framework specific to additive manufacturing using agrowaste. Safety standards, material protocols, and digital IP governance for decentralized fabrication remain vague and evolving (Insights 2023).

8.3 *Post-Processing and Surface Quality Limitations*

Agrowaste-based printed components frequently require post-processing to meet industrial standards. Issues such as surface roughness, porosity, and inconsistent layer bonding necessitate sanding, coating, or chemical treatment, adding time, cost, and complexity. PLA composites reinforced with rice husk and coffee husk exhibit good biodegradability but comparatively rough surface profiles unless meticulously post-treated (Silva et al. 2024). These post-processing demands can offset the ecological benefits of low-energy fabrication if not streamlined. Furthermore, rural makerspaces and decentralized labs often rely on manual finishing processes, which limit consistency and scalability compared to automated post-processing in industrial setups (Insights 2023).

8.4 Material Availability and Supply Chain Challenges

Although agricultural waste is abundant, its availability as a consistent and reliable feedstock is hampered by seasonality, geographic concentration, and preprocessing gaps. Rice husks and wheat straw, for example, are harvested cyclically and are often burned or disposed of before valorization, due to a lack of on-site grinding or drying infrastructure (Biswas 2022). Without decentralized preprocessing units, communities remain dependent on centralized facilities that increase transport emissions and logistics costs, undermining sustainability goals. Microbial degradation of stored residues, particularly under humid conditions, further limits shelf life and quality, demanding airtight and climate-controlled facilities, which may be inaccessible in rural settings (Karsoun et al. 2024).

8.5 Cultural Acceptance and Market Perception

Public perception continues to be a significant barrier to the adoption of agrowaste-derived products. In many regions, consumers associate "waste-based" materials with inferiority, impermanence, or sanitation concerns, especially in applications such as packaging, furniture, or biomedical goods. This cultural resistance can stifle market growth even when technical performance is verified (Pemas et al. 2024). Design innovation, storytelling, and sustainability certifications can help reposition biofabricated goods as aspirational and responsible choices. However, these strategies demand coordinated investment in marketing, policy, and education—elements still nascent across most agrowaste valorization initiatives (Commission n.d.). The landscape of agrowaste-based additive manufacturing is fraught with challenges, ranging from technical inconsistencies and supply chain inefficiencies to policy voids and cultural hesitations. Yet, these limitations also represent areas for targeted intervention and innovation. Interdisciplinary research, scalable equipment, policy reform, and community sensitization are vital to unlock the full potential of sustainable biofabrication.

9 Ethical, Societal, and Cultural Dimensions

9.1 *Ethical Considerations in Agrowaste Valorization*

The ethical landscape of agrowaste-based additive manufacturing is shaped by deeper questions of resource justice, environmental responsibility, and fair access to technology. Turning agricultural residues into functional products can be transformative—but it also raises critical concerns about who owns the raw materials, especially when they come from smallholder farms or indigenous communities. Without clear benefit-sharing frameworks, there's a real risk of extractive innovation—where local biomass is commercialized without returning value to the people who cultivate it (Liu et al. 2025). At the same time, the drive to reduce environmental harm must be weighed against the unintended consequences of scaling biofabrication. Bioplastics made from agrowaste may be biodegradable, but their production can still involve chemical treatments or energy-intensive steps that deserve closer scrutiny. That's why transparency in life cycle assessments and responsible sourcing protocols is essential—not just to back up sustainability claims, but to ensure they reflect real, measurable impact (Mohanta 2025).

9.2 *Societal Impact and Labor Redistribution*

Agrowaste-based additive manufacturing can reshape labor dynamics across the agricultural and manufacturing sectors. By decentralizing production and enabling rural biofabrication, it can create new employment opportunities in preprocessing, design, and fabrication. However, this shift may also displace traditional roles in centralized industries, necessitating reskilling and inclusive workforce development strategies. A stakeholder-driven framework for assessing social impacts of additive manufacturing identified over 40 distinct effects, ranging from job creation and skill enhancement to occupational hazards and digital exclusion (Huang et al. 2013; Naghshineh et al. 2021). In rural contexts, the empowerment of farmers and artisans through localized fabrication can foster economic resilience and reduce dependency on urban supply chains. However, equitable access to technology, training, and market linkages remains a challenge. Without deliberate inclusion strategies, there is a risk that biofabrication benefits will concentrate among tech-savvy or capital-rich actors, exacerbating existing inequalities.

9.3 *Cultural Acceptance and Design Sensitivity*

Cultural acceptance of agrowaste-derived products is a nuanced issue, shaped by perceptions of cleanliness, utility, and aesthetic value. In many cultures, the label "waste" carries a deep-rooted stigma—even when that waste is transformed into high-performance, sustainable products. Breaking through this perception requires more than just technical validation; it calls for design innovation and compelling storytelling that reframe waste as a valuable resource. Take bioplastics made from banana fibers, citrus peels, or pine needles, for example. In India and Europe, these materials have been successfully marketed by highlighting their natural origins, biodegradability, and alignment with cultural values like frugality and environmental stewardship (Mohanta 2025). Design sensitivity is equally crucial. For biofabricated products to gain traction, especially in tribal and agrarian communities, they must resonate with local aesthetics and functional expectations. Materials that echo traditional textures or familiar forms are often embraced more readily than those that feel synthetic or foreign. Participatory design—where users are actively involved in co-creating products—can deepen cultural relevance and foster a sense of ownership, making innovation not just inclusive but truly rooted in place.

9.4 Responsible Innovation and Governance

The governance of agrowaste valorization must evolve to meet ethical and societal challenges head-on. This means setting clear standards for material safety, biodegradability, and environmental impact; developing certification systems for biofabricated goods; and ensuring that intellectual property rights don't stifle open innovation. Responsible innovation frameworks call for anticipatory regulation, inclusive stakeholder engagement, and reflexivity in technology development—principles that are especially vital in emerging fields like agrowaste-based 3D printing (Prasad et al. 2024). To institutionalize ethical practices and scale inclusive models, we need a mix of public–private partnerships, community-led cooperatives, and targeted policy incentives. India's Biomanufacturing Framework and the EU's Circular Economy Action Plan offer compelling templates for integrating sustainability, equity, and innovation into material transitions. Ultimately, the ethical, societal, and cultural dimensions of agrowaste-based additive manufacturing aren't side notes—they're central to its credibility and long-term impact. Addressing these dimensions requires a holistic approach that blends technical rigor with social empathy, cultural intelligence, and governance foresight. By embedding values of justice, inclusivity, and respect into biofabrication systems, we can ensure that the journey from waste to wonder is not only sustainable—but also equitable.

10 Future Roadmap—Toward a Circular Bio-Economy

10.1 Reimagining Agrowaste Through Circularity

The future of agrowaste-based 3D printing is deeply intertwined with the rise of the circular bio-economy, a regenerative system where biological resources are continuously reused, recycled, and valorized to minimize waste and environmental harm. Unlike the traditional linear model of production, which follows a take-make-dispose trajectory, circular bio-economy frameworks seek to decouple economic growth from resource depletion by embedding sustainability into every stage of the value chain (Mohanta 2025; Pilafidis et al. 2022). When agricultural residues are transformed into bioplastics, construction materials, or packaging through additive manufacturing, they become powerful symbols of this paradigm shift, turning waste into wealth, shrinking carbon footprints, and strengthening local resilience. This isn't just a technical evolution; it's a reimagining of how we relate to materials, production, and the planet itself.

10.2 Strategic Pathways for Scaling Biofabrication

Mainstreaming agrowaste-based 3D printing calls for strategic pathways that bridge technological innovation with policy alignment, infrastructure support, and market readiness. A pivotal strategy is the creation of regional biofabrication hubs—decentralized production centers embedded in agrarian landscapes. These hubs can localize material sourcing, reduce logistical burdens, and empower rural economies. Studies by CSIRO highlight how such regional models enhance sustainability and resilience by tapping into local biomanufacturing potential while aligning with broader circular economy goals (CSIRO). Equally vital is the integration of advanced technologies to refine biofabrication workflows. The future of agrowaste-based manufacturing will be shaped by AI-driven design optimization, rheological modeling, and machine learning—tools that minimize trial-and-error, improve material consistency, and enable application-specific customization. ETH Zurich researchers advocate for a computationally guided approach that reduces resource consumption and environmental impact, offering a scalable and scientifically grounded model for sustainable production (ETH Zurich). Complementing these technological and infrastructural advances are circular design principles. These emphasize designing with the full product life cycle in mind, modular components for easy disassembly, biodegradable or compostable materials, and multifunctional designs that meet both ecological and social needs. Circularity in design doesn't just reduce waste, it redefines value by embedding utility, reusability, and environmental responsibility into the very DNA of manufactured goods.

10.3 Policy and Governance Frameworks

Robust policy frameworks are the backbone of any meaningful transition toward sustainable biofabrication. India's BioE3 Policy—launched by the Department of Biotechnology—offers a visionary roadmap for regenerative biomanufacturing. It places innovation, employment generation, and environmental stewardship at the heart of its agenda, signaling a national commitment to building a future-ready bioeconomy that is both inclusive and ecologically responsible. It proposes the creation of bio-AI hubs, biofoundries, and biomanufacturing clusters to scale sustainable technologies and foster inclusive growth. Internationally, the EU Circular Economy Action Plan and Quad-led biomanufacturing initiatives aim to harmonize standards, incentivize green innovation, and promote cross-border collaboration. These frameworks must be adapted to local contexts, ensuring that smallholder farmers, rural entrepreneurs, and community makerspaces are not left behind.

10.4 Research and Innovation Priorities

Future research must focus on material enhancement, primarily by improving mechanical, thermal, and biodegradation properties of agrowaste composites through nanotechnology, enzymatic treatments, and hybridization. The next focus is supposed to be on Life cycle assessments (LCA). This can be achieved by quantifying environmental impacts across production, usage, and disposal phases to validate sustainability claims and guide design decisions (Lizundia et al. 2022). The next necessity is Biorefinery integration, which can be achieved by developing closed-loop systems where agrowaste is converted into multiple value-added products—bioplastics, bioenergy, fertilizers—within a single facility (Yaashikaa and Kumar 2022). The other need is to emphasize social innovation by exploring participatory design, community-led fabrication, and cultural adaptation to ensure that biofabricated goods are accepted, valued, and sustained.

10.5 Vision for a Circular Bio-Economy

The roadmap ahead envisions a world where agricultural waste is no longer a liability but a cornerstone of sustainable development. Through strategic investments, interdisciplinary research, and inclusive governance, agrowaste-based 3D printing can evolve from niche experimentation to mainstream practice. It can catalyze a shift from extractive to regenerative economies, where production is localized, materials are biodegradable, and innovation is democratized. The fusion of digital craftsmanship and biological resources offers not just technical solutions, but a philosophical reorientation toward harmony with nature. Also, the circular bio-economy is not merely a destination—it is a journey of co-creation, resilience, and ecological intelligence.

References

Alawi A et al (2023) Eco-friendly geopolymer composites prepared from agro-industrial wastes: a state-of-the-art review. CivilEng 4(2):433–453

Albelo I, Raineri R, Salmon S (2024) Materials and methods for all-cellulose 3D printing in sustainable additive manufacturing 5(2):98–115

Alexander et al (2024) Life cycle assessment of polylactic acid (PLA) in 3D printing applications: evaluating environmental impacts and energy consumption. Int J Eng Res Technol 13(8):262–271

Ali Shafahat et al (2025) Development of sustainable polymer composite withagro-industrial residue for biomedical applications. Polymer Engineering and science, 1922–1933

Andanje MN et al (2023) Biocompatible and biodegradable 3D printing from bioplastics: a review 15(10):2355

Armstrong K, Makerspaces, fab labs and citizen-centric action: behind REFLOW's productive 'making' approach. 2021 27/07/2025]; Available from: https://reflowproject.eu/blog/makerspaces-fab-labs-and-citizen-centric-action/.

Ashok G, Kumar P, Prabhu TR (2025) Global Developments in additive manufacturing of polymer composite materials: a scientometric review. Archives of computational methods in engineering

Balu UR et al (2024) Linking EDC-laden food consumption and modern lifestyle habits with preeclampsia: a non-animal approach to identifying early diagnostic biomarkers through biochemical alterations. Food Chem Toxicol 194:115073

Barbara, Food waste as feedstock for 3D printed bioplastics, 2017–2020.

Bernardes F et al (2024) Global waste management outlook 2024:1–116

Birmingham, U.o. (2025) Bio-based resins could offer recyclable future for 3D printing

Biswas (2022) Sustainable efficacy of agriculture waste materials and the value of indigenous knowledge. Just Agric 2(11):1–28

Commission, E. Food waste as feedstock for 3D printed bioplastics. Projects, Success stories, All success stories [Project] 2020; Available from: https://projects.research-and-innovation.ec.europa.eu/en/projects/success-stories/all/food-waste-feedstock-3d-printed-bioplastics.

Darekar VS et al (2024) Transforming agriculture waste into useful filler for a sustainable epoxy-glass fabric composites: microwave absorbing applications. J Mater Sci: Mater Electron 35(29):1911

Darekar VS, GP, Goyal RK (2025) Rice Husk Ash an agro-waste recycled value added product in epoxy for microwave absorption. Polymer Compos

Das D et al (2025) Biocompatible cellulose derivatives: green chemistry and its sustainable applications. In: Kesari KK et al (eds) Functionalized cellulose materials: sustainable manufacturing and applications. Springer Nature Switzerland, Cham, pp 159–190

Dhar SK et al (2024) Novel Bacillus and Prestia isolates from Dwarf century plant enhance crop yield and salinity tolerance. Sci Rep 14(1):14645

FAO (2024) Global facts and figures 2024; Available from: https://www.fao.org/docs/foodlosswastelibraries/default-document-library/global-facts-and-figures_idaflw24_en.pdf?sfvrsn=72587608_1

Fotina NV et al (2024) Enhancement of wheat growth by plant growth-stimulating bacteria during phytopathogenic inhibition. Biocatal Agric Biotechnol 60:103294

Grozier T (2025) A manufacturer's guide to sustainability

Harika R et al (2025) Sweetener synapse: exploring non-nutritive sweeteners' role in zebrafish glucose homeostasis. Environ Technol Innov 39:104322

https://www.improvedinsights.com/insights/evaluatingfablabs. Evaluating Fab Labs and Makerspaces. 2023; Available from: https://www.improvedinsights.com/insights/evaluatingfablabs.
Huang SH et al (2013) Additive manufacturing and its societal impact: a literature review. Int J Adv Manuf Technol 67(5):1191–1203
Insights WBM, Agro-Waste startups tackle India's plastic waste and smog. 2023 27/07/2025; Available from: https://worldbiomarketinsights.com/agro-waste-startups-tackle-indias-plastic-waste-and-smog/
Jayaprakash M et al (2025) Collagen fusion with natural biopolymers in food packaging: an in-depth review from conceptualization to consumer application. In: Kesari KK et al (eds) Functionalized cellulose materials: sustainable manufacturing and applications. Springer Nature Switzerland, Cham, pp 65–87
Karsoun A et al (2024) Recycling agriculture waste for liquid deposition 3D printing. Innovative Infrastructure Solutions 9(5):161
Kasahun Tsegaye M et al (2024) Polylactic acid based biocomposite for 3D Printing: a review. Composite Materials 8(2):57–71
Kaur J et al (2022) An exopolysaccharide-producing novel Agrobacterium pusense strain JAS1 isolated from snake plant enhances plant growth and soil water retention. Sci Rep 12(1):21330
Kaur J et al (2023a) GC-MS validated phytochemical up-leveling with in vitro-raised Sansevieria trifasciata [Prain]: the Mother in Law's tongue gets more antibacterial. Current Plant Biology 35–36:100308
Kaur J et al (2023b) Reactive Black-5. Congo red and methyl orange: chemical degradation of Azo-dyes by agrobacterium 15(9):1664
Kaur S et al (2025a) Poisoned seas: chemical threats to marine life and human health. In: Prakash C, Kesari KK, Negi A (eds) Sustainable development goals towards environmental toxicity and green chemistry: environment and sustainability. Springer Nature Switzerland, Cham, pp 167–200
Kaur J, Mudgal G, Dhar SK (2025) Decoration plants and their role in remediation. In: Abdel Latef AAH, Zayed EM, Omar AA (eds) Sustainable remediation for pollution and climate resilience. Springer Nature Singapore: Singapore, pp 695–724
Liu J et al (2025) Valorization of agricultural residues: challenges and opportunities in the production of bio-based materials. BioResources 20(2):4798–4820
Lizundia E, Luzi F, Puglia D (2022) Organic waste valorisation towards circular and sustainable biocomposites. Green Chem 24(14):5429–5459
Machado TO et al (2024) A renewably sourced, circular photopolymer resin for additive manufacturing. Nature 629(8014):1069–1074
Mohanta YK et al (2025) Agro-waste-derived bioplastics: sustainable innovations for a circular economy. Waste and biomass valorization
Morales MA et al (2021) Development and characterization of rice husk and recycled polypropylene composite filaments for 3D printing 13(7):1067
Mudgal G et al (2025) Game changer: how cellulose-based bioleather is transforming the market. In: Kesari KK et al (eds) Functionalized cellulose materials: sustainable manufacturing and applications. Springer Nature Switzerland, Cham, pp 277–295
Mudgal G et al (2025) Strategic conservation plan, in sustainable remediation for pollution and climate resilience, Abdel Latef AAH, Zayed EM, Omar AA (eds). Springer Nature Singapore: Singapore, pp 139–206
Naghshineh B et al (2021) Social impacts of additive manufacturing: a stakeholder-driven framework. Technol Forecast Soc Chang 164:120368
Nath A, Das K, Dhal GC (2023) Global status of agricultural waste-based industries, challenges, and future prospects. In: Neelancherry R, Gao B, Wisniewski A (eds) Agricultural waste to value-added products: technical, economic and sustainable aspects. Springer Nature Singapore, Singapore, pp 21–45

Parashar M et al (2023) Two novel plant-growth-promoting Lelliottia Amnigena isolates from euphorbia Prostrata Aiton enhance the overall productivity of wheat and tomato 12(17):3081
Pemas S et al (2024) Valorization of tomato agricultural waste for 3D-printed polymer composites based on poly(lactic acid) 16(11):1536
Pilafidis S et al (2022) Valorization of agro-industrial wastes and residues through the production of bioactive compounds by macrofungi in liquid state cultures: growing circular economy 12(22):11426
Prasad G et al (2024) Advancing sustainable practices in additive manufacturing: a comprehensive review on material waste recyclability 16(23):10246
Project TB (2024) Agro-waste to biopolymers: upcycling with bio-additives. Plastics Engineering
Rahman AM et al (2023) Additive manufacturing using agriculturally derived biowastes: a systematic literature review 10(7):845
Rao PM et al (2025) Cellulose-based packaging: an emerging sustainable packaging option for future use. In: Kesari KK et al (eds) Functionalized cellulose materials: sustainable manufacturing and applications. Springer Nature Switzerland, Cham, pp 89–115
Sabana Ara B, Krishnan PSG, Kanny K (2023) Bio-based polymers: a review on processing and 3D printing. Polymer Sci Series A 65(5):421–446
Seetharaman B et al (2025) The piteous price of progress: the environmental and health costs of modern society. In: Prakash C, Kesari KK, Negi A (eds) Sustainable development goals towards environmental toxicity and green chemistry: environment and sustainability. Springer Nature Switzerland, Cham, pp 291–313
Sharma A et al (2021) Toxicity with waste-generated ionizing radiations: blunders behind the scenes. In: Kesari KK, Jha NK (eds) Free radical biology and environmental toxicity. Springer International Publishing, Cham, pp 305–325
Sharma R (2024) Sustainable agro-waste composite market- Data Intelo. Available from: https://dataintelo.com/report/sustainable-agro-waste-composite-market
Silva RGD et al (2024) Biodegradable hybrid PLA composites incorporating coffee husks and mineral fillers. Polímeros 34(3)
Singh GB et al (2024) Azo dye bioremediation: an interdisciplinary path to sustainable fashion. Environ Technol Innov 36:103832
Singh GB, Dey S, Mudgal G (2025) Microbial contributions to a circular economy. In: Kaur S, Dhiman S, Tripathi M (eds) Microbial metabolomics: recent developments, challenges and future opportunities. Springer Nature Singapore, Singapore, pp 419–439
Singh A et al (2025) Unveiling the transformative power of smart cellulosic nanomaterials: revisiting potential promises to sustainable future. In: Kesari KK et al (eds) Functionalized cellulose materials: sustainable manufacturing and applications. Springer Nature Switzerland, Cham, pp 1–42
Soriano-Cuadrado B et al (2024) Mechanical. Thermal and morphological study of bio-based PLA composites reinforced with lignin-rich agri-food wastes for their valorization in industry. 16(17):2462
Technology, F.I.f.P.S.a.D. Additive Manufacturing. 2025 17/07/2025].
Thakur H et al (2025) Smart sustainable materials: a blueprint for a better tomorrow. In: Kesari KK et al (eds) Functionalized cellulose materials: sustainable manufacturing and applications. Springer Nature Switzerland, Cham, pp 253–276
Ufitikirezi JDDM et al (2024) Agricultural waste valorization: exploring environmentally friendly approaches to bioenergy conversion. 16(9):3617
Vengadesan E et al (2025) Enhancement of polylactic acid (PLA) with hybrid biomass-derived rice husk and biocarbon fillers: a comprehensive experimental study. Discover Applied Sciences 7(3):161
Vedantam Sai Krishna et al (2025) Agricultural waste valorisation via activated carbon synthesis: a comprehensive study of material properties and environmental applications. Int J Agric Food Sci 7(3):28–35

Vijayan M, Vani D, Vyshnavi S (2024) Advancements in biofiller-reinforced PLA composites for 3D printing: a review. J Res Appl Sci Eng Technol

Vinayak A, Mudgal G, Singh GB (2021) Environment persistent free radicals: long-lived particles. In: Kesari KK, Jha NK (eds) Free radical biology and environmental toxicity. Springer International Publishing, Cham, pp 1–19

WWF-UK (2021) Driven to waste: The Global Impact of Food Loss and Waste on Farms. Woking, WWF-UK

Yaashikaa PRA, Kumar PS (2022) Valorization of agro-industrial wastes for biorefinery process and circular bioeconomy: a critical review. Munich personal ss archive, 1–11

Yadav VK et al (2025) Emerging trends in the valorization of agricultural waste and their utilization in agricultural, pharmaceuticals, and environmental cleanup. Waste Biomass Valorization 16(6):2779–2833

Yang C et al (2025) Additive manufacturing of fibre-reinforced. Polym Compos 17(12):1652

Yoha KS, Moses JA (2023) 3D printing approach to valorization of agri-food processing waste streams 12(1):212

Yoha KS, Moses JA (2023) 3D printing of agri-food processing wastes. In Encyclopedia

yourstoory.com. UKHI: Reinventing Plastic with Purpose and Agricultural Waste. 2025 17/07/ 2025]; Available from: https://yourstory.com/2025/06/ukhi-agrowaste-ecogran-bioplastics.

Recovery of Energy and Nutrients from Natural Resources, Biomass, and Solid Wastes

Shivani Sharma, Gaurav Mudgal, Vasantharekha Ramasamy, Sanjoy K. Dhar, Jaspreet Kaur, and Victor Atuchin

Abstract The rising tide of environmental degradation, resource scarcity, and organic waste accumulation underscores a global crisis in energy and nutrient management. Agriculture, urbanization, and industrial systems continue to generate vast quantities of biomass and solid waste, while depleting finite nutrients such as phosphorus and exacerbating climate change through fossil energy dependence. Conventional approaches—like isolated anaerobic digestion, composting, and chemical precipitation—offer partial remediation but remain fragmented, resource-intensive, and contextually inflexible. In response, this chapter investigates combined and dynamic recovery systems that valorise natural resources, biomass, and solid wastes through intelligent and sustainable engineering. It reveals current solutions

S. Sharma · G. Mudgal
University Institute of Biotechnology, Chandigarh University, Mohali, Punjab, India
e-mail: gauravmdgl@gmail.com

V. Ramasamy
Faculty of Engineering and Technology, Department of Biotechnology, School of Bioengineering, SRM Institute of Science and Technology, Kattankulathur, Tamil Nadu, India
e-mail: vasanthr@srmist.edu.in

S. K. Dhar (✉)
Banophool Adibashi Green Heart College, Mirpur, Dhaka, Bangladesh
e-mail: sanjoydharctg@gmail.com

J. Kaur
Kanya Maha Vidyalaya, Pathankot Chowk, Jalandhar, Punjab, India
e-mail: Kaurjaspreet2308@gmail.com

V. Atuchin
Laboratory of Optical Materials and Structures, Institute of Semiconductor Physics, SB RAS, Novosibirsk, Russia

Research and Development Department, Kemerovo State University, Kemerovo, Russia

Department of Industrial Machinery Design, Novosibirsk State Technical University, Novosibirsk, Russia

R&D Center "Advanced Electronic Technologies", Tomsk State University, Tomsk, Russia

V. Atuchin
e-mail: atuchin@isp.nsc.ru

K. K. Kesari et al. (eds.), *Smart Sustainable Materials Advancing Re-manufacturing and Technology*, World Sustainability Series, https://doi.org/10.1007/978-3-032-17907-4_4

that vary from thermochemical and biochemical conversion technologies to decentralized bio-CNG plants and microbial fuel cells, all designed to optimise resource intensities in agriculture and environmental infrastructure. Nutrient recovery technologies, including struvite precipitation and stripping off ammonia, are critically reviewed alongside nature-inspired systems such as constructed wetlands and bio-inspired platforms based on termite digestion. Digitalisation is presented as a transformational lever, facilitating AI-based waste diagnostics, nutrient mapping based on the Internet of Things, and simulation tools like Aspen Plus and SuperPro Designer to optimise scenarios. The socio-economic aspects are also examined, featuring community cooperatives, public–private collaboration, and facilitating policy regimes that encourage circular nutrient flows and regenerative development. Integrating ecological intelligence, technological innovation, and participation-based governance, the chapter provides a multi-disciplinary roadmap to future-proofed recovery systems. It conceives a shift from the linearity of waste disposal to modularity-based, data-driven ecosystems that regenerate energy, restore nutrients, and empower local economies. Based on in-depth case studies, modelling insights, and strategy analysis, this chapter provides actionable intelligence to the aim of climate-resilient, low-carbon infrastructure that aligns with the objectives of smart and sustainable infrastructure.

Keywords Regenerative design · Bio-CNG hubs · Modular biorefineries · Microbial fuel cells · Struvite precipitation · Ammonia stripping · Constructed wetlands · Digital diagnostics · AI-based optimization · Public–private partnerships

1 Introduction: Waste as Resource

The global trajectory of population growth, urbanization, and industrial expansion has precipitated a dramatic increase in both energy demand and waste generation. Municipal solid waste is projected to reach approximately 3.4 billion tonnes annually by 2050, escalating from 2.01 billion tonnes in 2016, as reported by the World Bank {https://www.worldbank} (Maalouf and Mavropoulos 2023). Simultaneously, the food and agriculture sector, while vital for human sustenance, contributes significantly to environmental degradation, including nutrient runoff and organic waste accumulation. Studies estimate that over 30% of global food production is lost or wasted, much of which ends up in landfills or incinerators (Upadhyay et al. 2023).

While fossil fuels continue to dominate the global energy mix, accounting for more than 80% of consumption, their environmental costs—from greenhouse gas emissions to land degradation—underscore an urgent need to pivot towards regenerative and decentralized energy systems (Wang and Azam 2024) Natural resources such as biomass, agricultural residues, urban sludge, and organic waste streams embody vast reservoirs of carbon, nitrogen, and phosphorus, which are often left untapped due to fragmented and outdated recovery infrastructures. This chapter aligns with the principles of smart and sustainable development, whereby systems are designed to extract value while minimizing ecological issues. Such change demands a move from linear resource consumption to circular synergies, the systems that integrate energy recovery with nutrient cycling to create self-sustaining ecological and agricultural loops (Liu et al. 2022).

Historically, energy and nutrient recovery have been worked out using siloed technologies. Anaerobic digestion is widely used for biogas generation, and similarly composting for organic fertilizer production, and incineration or gasification for energy recovery (Ahmed et al. 2019). Additionally, chemical precipitation techniques, such as struvite formation, are employed for selective nutrient recovery. However, these conventional approaches have revealed critical limitations. Anaerobic digestion often exhibits low nutrient recovery efficiency and risks methane leakage. Composting processes are slow and yield low-concentration fertilizers; incineration carries high capital costs and pollution risks, and chemical recovery processes often depend on external chemical inputs and fail to address comprehensive nutrient loops (Midden et al. 2023). These technologies, while technically viable, typically lack modularity, digital adaptability, and multi-output configurations—features essential for scalability in rural and peri-urban contexts. Moreover, existing systems rarely integrate with nature-based or bio-inspired solutions that could enhance performance through ecological mimicry. Policy instruments supporting such integrated systems remain underdeveloped, and public awareness of circular resource valorization is limited. Many of the processes involved are depicted in Fig. 1.

This chapter, therefore, addresses these gaps by exploring advanced, decentralized, and hybrid frameworks that combine thermochemical, biochemical, and data-driven innovations. It examines the valorization of diverse biomass types, including lignocellulosic, aquatic, and algal feedstocks; modular biorefineries and bio-CNG hubs; nutrient extraction from urban wastewater and landfill leachate; and the role of smart technologies—such as AI, sensor networks, and predictive modeling—in optimizing resource recovery. The broader socio-economic and policy contexts are also assessed, emphasizing the role of interdisciplinary collaboration and governance in steering a transition toward circular resource culture.

2 Biomass Valorization: From Feedstock to Function

Biomass valorization has emerged as a cornerstone of sustainable resource management, offering a pathway to transform organic residues into energy, biofuels, and nutrient-rich byproducts. Globally, biomass accounts for approximately 14% of total energy consumption, with developing nations utilizing nearly 38% of this energy, particularly in rural areas where access to conventional energy remains limited (Thaha et al. 2025). Despite its ubiquity, biomass remains underutilized due to technological, logistical, and policy constraints. The valorization of biomass—defined as the systematic conversion of biological feedstocks into value-added products—addresses these gaps by integrating biochemical, thermochemical, and hybrid approaches.

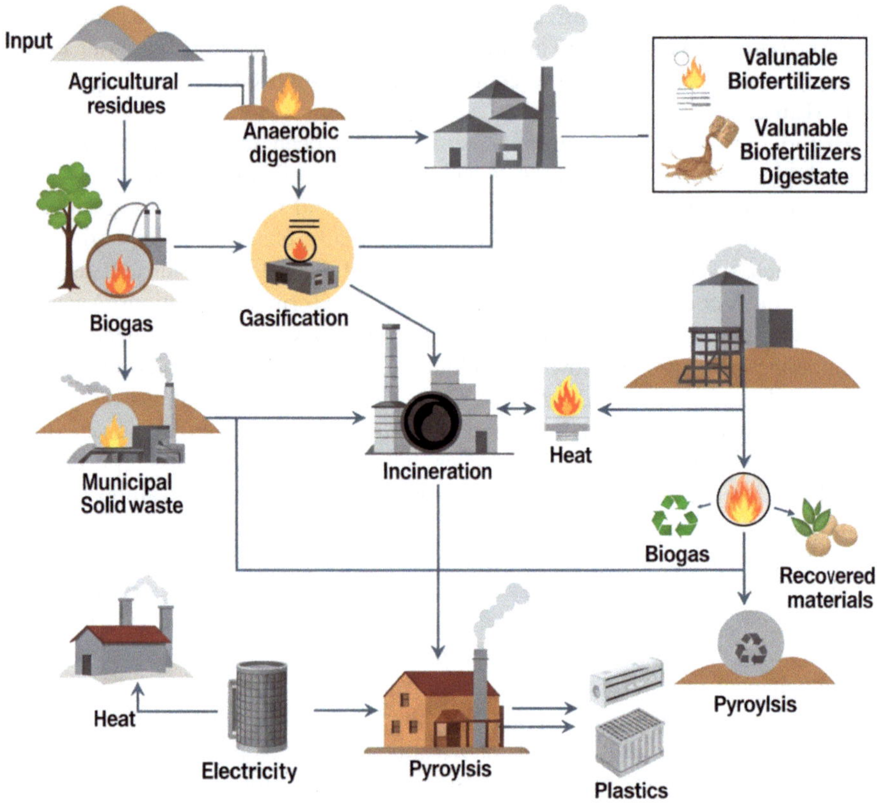

Fig. 1 Conceptual overview of an integrated framework for energy and nutrient recovery from biomass and solid wastes. This diagram illustrates the key inputs, conversion technologies, and valuable outputs of a comprehensive waste valorization system

Biomass can be broadly categorized into three generations based on feedstock origin and sustainability. First-generation biomass includes edible crops such as corn, sugarcane, and wheat, which are primarily used for bioethanol and biodiesel production. However, their use raises concerns over food security and land competition. Second-generation biomass comprises lignocellulosic residues, including crop stubble, forest waste, and agro-industrial byproducts. These feedstocks are non-edible and abundant, making them ideal for sustainable biofuel production (Ning et al. 2021). Third-generation biomass, such as microalgae and cyanobacteria, offers high productivity and minimal land use, though its commercial viability is still emerging (Raina et al. 2024).

Valorization routes are usually categorized as thermochemical and biochemical processes. Thermochemical processes, such as pyrolysis, gasification, and hydrothermal liquefaction, occur at high temperatures and transform biomass into syngas, bio-oil, and biochar. These processes are most efficient for lignocellulosic biomass, high in cellulose, hemicellulose, and lignin. Fast pyrolysis, for example, at temperatures around 450–600 °C produces high-energy–density bio-oil, while gasification generates syngas that can be used to produce electricity or chemicals (Danso-Boateng and Achaw 2022). Hydrothermal liquefaction, carried out under subcritical or supercritical conditions, allows wet biomass to be converted without drying, so it is best suited to sewage sludge and sea stocks (Hidalgo et al. 2025).

Biochemical valorization, however, depends on microbial and enzymatic conversions to produce biomass-based bioethanol, biogas, and organic acids. Anaerobic digestion is extensively practiced to produce biogas from food scraps and animal dung, while fermentation technologies allow sugars to be converted to ethanol and butanol. Improvements brought about by enzymatic hydrolysis and genetic engineering enhanced the yield of lignocellulosic ethanol, as evident through case studies that were successful at converting vegetable and fruit wastes to bioethanol yields higher than 6% volumetrically (Morales-Vera et al. 2022). In addition, integrated biorefineries nowadays integrate both the thermochemical and the biochemical modules to optimize the recovery of resources as much as possible as well as minimize wastes.

Nevertheless, biomass valorization suffers some setbacks. Feedstock inconsistency, pretreatment expense, and conversion effectiveness are the notable ones. For instance, the lignocellulosic biomass has to be highly pretreated to disable the stubborn structure, sometimes requiring chemical solvents or high-temperature steam (Raina et al. 2024). Moreover, the viability of bioethanol production based on waste biomass hinges on the supply value chain, policy support, and market integration (Morales-Vera et al. 2022). Strategic measures—but like the modular biorefinery, decentralized processing sites, as well as remote monitoring networks—are, nevertheless, offsetting these constraints.

In this chapter, we shall discuss the valorization of various biomass feedstocks, such as lignocellulosic wastes, aquatic biomass, and urban organic wastes. We shall analyze conversion technologies within the thermochemical as well as biochemical spheres, evidencing innovations like microbial fuel cells, AI-based process intensification, and nature-based solutions. Case studies from India, Brazil, and the Netherlands shall be presented based on actual applications, while modeling instruments shall be employed to estimate scalability as well as environmental influence. Integrating the technical, the ecological, as well as socio-economy viewpoints, this section shall contribute a thorough insight of biomass valorization as a support column of smart and sustainable growth.

3 Energy Recovery Technologies: Innovations in Decentralized Systems

Energy recovery from organic waste streams has undergone tremendous transformation during the last twenty years, both due to the pressure from climate mitigation as well as rural energy access. Among the most effective technologies is anaerobic digestion (AD), where microbial consortia degrade organic substrates under oxygen-free conditions, yielding biogas—a gaseous mixture mainly consisting of methane as well as carbon dioxide. AD plants are best suited to rural settings, where there is a high availability of livestock manure, crop wastes, as well as food processing wastes. Livestock farms alone, in Canada, produce around 150 million tonnes manure each year, where cattle contribute to almost 84% (Alengebawy et al. 2024). If left unmanaged, the waste streams release methane, a green house gas having a global warming potential 28 times higher than CO_2, where it prevails over a period of 100 years.

Anaerobic digestion captures not only the methane that would otherwise be lost into the atmosphere but also generates digestate, a rich nutrient slurry that can be used as a substitute for synthetic fertilizers. Research has demonstrated that applications of digestate enhance the structure and fertility of soils while minimum-pathogen odor masses are decreased (Alengebawy et al. 2024). Decentralized biogas systems have been installed in rural areas around India to support irrigation pumps, cooking ranges, as well as small-scale electricity generation. These systems are being combined increasingly with the modular biorefineries that integrate AD, composting, biochar production, and nutrient recovery, generating multi-product platforms that improve the efficiency of resources (Rao et al. 2010).

In addition to AD, microbial fuel cells (MFCs) are an emerging-technology that treats waste-water while generating electricity. MFCs harness the electrogenic properties of certain microbes to oxidize organic substrates, conveying electrons to an external circuit. It produces low-voltage electricity while liberating chemical oxygen demand (COD) in waste-water streams. MFCs are experimented upon the substrates like brewery effluent, food wastes, as well as agricultural runoff, wherein the substrates' power densities up to 2,668 mW/m^2 are achieved through graphene-based, stainless steel anodes that are modified. While, up to now, commercialization is restricted due to the limitations on scalability as well as cost, research into nano structure-based electrodes, the biocathodes, as well as competing membranes, is enhancing the performance as well as the economic viability (Zhao et al. 2017; Obileke et al. 2021).

In rural India, advanced waste-to-energy systems are being increasingly fitted with IoT-capable sensors and AI-based monitoring platforms. These help provide real-time diagnostics of waste stream composition, biogas production, and nutrient levels, facilitating predictive maintenance as well as adaptive control. For instance, in Jharkhand, Schneider Electric's Climate Smart Village project installed sun-based micrograms with smart algorithms for energy allocation to optimize irrigation as well as agro-processing loads, leading to the doubling of household income as well as a savings annually of up to 60,000 kg in carbon emissions (Dhaliwal et al. 2019).

Similarly, startups like WeVOIS are integrating IoT with decentralized waste collection and processing hubs, transforming organic waste into biogas and compost while empowering sanitation workers and reducing urban–rural disparities (Devi et al. 2024).

Despite these advances, several challenges persist. AD systems require consistent feedstock quality and temperature control, which can be difficult in remote or variable climates. MFCs, while promising, still suffer from low energy conversion efficiencies and high material costs. Moreover, policy frameworks often lag behind technological innovation, limiting incentives for decentralized energy recovery. Addressing these gaps will require integrated strategies that combine technical innovation with community engagement, capacity building, and supportive governance (Sharma et al. 2024).

4 Nutrient Recovery: Closing the Loop in Agricultural Systems

Nutrient recovery from organic waste and wastewater streams is increasingly recognized as a cornerstone of sustainable development, particularly in the context of agriculture and food security (Tables 1 and 2). Phosphorus and nitrogen, two essential macronutrients, are often lost through conventional waste management practices, contributing to environmental degradation and resource inefficiency. For instance, excessive phosphorus discharge into water bodies leads to eutrophication, while nitrogen losses contribute to greenhouse gas emissions and groundwater contamination (Guerra-Rodríguez et al. 2020). According to Santos et al. (2024), struvite precipitation can remove up to 80–90% of soluble reactive phosphorus and 20–30% of nitrogen from wastewater under optimized conditions (Santos et al. 2024).

Phosphorus recovery is particularly urgent given its classification as a critical raw material by the European Union, due to geopolitical supply risks and the finite nature of phosphate rock reserves (Cordell et al. 2011). One of the most widely adopted technologies for phosphorus recovery is struvite precipitation, which involves the crystallization of magnesium ammonium phosphate ($MgNH_4PO_4 \cdot 6H_2O$) under controlled pH and molar ratios. This process is typically applied to anaerobic digestate, sewage sludge liquor, or industrial effluents. Studies have shown that maintaining a pH around 9.5 and a Mg: PO_4 ratio of at least 1:1 significantly enhances recovery efficiency (Lorick et al. 2020). However, the presence of competing ions such as calcium can reduce struvite purity and crystallinity, necessitating pretreatment or seeding strategies to improve product quality (Enyemadze et al. 2021).

In the Netherlands, urban sewage sludge is increasingly valorized as a fertilizer base, with over 95% of sludge treatment residues repurposed as reusable raw materials (Sugurbekova et al. 2023). The concept of transforming wastewater treatment plants into "Energy and Resource Factories" has enabled the recovery of phosphorus, nitrogen, cellulose, and bioplastics from municipal waste streams. Technologies such as crystallization reactors, fluidized bed systems, and electrochemical struvite recovery are being deployed to enhance nutrient capture and reduce environmental impact (Hussain et al. 2023).

Nitrogen recovery, while more complex due to its volatile nature, is also gaining traction through methods such as ammonia stripping, membrane distillation, and bioelectrochemical systems. Ammonia stripping involves elevating pH and temperature to convert ammonium ions into gaseous ammonia, which is then absorbed

Table 1 Comparative Overview of Nutrient Recovery Techniques from Wastewater Shown here is an evaluation of five key nutrient recovery strategies—struvite precipitation, ammonia stripping, membrane distillation, bioelectrochemical systems, and constructed wetlands—based on their target nutrients, operating conditions, recovery efficiency, technological challenges, and supporting literature. The comparative format offers a practical lens for assessing the ecological, technical, and economic feasibility of integrating these methods into circular agricultural systems and sustainable wastewater management frameworks

Method	Target Nutrients	Operating conditions	Recovery efficiency	Challenges	References
Struvite precipitation	Phosphorus, Ammonium	pH ~ 9.5, Mg:PO_4 ratio ≥ 1:1	80–90% P, ~30% N	Calcium interference, Mg cost	Santos et al. (2024), Otieno et al. (2023)
Ammonia stripping	Ammonium	High pH (≥11), elevated temperature	Up to 95% ammonia recovery	Energy-intensive, acid absorption is needed	Zhu et al. (2023), Kinidi et al. (2018)
Membrane distillation	Ammonia	Hydrophobic membranes, temp gradient	70–85% ammonia recovery	Fouling, membrane wetting	Rivera et al. (2022), Guo et al. (2019)
Bioelectrochemical systems	Nitrogen (NH_4^+, NO_3^-)	Electrochemical setup, microbial fuel	30–70% N recovery	Scalability, low voltage, complex setup	Cerrillo et al. (2023), Rodríguez Arredondo et al. (2015), Liang et al. (2019)
Constructed wetlands	Nitrogen, Phosphorus	Nature-based, seasonal variability	Up to 90% N, 60–85% P	Land use, climate sensitivity	Paula et al. (2024), Yang et al. (2023), Land et al. (2016)

into acid solutions to produce ammonium sulfate, a valuable fertilizer. Membrane-based technologies, including direct contact membrane distillation (DCMD) and liquid–liquid membrane contactors (LLMC), offer high selectivity and efficiency for ammonia recovery from digestate and landfill leachate (Zhu et al. 2024). Electrically conducting gas-stripping membranes have also demonstrated promising results, achieving recovery rates of 68.8 ± 8.0 g N/m^2/day with energy consumption as low as 7.1 ± 1.1 kWh/kg N (Iddya et al. 2020).

Table 2 Comparative Analysis of Biomass Conversion Technologies for Energy and Resource Recovery. This table presents a structured comparison of five key biomass conversion technologies—pyrolysis, anaerobic digestion, fermentation, microbial fuel cells, and hydrothermal liquefaction. It outlines their feedstock compatibility, primary outputs, operational advantages, limitations, and references to current scientific literature for evaluating recovery pathways aligned with sustainable development goals

Conversion method	Feedstock compatibility	Key outputs	Advantages	Limitations	References
Pyrolysis	Lignocellulosic, agrowaste	Bio-oil, syngas, biochar	Fast processing, biochar benefits	Needs dry feedstock, energy input	Jerzak et al. (2024), Nguyen et al. (2022), Huber and Brown (2017)
Anaerobic digestion	Food waste, manure, crop waste	Biogas, digestate	Renewable energy, soil enhancer	Methane leakage risk, slow rate	Dhull (2024), Uddin and Wright (2023)
Fermentation	Sugars, fruit/vegetable waste	Ethanol, organic acids	Biochemical precision, scalable	Complex pre-treatment for residues	Wall et al. (2024), Julie Chandra and Thomas (2023), Garba (2020)
Microbial fuel cells	Wastewater, organic effluents	Electricity, treated water	Simultaneous treatment and energy	Low voltage output, costly setup	Moradian et al. (2021), Barakat (2023)
Hydrothermal liquefaction	Aquatic biomass, sludge	Bio-crude, gases	No drying needed, high energy yield	High pressure required	Mathanker et al. (2021), Qiu et al. (2025), Yang et al. (2020)

Despite these advancements, several challenges remain (Tables 1 and 2). Struvite precipitation requires external magnesium sources, which can increase operational costs. Ammonia stripping and membrane distillation are energy-intensive and sensitive to feed composition. Moreover, regulatory barriers—such as the lack of end-of-waste status for recovered products—limit market integration and scalability (Farghali et al. 2024). Addressing these issues will require coordinated efforts in technology development, policy reform, and stakeholder engagement.

5 Hybrid and Modular Systems: Resilience Through Design

Hybrid and modular systems represent a transformative approach to resource recovery, offering decentralized, scalable, and multifunctional solutions tailored to diverse agro-environmental contexts (Fig. 2). Unlike conventional centralized facilities, these systems are designed to operate efficiently in rural and peri-urban regions, where infrastructure limitations and feedstock variability often hinder the deployment of large-scale technologies. Modular biorefineries, in particular, have gained traction as compact, plug-and-play units capable of processing a wide range of biomass types—including crop residues, livestock manure, and food waste—into energy, biochar, compost, and nutrient-rich outputs (Capodaglio 2017).

The concept of modularity enables flexibility in system configuration, allowing operators to adjust processing capacity, output streams, and technology combinations based on local needs and resource availability. For instance, the EU-funded MANUREFINERY project demonstrated the viability of mobile biorefining units that convert livestock manure into microbial protein, caproic acid, and biofertilizer ingredients, while integrating digital twins and data science for process optimization (Johnson et al. 2021). Similarly, the Rural BioReFarmeries initiative in Europe showcased decentralized green biorefineries that transform grassland biomass into human-grade protein, bio-based packaging, and clean energy, thereby enhancing farmer competitiveness and regional resilience (Skondras et al. 2024).

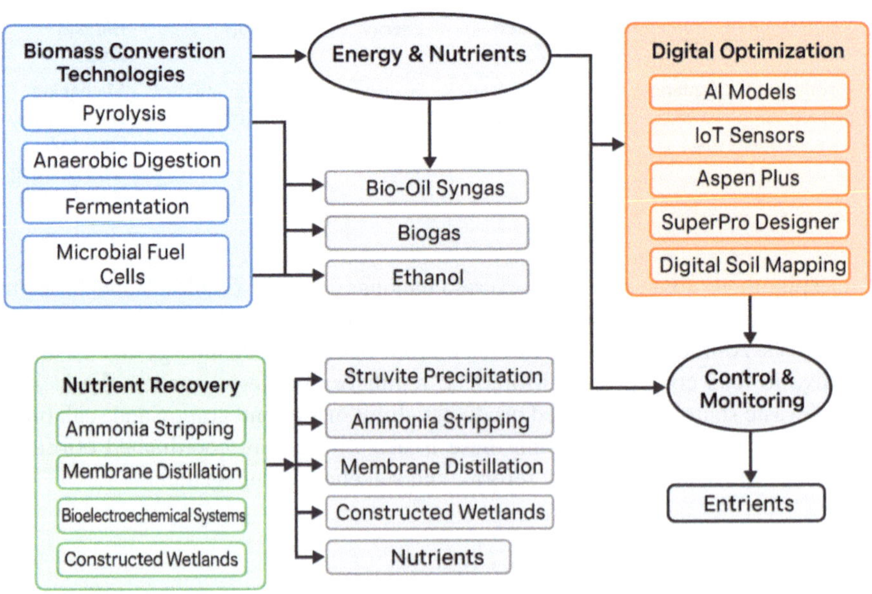

Fig. 2 Modular biorefinery integration for decentralized energy and nutrient recovery

In India, sugar mills are increasingly being repurposed into bio-CNG hubs, leveraging press mud and spent wash as feedstocks for anaerobic digestion and biogas upgrading. These multi-output platforms not only generate renewable fuel but also produce enriched compost and recover nutrients such as phosphorus and potassium from digestate streams (Kumar et al. 2022). The integration of such systems with smart monitoring tools—IoT sensors, AI-based diagnostics, and predictive maintenance algorithms—further enhances operational efficiency and environmental performance. Biochar-based recovery platforms also exemplify hybrid functionality. When combined with composting systems, biochar improves aeration, microbial activity, and heat retention, thereby accelerating biodegradation and enabling heat recovery for domestic or agricultural use. Studies have shown that adding 10% biochar to composting reactors can increase biodegradation yields by 50% and improve heat extraction rates by up to 0.3 kW/m^3 (Jindo et al. 2012). Moreover, engineered biochars derived from organic waste have been successfully used to immobilize heavy metals, recover phosphate from wastewater, and enhance soil fertility, demonstrating their versatility across energy, environmental, and agricultural domains (Sharma et al. 2024).

Despite their promise, hybrid and modular systems face challenges related to capital investment, technical integration, and regulatory compliance. Ensuring feedstock consistency, managing multi-stream outputs, and aligning with local governance frameworks require careful planning and stakeholder engagement. Nonetheless, the adaptability and resilience of these systems make them ideal for regions with fragmented waste streams, limited infrastructure, and high demand for sustainable resource solutions.

Schematic representation of a hybrid resource recovery platform combining anaerobic digestion, biochar composting, and nutrient extraction technologies (e.g., struvite precipitation, ammonia stripping). Illustrated pathways demonstrate feedstock inputs, multi-stream outputs (bio-CNG, compost, recovered nutrients), and digital optimization layers (IoT sensors, AI diagnostics). Designed to function in rural and peri-urban settings, the system reflects regenerative design principles, multi-output resilience, and circular synergy.

6 Bio-Inspired and Nature-Based Solutions

Nature-based and bio-inspired solutions are increasingly recognized as powerful frameworks for designing regenerative systems that recover energy and nutrients while enhancing ecological resilience (Dhar et al. 2024; Mudgal et al. 2025; Kaur et al. 2025a, 2023a, 2023b, 2022; Mudgal and Kesari 2025). Wetlands, for instance, serve as natural bioreactors that filter pollutants, sequester nutrients, and support biomass cultivation. Constructed wetlands and wetland buffer zones (WBZs) have demonstrated high efficiency in removing nitrogen and phosphorus from agricultural runoff and municipal wastewater. In a study conducted in Poland, vegetation harvesting in WBZs achieved nitrogen removal rates of up to 92% and phosphorus removal between 17–63%, depending on site conditions and management practices (Jabłońska et al. 2021). The removal of aboveground biomass after mowing was found to be critical, as it prevented nutrient leaching during decomposition and enhanced long-term purification capacity.

Beyond nutrient capture, wetlands also offer opportunities for biomass valorization. Harvested wetland plants such as cattail (*Typha spp.*) and reed (*Phragmites spp.*) can be processed into compost, bioenergy feedstocks, or industrial raw materials, contributing to circular bioeconomy goals. Floating treatment wetlands (FTWs), a variant of constructed wetlands, have been successfully deployed in eutrophic lakes and wastewater lagoons, where they facilitate nutrient uptake through dense root systems and microbial biofilms. These systems are particularly effective in shallow water bodies with fluctuating hydrology, offering scalable and low-maintenance remediation options (White 2021).

Bio-inspired design also draws from the digestive and metabolic strategies of organisms such as termites, which possess highly efficient lignocellulolytic systems. Termite gut symbiosis—comprising bacteria, fungi, and protists—enables the breakdown of complex biomass into fermentable sugars and bioenergy intermediates. Termites utilize a stepwise enzymatic process to hydrolyze lignin, cellulose, and hemicellulose, offering a model for biomimetic biorefineries that valorize lignocellulosic waste (Dar et al. 2024). These systems can be emulated through engineered microbial consortia and compartmentalized reactors that mimic termite gut physiology, potentially enhancing yields of biohydrogen, methane, and organic acids. In addition, termite mound construction has influenced passive ventilation systems among building and reactor designs, making use of the principles of airflow optimization and thermoregulation (Ramasam et al. 2024). Termite mounds' porous, stratified structure allows for moisture storage as well as gas exchange, which may be augmented into modular composting units or improved anaerobic digesters having enhanced thermal stability as well as aeration. Ecosystem modeling also assists with the optimization of energy flows within nature-based systems. Ecosystems like the Y-shaped energy flow paradigm and universal trophic models assist with the measurement of nutrient cycling, biomass storage, and energy dissipation throughout the trophic levels (ODUM 2015). Scenario planning and system optimization that conform to the principles of ecology allow the designed recovery systems to be retained within functional integrity levels as they exhibit resilience.

7 Digitalization and Smart Monitoring

Digitalization is fast transforming the landscape of energy and nutrient recovery, making it possible to make systems more adaptive, efficient, and data-based. The convergence of artificial intelligence (AI), Internet of Things (IoT) sensors, and predictive analytics uncovers the capability to conduct real-time diagnostics of waste streams, nutrient flows, and the performance of the systems. These technologies improve, not only the control of the systems during operations, but also decision-making in environmental as well as agrarian spheres (Alzahrani et al. 2023).

Artificially intelligent waste stream diagnostics have shown great promise in the name of automating the classification and identification of disparate waste materials. Convolutional neural networks (CNNs) and other object detection algorithms like YOLO (You Only Look Once) are being implemented to classify plastic waste in real time, enhancing the recycling rate and minimizing the contaminating rate (Giel et al. 2025). They are capable of recognizing multiple wastes within the same frame, considering the size, color, and orientation variation. Combined with sensor fusion platforms, they offer overarching insights into the composition of wastes, facilitating specific recovery measures and aiding circular economy objectives.

In smart farming, data-based nutrient mapping is transforming precision cultivation. Digital soil mapping (DSM) applications, many based on machine algorithms, produce high-resolution maps of the nutrient profile, moisture level, and acidity/basicity (pH) of the soils.

Applications such as GeaGrow use artificial neural networks to predict soil properties and recommend crop-specific fertilizer regimes, achieving mean absolute errors as low as 1.9750 for NPK prediction and over 99% accuracy in soil texture classification (Folorunso et al. 2025). These tools empower farmers to make informed decisions, reduce input costs, and minimize environmental impacts associated with over-fertilization. Sensor networks also play a pivotal role in monitoring nutrient dynamics and system health. Ultrasonic, capacitive, and electrochemical sensors are used to track fill levels, moisture content, and nutrient concentrations in composting units, digesters, and soil beds. IoT-enabled platforms transmit this data to cloud-based dashboards, where AI algorithms analyze trends and trigger alerts for maintenance or intervention. For example, smart waste segregation systems equipped with capacitive moisture sensors and inductive proximity detectors have achieved classification accuracies exceeding 90% for metallic and wet waste types, with response times under one second (Lakhouit 2025).

Predictive modeling software like Aspen Plus and SuperPro Designer advance system design as well as optimization. Aspen Plus is extensively used to model thermochemical as well as biochemical conversion processes, such as pyrolysis, gasification, and fermentation. It also allows integration with external applications as well as rigorous mass as well as energy balance calculations (Gosling 2005). SuperPro Designer, meanwhile, is strong on batch as well as continuous process modeling, with capabilities like equipment scheduling, cost estimation, and environmental impact analysis (Canizales et al. 2020). These programs enable researchers as well as engineers to analyze the performance of systems under multiple scenarios, determine the site of bottlenecks, as well as optimize the channels for resource recovery. Despite their potential, digital technologies also come up against data quality, interoperability, and user access issues. Most AI algorithms are learnt on small-scale data that lacks real-world variation, while sensor networks demand strong connectivity as well as calibration. It will address the limitations through multi-stakeholder collaboration on data standardization, open source platforms, and capacity building.

8 Socio-Economic Integration and Policy Levers

Socio-economic integration is an indispensable facilitator of the scaling up and diffusion of energy and nutrient recovery technologies. Although technological advancement is the core component of the technologies, their effective functioning depends on inclusive governments, community engagement, and conducive policy settings. In the agricultural and peri-urban settings, decentralized energy technologies like solar microgrids, biogas plants, and small-scale modifiable biorefineries have presented substantial capability to empower farmer co-ops as well as self-help groups. According to a 2025 study by the Council on Energy, Environment and Water (CEEW), decentralized renewable energy (DRE) technologies have led to a 33% increase in annual income among rural users, with over 90% reporting improved livelihoods and 81% noting enhanced confidence and skills, particularly among women-led collectives (Boostani et al. 2024).

Farming cooperatives and producer organizations play a pivotal role in mainstreaming circular resource systems. By pooling resources and sharing infrastructure, these groups can collectively invest in waste-to-energy platforms, composting units, and nutrient recovery technologies. For example, agrivoltaic cooperatives in Germany have successfully integrated solar energy generation with crop cultivation, enabling villages to become electricity self-sufficient while maintaining agricultural productivity (Bauknecht 2024). In India, decentralized solar setups have facilitated innovations such as solar-powered irrigation, fencing, and food processing, reducing operational costs and improving yield quality (Mueller et al. 2016).

Policy frameworks are essential to incentivize and regulate nutrient recycling from municipal and agricultural waste. The Solid Waste Management Rules (SWM), 2016 in India mandate source segregation, decentralized processing, and recovery of commercially recyclable and nutrient-rich fractions from urban waste streams (MOEFCC 2016). These rules empower urban local bodies to levy user fees, promote composting and biomethanation, and engage private partners for scientific waste management. However, implementation remains uneven, with only 50% of waste being treated and segregation rates below 50% in major cities like Delhi and Mumbai (Kala and Bolia 2024).

Public–private partnerships (PPPs) offer a strategic mechanism to scale circular resource systems. By aligning private sector innovation with public sector mandates, PPPs can mobilize capital, enhance technical capacity, and ensure long-term sustainability. In Ajman, UAE, PPPs have been shown to positively impact economic sustainability in industrial waste management, contributing to circular industry ecosystems and advancing SDGs 11, 12, and 17 (Alhosani 2025). Similarly, Thailand's Bio-Circular-Green (BCG) economy model integrates PPPs in renewable energy, agriculture, and infrastructure projects, promoting resource efficiency and climate resilience. Despite these promising models, several barriers persist. These include fragmented policy landscapes, lack of market linkages for recovered products, limited access to working capital, and insufficient stakeholder coordination. Addressing these challenges requires multi-level governance, capacity building, and inclusive financing mechanisms. Targeted subsidies, carbon credits, and performance-based incentives can accelerate adoption, while participatory planning and community ownership ensure social legitimacy and long-term engagement.

9 Future Perspectives and Emerging Frontiers

As the global sustainability agenda intensifies, energy and nutrient recovery systems are poised to evolve beyond circularity toward regenerative and carbon-negative paradigms. Emerging technologies and integrated frameworks are redefining what is possible, not only in terms of resource efficiency but also in climate mitigation, ecosystem restoration, and socio-economic transformation.

One of the most promising frontiers is the development of carbon-negative recovery processes, which actively remove more carbon dioxide from the atmosphere than they emit. Enhanced oil recovery (EOR) using CO_2 captured from biomass combustion or direct air capture has demonstrated potential to produce net carbon-negative oil, provided that the injected CO_2 is permanently sequestered underground and sourced from biogenic or atmospheric origins (Godec et al. 2013). Lifecycle analyses of CO_2-EOR operations show that with optimized injection strategies—such as water-alternating-gas (WAG) and continuous gas injection (CGI)—carbon storage can offset emissions from oil production and combustion, especially when coupled with monitoring and verification protocols (Nuñez-Lopez et al. 2019). These approaches not only support climate goals but also offer a lower-cost pathway to scale carbon capture, utilization, and storage (CCUS) technologies.

Parallel to terrestrial innovations, the blue economy is emerging as a vital domain for energy and nutrient recovery. Marine biomass—including macroalgae, microalgae, and aquatic plants—offers high productivity, minimal land use, and multifunctional benefits. Algae cultivation, for instance, can absorb significant amounts of CO_2, treat wastewater, and produce biofuels, bioplastics, and biofertilizers (Nuñez-Lopez et al. 2019). Marine biorefineries are being developed to valorize seaweed and fishery byproducts into high-value compounds, contributing to food security, clean energy, and coastal resilience (Veríssimo et al. 2021). Integrating marine biomass into circular systems not only diversifies feedstocks but also enhances ecosystem services such as nutrient cycling, carbon sequestration, and biodiversity conservation.

Looking ahead to 2035, scenario modeling is becoming an essential tool for forecasting and designing resilient recovery systems. Models such as Aspen Plus, SuperPro Designer, and dynamic lifecycle assessments are being used to simulate energy flows, nutrient recovery efficiencies, and environmental impacts under various policy and technological trajectories. For example, the Fuel Portfolio Scenario Model (FPSM) developed in California evaluates low-carbon fuel standards and recovery targets for 2030 and 2035, helping policymakers balance credit markets and optimize biofuel integration (Nuñez-López et al. 2017). Similarly, participatory disaster recovery simulation models like DESaster enable communities to co-design recovery pathways that align with resilience goals and local priorities (Miles 2018).

These modeling frameworks support adaptive governance, allowing stakeholders to test interventions, anticipate trade-offs, and co-create strategies that are robust under uncertainty. By 2035, it is expected that recovery systems will be increasingly modular, data-driven, and integrated across land and sea interfaces. Innovations such as bioelectrochemical nutrient recovery, AI-optimized biorefineries, and marine carbon farming will likely become mainstream, supported by policy instruments that reward regeneration and climate-positive outcomes.

10 Conclusion: Toward a Regenerative Resource Culture

The journey through energy and nutrient recovery from natural resources, biomass, and solid wastes reveals a dynamic and interdisciplinary landscape—one that is increasingly shaped by ecological design, technological innovation, and socio-economic integration. As global pressures mount from climate change, resource scarcity, and waste accumulation, the imperative to transition from linear to circular systems becomes not just a sustainability goal but a survival strategy (Mudgal et al. 2025; Mudgal and Kesari 2025; Kaur et al. 2023a, 2025b; Harika et al. 2025; Singh et al. 2025, 2024; Singh and Kesari 2025; Thakur and Kesari 2025; Rao and Kesari 2025; Seetharaman et al. 2025; Balu et al. 2024; Fotina et al. 2024; Sharma et al. 2021; Vinayak et al. 2021). This chapter has explored how diverse recovery pathways—ranging from thermochemical valorization and biochemical conversion to nature-based solutions and digital monitoring—can be orchestrated into synergistic frameworks that regenerate rather than deplete. A recurring theme across the sections is the need for modularity, decentralization, and adaptability. Hybrid systems and modular biorefineries offer scalable solutions tailored to local contexts, especially in rural and peri-urban regions where centralized infrastructure is lacking. These platforms enable multi-output recovery—energy, biochar, compost, and nutrients—while fostering resilience through design flexibility and community ownership. The integration of bio-inspired strategies, such as termite gut bioconversion and wetland-based nutrient capture, further enriches the technological palette with ecological intelligence and low-impact remediation.

Digitalization becomes an effective enabler, making recovery systems smart, adaptable networks. AI-based diagnostics, real-time monitoring through sensors, and predictive modeling software such as Aspen Plus and SuperPro Designer enable real-time optimization, scenario analysis, and lifecycle analysis. These enable not only enhanced operational effectiveness but also transparency, accountability, and adaptable governance. No less important is the socio-economic and policy angle. Cooperative schemes, public–private collaborations, and specific policy instruments—like India's Solid Waste Management Rules (2016) and the EU's strategic raw materials plan—is the institutional framework for the circular management of resources.

Tax incentives, decentralized energy, nutrient cycling, and regenerative agriculture need to be upscaled based on inclusive financing, strengthened capacities, and participatory planning.

Towards the future, energy and nutrient recovery hinges on regenerative resource cultures—sustainabilities that extend beyond mere sustenance to positively restore ecological integrity, enrich biodiversity, and enable communities. Carbon-negative technologies, marine biomass valorisation, and scenario modelling to 2035 indicate transformational avenues that meet climate objectives as well as socio-economic justice. As emphasised recently, regenerative measures need to be systemic, cooperative, and situational, embracing complexity rather than sanitizing it. It is therefore time to call for a paradigm shift: from fragmented, reactive waste management to integrated, anticipatory resource recovery. Bridging theory and practice—to interdisciplinary innovation, ecological design, as well as digital intelligence—we are capable of creating resilient systems that not only recover energy and nurture nutrients, but also regenerate landscape, livelihood, as well as local economies. Stating the challenge now is scaling the solutions, ingraining the latter into policy as well as practice, as well as engendering a culture of regeneration that is as grandiose as it is inclusive.

References

Ahmed M et al (2019) Innovative processes and technologies for nutrient recovery from wastes: a comprehensive review 11(18):4938

Alengebawy A et al (2024) Anaerobic digestion of agricultural waste for biogas production and sustainable bioenergy recovery: a review. Environ Chem Lett 22(6):2641–2668

Alhosani K (2025) Leveraging public–private partnerships for a circular industry economy: advancing economic sustainability in industrial waste management in the emirate of Ajman, UAE. Challenges, 16. https://doi.org/10.3390/challe16030031

Alzahrani AIA, Chauhdary SH, Alshdadi AA (2023) Internet of Things (IoT)-based wastewater management in smart cities. Electronics, 12. https://doi.org/10.3390/electronics12122590

Balu UR et al (2024) Linking EDC-laden food consumption and modern lifestyle habits with preeclampsia: a non-animal approach to identifying early diagnostic biomarkers through biochemical alterations. Food Chem Toxicol 194:115073

Barakat NAM et al (2023) Synergistic advancements in sewage-driven microbial fuel cells: novel carbon nanotube cathodes and biomass-derived anodes for efficient renewable energy generation and wastewater treatment. Frontiers Chem 11

Bauknecht M (2024) With agrivoltaics cooperatives towards electricity self-sufficient villages in Germany? 74(4):74–98

Boostani P, Pellegrini-Masini G, Klein J (2024) The role of community energy schemes in reducing energy poverty and promoting social inclusion: a systematic literature review. Energies, 17. https://doi.org/10.3390/en17133232

Canizales L et al (2020) SuperPro Designer®. User-oriented software used for analyzing the techno-economic feasibility of electrical energy generation from sugarcane vinasse in Colombia 8(9):1180

Capodaglio AG (2017) Integrated, decentralized wastewater management for resource recovery in rural and peri-urban areas. Resources, 6. https://doi.org/10.3390/resources6020022

Cerrillo M, Riau V, Bonmatí A (2023) Recent advances in bioelectrochemical systems for nitrogen and phosphorus recovery using membranes. Membranes 13(2):186

Cordell D et al (2011) Towards global phosphorus security: a systems framework for phosphorus recovery and reuse options. Chemosphere 84(6):747–758

Danso-Boateng E, Achaw O-W (2022) Bioenergy and biofuel production from biomass using thermochemical conversions technologies—a review. AIMS Energy 10(4):585–647

Dar MA et al (2024) Termite microbial symbiosis as a model for innovative design of lignocellulosic future biorefinery: current paradigms and future perspectives 4(1):180–201

de Paula RZ, Fontana L, de Jesus TA (2024) Phosphorus removal by free water surface constructed wetlands for the wastewater treatment: bibliometric and bibliographic review. Discover Water 4(1):49

Devi R et al (2024) Development of technologies for municipal solid waste management: current status, challenges, and future perspectives. In: Gupta A, Kumar R, Kumar V (eds) Integrated waste management: a sustainable approach from waste to wealth. Springer Nature Singapore, Singapore, pp 37–62

Dhaliwal A, Sahay A (2019) Schneider electric india private limited: the CSR Dilemma. In: Prange C, Kattenbach R (eds) Management practices in Asia: case studies on market entry, CSR, and coaching. Springer International Publishing, Cham, pp 187–201

Dhar SK et al (2024) Novel Bacillus and Prestia isolates from Dwarf century plant enhance crop yield and salinity tolerance. Sci Rep 14(1):14645

Dhull P et al (2024) A comprehensive review on anaerobic digestion with focus on potential feedstocks, limitations associated and recent advances for biogas production. Environmental science and pollution research

Enyemadze I et al (2021) Phosphorus recovery by struvite precipitation: a review of the impact of calcium on struvite quality. J Water Sanitation Hygiene Dev 11(5):706–718

Farghali M et al (2024) Strategies for ammonia recovery from wastewater: a review. Environ Chem Lett 22(6):2699–2751

Folorunso O et al (2025) GeaGrow: a mobile tool for soil nutrient prediction and fertilizer optimization using artificial neural networks, Vol 9

Fotina NV et al (2024) Enhancement of wheat growth by plant growth-stimulating bacteria during phytopathogenic inhibition. Biocatal Agric Biotechnol 60:103294

Garba A (2020) Biomass conversion technologies for bioenergy generation: an introduction. In: Peixoto Basso T, Basso TO, Basso LC (eds) Biotechnological applications of biomass. IntechOpen: Rijeka

Giel R, Fiedeń M, Dąbrowska A (2025) Real-time automatic identification of plastic waste streams for advanced waste sorting systems 17(5):2157

Godec ML, Kuuskraa VA, Dipietro P (2013) Opportunities for using anthropogenic CO_2 for enhanced oil recovery and CO_2 storage. Energy Fuels 27(8):4183–4189

Gosling I (2005) Process simulation and modeling for industrial bioprocessing: tools and techniques. Ind Biotechnol 1(2):106–109

Guerra-Rodríguez S et al (2020) Towards the implementation of circular economy in the wastewater sector: challenges and opportunities. Water, 12. https://doi.org/10.3390/w12051431

Guo J et al (2019) Enhanced ammonia recovery from wastewater by Nafion membrane with highly porous honeycomb nanostructure and its mechanism in membrane distillation. J Membr Sci 590:117265

Harika R et al (2025) Sweetener synapse: exploring non-nutritive sweeteners' role in zebrafish glucose homeostasis. Environ Technol Innov 39:104322

Hidalgo D et al (2025) Integrated approach for biomass conversion using thermochemical routes with anaerobic digestion and syngas fermentation 17(8):3615

Huber GW, Brown RC (2017) Prospects and challenges of pyrolysis technologies for biomass conversion. Energ Technol 5(1):5–6

Hussain MNN et al (2023) Nitrogen and phosphorus recovery as struvite from nutrient rich wastewater using electrochemical method: a review. AIP Conf Proc 2427(1)

Iddya A et al (2020) Efficient ammonia recovery from wastewater using electrically conducting gas stripping membranes. Environ Sci Nano 7(6):1759–1771

Jabłońska E et al (2021) Impact of vegetation harvesting on nutrient removal and plant biomass quality in wetland buffer zones. Hydrobiologia 848(14):3273–3289

Jerzak W, Acha E, Li B (2024) Comprehensive review of biomass pyrolysis: conventional and advanced technologies, reactor designs, product compositions and yields, and techno-economic analysis. Energies 17(20):5082

Jindo K et al (2012) Biochar influences the microbial community structure during manure composting with agricultural wastes. Sci Total Environ 416:476–481

Johnson C et al (2021) The Bio-based industries joint undertaking as a catalyst for a green transition in Europe under the European green deal. EFB Bioeconomy J 1:100014

Julie Chandra CS et al (2023) Modern biomass conversion technologies. In: Thomas S et al (eds) Handbook of biomass. Springer Nature Singapore, Singapore, pp 1–31

Kala K, Bolia NB (2024) Empowering the informal sector in urban waste management: towards a comprehensive waste management policy for India. Environ Dev 49:100968

Kaur J et al (2022) An exopolysaccharide-producing novel Agrobacterium pusense strain JAS1 isolated from snake plant enhances plant growth and soil water retention. Sci Rep 12(1):21330

Kaur J et al (2023a) GC-MS validated phytochemical up-leveling with in vitro-raised Sansevieria trifasciata [Prain]: the Mother in Law's tongue gets more antibacterial. Current Plant Biology 35–36:100308

Kaur J et al (2023b) Reactive Black-5. Congo red and methyl orange: chemical degradation of azo-dyes by agrobacterium 15(9):1664

Kaur S et al (2025b) Poisoned seas: chemical threats to marine life and human health. In: Prakash C, Kesari KK, Negi A (eds) Sustainable development goals towards environmental toxicity and green chemistry: environment and sustainability. Springer Nature Switzerland, Cham, pp 167–200

Kaur J, Mudgal G, Dhar SK (2025) Decoration plants and their role in remediation. In: Abdel Latef AAH, Zayed EM, Omar AA (eds) Sustainable remediation for pollution and climate resilience. Springer Nature Singapore: Singapore, pp 695–724

Kinidi L et al (2018) Recent development in ammonia stripping process for industrial wastewater treatment 2018(1):3181087

Kumar A, Rao AB (2022) A historical perspective on the biofuel policies in India. In: Agarwal AK, Valera H (eds) Greener and scalable E-fuels for decarbonization of transport. Springer Singapore, Singapore, pp 33–64

Lakhouit A (2025) Revolutionizing urban solid waste management with AI and IoT: a review of smart solutions for waste collection, sorting, and recycling. Results Eng 25:104018

Land M et al (2016) How effective are created or restored freshwater wetlands for nitrogen and phosphorus removal? A systematic review. Environ Evid 5(1):9

Liang Y, Feng H (2019) Removal and recovery of nitrogen pollutants in bioelectrochemical system. In: Wang A-J et al (eds) Bioelectrochemistry stimulated environmental remediation: from bioelectrorespiration to bioelectrodegradation. Springer Singapore, Singapore, pp 157–203

Liu Z et al (2022) A review on the treatment of dyes in printing and dyeing wastewater by plant biomass carbon. Biores Technol 354:127168

Lorick D et al (2020) Effectiveness of struvite precipitation and ammonia stripping for recovery of phosphorus and nitrogen from anaerobic digestate: a systematic review. Environ Evid 9(1):27

Maalouf A, Mavropoulos A (2023) Re-assessing global municipal solid waste generation. Waste Manag Res 41(4):936–947

Mathanker A et al (2021) A review of hydrothermal liquefaction of biomass for biofuels production with a special focus on the effect of process parameters, co-solvents, and extraction solvents. Energies 14(16):4916

Miles SB (2018) Participatory disaster recovery simulation modeling for community resilience planning. Int J Disaster Risk Sci 9(4):519–529

MOEFCC (2016) The solid waste management rules

Moradian JM, Fang Z, Yong Y-C (2021) Recent advances on biomass-fueled microbial fuel cell. Bioresources and Bioprocessing 8(1):14
Morales-Vera R et al (2022) Life cycle assessment of bioethanol production: a case study from poplar biomass growth in the U.S. pacific northwest. 8(12):734
Mudgal G et al (2025) Game changer: how cellulose-based bioleather is transforming the market. In: Kesari KK et al (eds) Functionalized cellulose materials: sustainable manufacturing and applications. Springer Nature Switzerland, Cham, pp 277–295
Mudgal G et al (2025) Strategic conservation plan, in sustainable remediation for pollution and climate resilience. In: Abdel Latef AAH, Zayed EM, Omar AA (eds). Springer Nature Singapore: Singapore, pp 139–206
Mueller A et al (2016) Evolution and feasibility of decentralized concentrating solar thermal power systems for modern energy access in rural areas. MRS Energy Sustain 3:E4
Nguyen HN, Khuong DA (2022) New trends in pyrolysis methods: opportunities, limitations, and advantages. In: Ramola S et al (eds) Engineered biochar: fundamentals, preparation, characterization and applications. Springer Nature Singapore, Singapore, pp 105–126
Ning P et al (2021) Recent advances in the valorization of plant biomass. Biotechnol Biofuels 14(1):102
Nuñez-López V et al (2017) Carbon balance of CO_2-EOR for NCNO classification. Energy Procedia 114:6597–6603
Nuñez-Lopez V et al (2019) Carbon life cycle analysis of CO_2-EOR for net carbon negative oil (NCNO) classification (Final Report). United States
Obileke K et al (2021) Microbial fuel cells, a renewable energy technology for bio-electricity generation: a mini-review. Electrochem Commun 125:107003
ODUM EP (2015) Energy flow in ecosystems: a historical review. American Zoologist 8(1):11–18
Otieno B et al (2023) Struvite recovery from anaerobically digested waste-activated sludge: a short review. J Mater Res 38(16):3815–3826
Qiu B et al (2025) Hydrothermal liquefaction for producing liquid fuels and chemicals from biomass-derived platform compounds: a review. Environ Chem Lett 23(1):81–115
Raina N et al (2024) Biomass valorization for bioenergy production: current techniques, challenges, and pathways to solutions for sustainable bioeconomy. BioEnergy Res 17(4):1999–2028
Ramasam NG et al (2024) Termite mould building—a comparative study on bio-memetics model. AIP Conf Proc 3187(1)
Rao PV et al (2010) Biogas generation potential by anaerobic digestion for sustainable energy development in India. Renew Sustain Energy Rev 14(7):2086–2094
Rao PM et al (2025) Cellulose-based packaging: an emerging sustainable packaging option for future use. In: Kesari KK et al (eds) Functionalized cellulose materials: sustainable manufacturing and applications. Springer Nature Switzerland, Cham, pp 89–115
Rivera F et al (2022) A systematic study of ammonia recovery from anaerobic digestate using membrane-based separation 12(1):19
Rodríguez Arredondo M et al (2015) Bioelectrochemical systems for nitrogen removal and recovery from wastewater. Environ Sci: Water Res Technol 1(1):22–33
Santos AF et al (2024) Nutrient recovery via struvite precipitation from wastewater treatment plants: influence of operating parameters. Coexisting Ions, and Seeding 16(12):1675
Seetharaman B et al (2025) The piteous price of progress: the environmental and health costs of modern society. In: Prakash C, Kesari KK, Negi A (eds) Sustainable development goals towards environmental toxicity and green chemistry: environment and sustainability. Springer Nature Switzerland, Cham, pp 291–313
Sharma A et al (2021) Toxicity with waste-generated ionizing radiations: blunders behind the scenes. In: Kesari KK, Jha NK (eds) Free radical biology and environmental toxicity. Springer International Publishing, Cham, pp 305–325
Sharma HB et al (2024) Engineered biochar/hydrochar derived from organic wastes for energy, environmental, and agricultural applications. Clean Technol Environ Policy 26(12):4059–4093

Singh GB et al (2024) Azo dye bioremediation: an interdisciplinary path to sustainable fashion. Environ Technol Innov 36:103832

Singh GB, Dey S, Mudgal G (2025) Microbial contributions to a circular economy. In: Kaur S, Dhiman S, Tripathi M (eds) Microbial metabolomics: recent developments, challenges and future opportunities. Springer Nature Singapore, Singapore, pp 419–439

Singh A et al (2025) Unveiling the transformative power of smart cellulosic nanomaterials: revisiting potential promises to sustainable future. In: Kesari KK et al (eds) Functionalized cellulose materials: sustainable manufacturing and applications. Springer Nature Switzerland, Cham, pp 1–42

Skondras A et al (2024) Governance strategies for sustainable circular bioeconomy development in Europe: insights and typologies. Sustainability, 16. https://doi.org/10.3390/su16125140

Sugurbekova G et al (2023) Sewage sludge management and application in the form of sustainable fertilizer. Sustainability, 15. https://doi.org/10.3390/su15076112

Thaha AN et al (2025) High impact biomass valorization for second generation biorefineries in India: recent developments and future strategies for sustainable circular economy. Biomass, 5. https://doi.org/10.3390/biomass5010016

Thakur H et al (2025) Smart sustainable materials: a blueprint for a better tomorrow. In: Kesari KK et al (eds) Functionalized cellulose materials: sustainable manufacturing and applications. Springer Nature Switzerland, Cham, pp 253–276

Uddin MM, Wright MM (2023) Anaerobic digestion fundamentals, challenges, and technological advances. De Gruyter 8(9):2819–2837

Upadhyay S et al (2023) Utilization of food waste for the development of composite bread 15(17):13079

van Midden C et al (2023) The impact of anaerobic digestate on soil life: a review. Appl Soil Ecol 191:105066

Veríssimo NV et al (2021) From green to blue economy: Marine biorefineries for a sustainable ocean-based economy. Green Chem 23(23):9377–9400

Vinayak A, Mudgal G, Singh GB (2021) Environment persistent free radicals: long-lived particles. In: Kesari KK, Jha NK (eds) Free radical biology and environmental toxicity. Springer International Publishing, Cham, pp 1–19

Wall AAJ, Rabemanolontsoa H, Venus J (2024) Bioprocessing and fermentation technology for biomass conversion. Appl Sci 14(1):5

Wang J, Azam W (2024) Natural resource scarcity, fossil fuel energy consumption, and total greenhouse gas emissions in top emitting countries. Geosci Front 15(2):101757

White SA (2021) Plant nutrient uptake in full-scale floating treatment wetlands in a Florida Stormwater pond: 2016–2020. Water, 13. https://doi.org/10.3390/w13040569.

Yang C et al (2020) Hydrothermal liquefaction and gasification of biomass and model compounds: a review. Green Chem 22(23):8210–8232

Yang L et al (2023) Technical structure and influencing factors of nitrogen and phosphorus removal in constructed wetlands. Water Sci Technol 89(2):271–289

Zhao C-E et al (2017) Nanostructured material-based biofuel cells: recent advances and future prospects. Chem Soc Rev 46(5):1545–1564

Zhu S, Ye G, Kang X (2023) Combined stripping-struvite precipitation process for treating high ammonia-nitrogen wastewater generated by the craft of vanadium extraction. Environ Sci Water Res Technol 9(11):3049–3061

Zhu Y et al (2024) Review of ammonia recovery and removal from wastewater using hydrophobic membrane distillation and membrane contactor. Sep Purif Technol 328:125094

Wood and Harvested Crop Straws for 3D Printing and Greener Solutions

Sanjoy K. Dhar, Jaspreet Kaur, Abhijeet Singh, Kavya Goyal, Vasantharekha Ramasamy, Sonia Kumari Shishodia, and Gaurav Mudgal

Abstract The progressive integration of lignocellulosic biomass—specifically post-harvest wood and straw residues—into additive manufacturing platforms marks a paradigm shift in sustainable design, ecological engineering, and biofabrication science. This chapter critically investigates the physicochemical attributes of wood and straw-based biomaterials, elucidating their cellulose-dominant microstructures, hierarchical porosity, and thermochemical thresholds that govern rheological behavior and printability in additive manufacturing environments. Through a comparative analysis of Fused Deposition Modeling, Binder Jetting, and Direct Ink Writing, the work examines biomass-processing compatibilities, binder-particle interactions, and multi-material layering protocols to address anisotropic challenges and extrusion variability. Embedding circularity into the design logic, this discourse foregrounds life cycle assessment, biodegradability metrics, and carbon emission deltas across diverse application verticals. Special emphasis is placed on functional deployments in smart agriculture and environmental sensing—where straw-derived sensor housings, moisture-responsive actuators, and wood-pulp probes enable precision diagnostics and adaptive control. Concurrently, the emergence of programmable bio-composites and smart surface interfaces underscores a convergence of material science, biomimicry, and AI-enabled analytics, facilitating 4D transformations and

S. K. Dhar (✉)
Banophool Adibashi Green Heart College, Mirpur, Dhaka, Bangladesh
e-mail: sanjoydharctg@gmail.com

J. Kaur (✉)
Kanya Maha Vidyalaya, Jalandhar, Punjab, India
e-mail: kaurjaspreet2308@gmail.com

A. Singh
School of Bio Sciences and Technology (SBST), VIT University, Vellore, Tamil Nadu, India

K. Goyal · S. K. Shishodia · G. Mudgal (✉)
University Institute of Biotechnology, Chandigarh University, Mohali, Punjab, India
e-mail: gauravmdgl@gmail.com

V. Ramasamy
Department of Biotechnology, School of Bioengineering, Faculty of Engineering and Technology, SRM Institute of Science and Technology, Kattankulathur, Chengalpattu District, Tamil Nadu, India

K. K. Kesari et al. (eds.), *Smart Sustainable Materials Advancing Re-manufacturing and Technology*, World Sustainability Series, https://doi.org/10.1007/978-3-032-17907-4_5

stimulus-responsive behavior in printed constructs. The chapter culminates by articulating the policy, commercialization, and scaling imperatives underpinning mainstream adoption. It proposes a bio-inspired design manifesto wherein waste valorization is reimagined as systemic innovation, and every biodegradable layer embodies regenerative intent. This work reimagines lignocellulosic additive manufacturing not just as a technical innovation, but as part of a larger, eco-conscious movement. By embedding it within a techno-ecological and transdisciplinary framework, we shift the focus from novelty to necessity—championing sustainable fabrication that respects material cycles, embraces ecological wisdom, and upholds ethical design. It's a call to build not just smarter, but more responsibly.

Keywords Biopolymers · Additive manufacturing · Circular economy · Sensor integration · Smart surfaces · Lifecycle assessment · Programmable materials · Eco-design · 4D printing · Biomimicry

1 Introduction: The Rise of Bio-based Additive Manufacturing

Amid escalating environmental degradation, dwindling resources, and climate instability, the global imperative to rethink waste has never been more urgent (Mudgal et al. 2025b; Harika et al. 2025; Singh et al. 2025b; Thakur and Kesari 2025; Kaur et al. 2023a, b, 2025b; Balu et al. 2024; Fotina et al. 2024; Sharma et al. 2021; Vinayak et al. 2021). The traditional linear model of production—extract, manufacture, consume, dispose—has proven unsustainable, producing over 2 billion tonnes of municipal solid waste annually, with projections rising to 3.4 billion tonnes by 2050 (Nature 2022). This challenge demands a paradigm shift towards circularity, where waste is reframed not as an endpoint but as a resource for regeneration. Among the most promising technologies enabling this transition is additive manufacturing (AM), more widely known as 3D printing. By fabricating objects layer by layer, AM minimises material waste, allows for the creation of intricate geometries, and supports decentralised production. When combined with bio-based feedstocks such as agricultural residues—particularly wood and straw—AM emerges as a powerful catalyst for sustainable innovation. These biomaterials, rich in lignocellulosic content, are abundant, renewable, and biodegradable, making them highly suitable for eco-conscious design and fabrication (Garima and Gupta 2024).

1.1 Agricultural Waste: A Hidden Resource

Globally, between 10–50% of agricultural output is discarded each year, creating substantial environmental and economic burdens (Konan et al. 2025; Kottaridis et al.

2025; Narra et al. 2025). Crop straws and wood residues, often burned or landfilled, not only emit greenhouse gases but also waste valuable biomass. However, these materials contain key structural components—cellulose, hemicellulose, and lignin—that can be utilised for biocomposite development and 3D printing applications (Kantaros et al. 2025). Their fibrous architecture provides mechanical strength, while their organic origin ensures biodegradability and carbon neutrality.

Recent research has confirmed the feasibility of transforming straw and wood waste into printable filaments and pastes through pre-treatment methods such as extrusion-biodelignification, enzymatic hydrolysis, and mechanical fibrillation (Konan et al. 2025; Kottaridis et al. 2025; Narra et al. 2025). These processes improve flowability, reduce brittleness, and enhance interlayer adhesion—critical factors for successful additive manufacturing.

1.2 Biomimicry Meets Additive Manufacturing

Nature has long mastered the art of sustainable design through principles like hierarchical structuring, multifunctionality, and resource efficiency. These time-tested strategies now serve as a blueprint for engineering solutions that are both innovative and environmentally conscious. Biomimetic additive manufacturing (BAM) brings these natural principles into the realm of advanced fabrication, merging them with additive manufacturing (AM) technologies to produce structures that mirror biological systems in both shape and behavior (Kantaros et al. 2025). Take, for instance, the layered toughness of nacre or the remarkable tensile strength of spider silk—these biological marvels inspire the creation of multi-material prints that balance strength with minimal material usage. This fusion of biomimicry and AM unlocks the ability to fabricate smart surfaces, responsive components, and self-morphing designs. It's a leap toward 4D printing, where printed objects can transform over time in response to environmental cues (Garima and Gupta 2024). Such breakthroughs aren't just technically impressive—they carry transformative potential across fields like agriculture, architecture, and environmental monitoring.

1.3 Applications Across Sectors

The use of wood and straw-based biomaterials in additive manufacturing (AM) goes far beyond technical innovation—it represents a strategic shift with wide-reaching impact across key sectors:

- **Agriculture**: By transforming crop residues into printed components like sensor housings, irrigation nozzles, and soil probes, AM reduces dependence on petroleum-based plastics and introduces biodegradable solutions that perform effectively in field conditions (Garima and Gupta 2024).

- **Architecture**: Straw-infused biocomposites bring more than sustainability—they offer thermal insulation, sound dampening, and design flexibility. These materials are ideal for interior panels, modular walls, and lightweight structural elements, blending function with aesthetic appeal (Mahajan et al. 2024).
- **Consumer Goods**: From compostable packaging to customizable eco-furniture, AM empowers rapid prototyping and personalization using renewable feedstocks. This aligns perfectly with the growing consumer appetite for environmentally responsible products (Sacchi 2025).
- **Environmental Monitoring**: Printed devices embedded with sensors can monitor soil health, air quality, and water purity. The porous and reactive nature of bio-based materials enhances sensitivity, making them well-suited for real-time environmental diagnostics (Yildiz and Pulatsü 2022).

1.4 Futuristic Horizons and Strategic Vision

Looking ahead, the fusion of AI, machine learning, and bio-composites will unlock new dimensions in sustainable design. Imagine programmable straw-based materials that respond to moisture levels or wood-derived smart surfaces that adjust thermal conductivity based on ambient temperature. These innovations are not speculative—they are emerging realities in labs and pilot projects worldwide (Kantaros et al. 2025).However, challenges remain. Material standardization, scalability, and regulatory frameworks must evolve to support widespread adoption. Industry-academic collaborations, government incentives, and open-source platforms will be pivotal in bridging these gaps (Nature 2022).

1.5 A Call to Action

This chapter advocates for a bio-inspired design culture—one that reclaims agricultural waste as a resource, reimagines fabrication through additive technologies, and reprints the future with sustainability at its core. By embracing wood and straw waste as viable inputs for 3D printing, we not only reduce environmental impact but also catalyze innovation across disciplines. As the world grapples with climate change, resource depletion, and ecological fragility, biomaterials in bloom represent more than a technical solution—they embody a philosophical shift toward harmony with nature. Let us dig deep into the fertile ground of waste and cultivate a future where every stalk and splinter becomes a building block for regeneration (Fig. 1).

Here, the workflow describes first the key processes or components, detailing the transformation of materials into 3D printable filaments and various object creations. Next, it highlights the primary benefits or outcomes emphasized in the illustration,

Fig. 1 Multifaceted applications of wood and harvested crop straws within the realm of 3D printing and more sustainable, eco-friendly solutions

such as eco-friendly advantages like reduced plastic waste and carbon sequestration. Finally, the workflow concludes by stating the overall purpose of promoting sustainable solutions in additive manufacturing for a greener future.

2 Biomaterial Properties: From Lignocellulosics to Printable Pulp

2.1 Structural and Chemical Attributes of Wood and Straw

Wood and straw are lignocellulosic biomaterials, composed primarily of cellulose, hemicellulose, and lignin, along with trace minerals and extractives. These components define their mechanical behavior, biodegradability, and suitability for 3D printing (Table 1). More precisely, cellulose is a linear polysaccharide providing tensile strength and rigidity. It forms microfibrils that reinforce the cell wall structure. Hemicellulose, a branched polymer that binds cellulose and lignin, contributes to flexibility and water retention (Li et al. 2023a). Other than these two, Lignin, a complex aromatic polymer that imparts compressive strength and hydrophobicity. Its structure varies across species and affects thermal behavior (Yang et al. 2011).

Straws from crops like wheat, sorghum, and reed exhibit hollow internodes, vascular bundles, and anisotropic mechanical properties. Their node-containing architecture enhances shear and compression resistance, making them viable for lightweight structural applications (Song et al. 2024). MDPI Biomimetics. The elemental composition bears a high content of carbon (50–53%) and oxygen (45–49%), with trace minerals like K, Si, and Cl influencing thermal and mechanical behavior (Song et al. 2024). IN context of microstructure: Straw walls consist of outer epidermal layers, mechanical tissues, and vascular bundles, which vary by species and growth stage (Li et al. 2023a).

Wood exhibits hierarchical porosity from nanoscale fibrils to macroscale lumina. Its anisotropic structure—with cellulose fibrils aligned along the grain—supports efficient load-bearing and nutrient transport (Chen et al. 2020). Its density range from ~150 kg/m^3 (balsa) to ~1,200 kg/m^3 (pockwood), depending on species and cell wall thickness (Chen et al. 2020).Directional channels and fibrils create structural anisotropy, influencing print orientation and mechanical fidelity.

2.2 Pre-treatment Technologies for Enhanced Printability

To convert wood and straw into printable pulp or filament, several pre-treatment strategies are employed. These aim to improve flowability, adhesion, and thermal stability during additive manufacturing.

2.2.1 Mechanical Pre-treatments

- **Grinding and Milling**: Reduces particle size to improve dispersion and extrusion compatibility.

Table 1 A comprehensive comparison of wood and straw residues as feedstocks for bio-based 3D printing

Attribute	Wood residue	Straw residue	Implications for 3D printing
Cellulose content (%)	40–50	30–40	Higher cellulose improves tensile strength and interlayer bonding
Lignin content (%)	20–30	15–20	Affects rigidity and thermal degradation threshold
Hemicellulose content (%)	15–25	20–30	Influences flexibility and water absorption
Moisture retention	Low to moderate	High	Impacts drying time and extrusion stability
Density (kg/m^3)	300–1200 (species-dependent)	80–150	Affects part weight and structural load capacity
Thermal degradation threshold (°C)	~250	~200	Dictates safe processing temperature range
Structural anisotropy	Strong (grain-aligned microfibrils)	Moderate (hollow stems, vascular bundles)	Influences print orientation and mechanical fidelity
Porosity and surface area	Moderate to low	High	Affects binder absorption and surface functionalization
Elemental composition (C/O ratio)	~50–53% C/~45–47% O	~50–51% C/ ~48–49% O	Determines combustion behavior and biodegradability
Microstructure complexity	Layered cell walls, lumina, tracheids	Epidermal layers, vascular bundles, parenchyma	Impacts mechanical modeling and print resolution
Availability and cost	Widely available; moderate cost	Abundant post-harvest; low cost	Influences scalability and economic feasibility
Environmental impact	Renewable; moderate biodegradability	Renewable; high biodegradability	Supports circular economy and compostable applications

The table highlights the key physical, chemical, and thermal properties that influence printability, mechanical performance, and environmental behavior, serving as a foundational reference for material selection and design optimization

- **Ball Milling**: Enhances surface area and disrupts crystalline cellulose for better bonding.

2.2.2 Chemical Pre-treatments

- **Delignification**: Removes lignin using alkaline or enzymatic methods to reduce brittleness and improve flexibility (Yang et al. 2011).
- **Hydrolysis**: Acid or enzymatic hydrolysis breaks down hemicellulose and cellulose chains for smoother extrusion.
- **Surface Functionalization**: Plasma or chemical treatments introduce polar groups to enhance ink or binder adhesion (Olariu et al. 2023).

2.2.3 Plasma and Thermal Treatments

- **Plasma Pre-Treatment**: Alters surface energy and wettability, improving ink adhesion and layer bonding (Hassabo et al. 2022).
- **Thermal Conditioning**: Pre-heating or controlled drying reduces moisture content and stabilizes feedstock behavior during printing.

2.2.4 Binder Integration

- **Biopolymer Blending**: Mixing straw or wood particles with PLA, PHA, or starch-based binders enhances printability and mechanical performance.
- **Crosslinking Agents**: Additives like citric acid or silicates improve interlayer adhesion and reduce warping.

The printability of lignocellulosic materials hinges on balancing mechanical integrity, thermal behavior, and flow dynamics. Pre-treatment is not merely a technical step—it is a design intervention that transforms agricultural waste into functional, printable matter. As additive manufacturing evolves, programmable pre-treatment protocols—guided by AI and material informatics—will enable tailored pulp formulations for specific applications, from biodegradable sensors to architectural panels.

3 3D Printing Techniques Compatible with Agricultural Residue

3.1 Fused Deposition Modeling (FDM): A Versatile Platform for Biomaterials

Fused Deposition Modeling (FDM) stands out as one of the most popular additive manufacturing methods, thanks to its straightforward setup, cost-effectiveness, and ability to work seamlessly with thermoplastic biocomposites. The process involves feeding a filament—usually composed of a polymer or a polymer blend—into a heated nozzle, which then extrudes the material layer by layer to gradually build the desired object (Siyoum et al. 2025).

3.1.1 Compatibility with Agricultural Residue

Agricultural residues such as wood flour, straw fibers, and lignocellulosic blends can be incorporated into FDM filaments by blending them with biodegradable polymers like PLA, PHA, or starch-based thermoplastics (Iqbal et al. 2024). These biocomposites offer:

- **Improved stiffness and texture** due to fibrous reinforcement
- **Reduced carbon footprint** compared to petroleum-based filaments
- **Enhanced biodegradability**, especially in soil or composting environments

However, challenges such as nozzle clogging, inconsistent extrusion, and interlayer adhesion must be addressed through pre-treatment and binder optimization (Khazaee 2025).

3.1.2 Innovations in FDM for Biomaterials

Recent studies have explored multi-material FDM, dual-nozzle systems, and AI-guided slicing algorithms to optimize print quality and material compatibility. For example, wax-based binders and stearic acid additives have shown promise in improving flowability and mechanical strength of biomass-infused filaments (Khazaee 2025).

3.1.3 Binder Jetting: Powder-Based Precision for Biomass

Binder Jetting is a powder-bed-based additive manufacturing technique where a liquid binder selectively fuses powder particles layer by layer. It is particularly suited for porous and fibrous materials, making it ideal for agricultural residues processed into fine powders (Kafle et al. 2021).

3.1.4 Biomass Integration

Wood and straw powders can be used as the base material, with binders such as:

- **Water-based adhesives**
- **Biopolymer solutions (e.g., alginate, gelatin)**
- **Enzymatic or reactive binders for crosslinking**

Binder jetting allows for low-temperature processing, preserving the organic integrity of biomass and enabling functional porosity for applications like soil sensors, air filters, and biodegradable packaging (Mohanty et al. 2018).

3.1.5 Emerging Trends

- **Post-processing sintering** and **infiltration techniques** are being adapted for bio-based powders to enhance strength and water resistance.
- **Hybrid binder systems** combining natural and synthetic agents are under investigation to balance **eco-performance** and **mechanical fidelity** (Khazaee 2025).

3.1.6 Biocomposites and Hybrid Printing Innovations

The fusion of **natural fibers** with **synthetic or bio-based polymers** has led to a new class of **hybrid biocomposites** that are tailor-made for additive manufacturing.

3.1.7 Hybridization Strategies

- **Dual-fiber reinforcement**: Combining straw with flax, hemp, or coir to balance tensile and flexural properties (Bahrami et al. 2020)
- **Matrix blending**: Mixing PLA with PHA or starch to tune biodegradability and printability
- **Layered printing**: Alternating material layers to create gradient properties or functional zones

These strategies enable customization of mechanical, thermal, and aesthetic properties, making them suitable for agriculture, architecture, and consumer goods.

3.1.8 Smart and Sustainable Applications

- **Self-healing composites** using encapsulated bio-resins
- **Moisture-responsive structures** for irrigation control
- **Biodegradable electronics casings** for environmental monitoring

Hybrid biocomposites are also being explored for 4D printing, where printed objects change shape or function over time in response to stimuli like humidity or temperature (Kafle et al. 2021). The compatibility of agricultural residues with FDM and binder jetting opens a gateway to low-cost, sustainable, and decentralized manufacturing (Table 2). As material informatics, AI-driven design, and bio-inspired engineering converge, the future of additive manufacturing will be defined not just by performance—but by planetary stewardship.

4 Eco-processing and Material Science: Life Cycle Insights

4.1 Circularity, Biodegradability, and Carbon Footprint Comparisons

The integration of agricultural residues into additive manufacturing aligns with the principles of the circular economy, which emphasizes resource regeneration, waste valorization, and reduced environmental impact. Unlike traditional petroleum-based polymers, materials derived from biomass, like wood flour and straw fibers, bring added environmental benefits. They naturally break down over time, come from renewable sources, and even help capture and store carbon, making them a compelling choice for sustainable design.

4.1.1 Circularity in Biomaterials

Circularity in biomaterials isn't just a technical concept, it's a cornerstone of sustainable design and regenerative manufacturing, especially when additive technologies tap into agricultural residues. At its heart, circularity means giving materials a second life: reusing, recycling, or allowing them to biodegrade in ways that prevent harmful end-of-life outcomes like landfilling or incineration. This approach aligns seamlessly with the principles of the circular economy, which aims to break the link between material use and environmental harm by fostering closed-loop systems of resource flow (Scheers 2024).

Agricultural byproducts—such as straw, husks, and wood waste—naturally embody this circular ethos. Sourced from renewable biological origins, they can be reintegrated into ecological cycles after use. With proper processing, these materials can be composted or returned to the soil, enriching nutrient cycles and supporting soil

Table 2 Compatibility of 3D printing techniques with wood and straw-based biomass materials

Printing technique	Biomass form used	Technical parameters	Advantages	Limitations	Strategic applications
Fused deposition modeling (FDM)	Straw/wood blended filament	Temp: 180–220 °C Layer height: 0.1–0.3 mm	Widely accessible Low-cost prototyping PLA compatibility	Nozzle clogging Requires binder optimization Limited resolution	Furniture parts Sensor housings Eco-packaging
Binder jetting	Fine straw/wood powder	Binder viscosity: 10–100 mPa·s Layer thickness: ~ 100 μm	Low-temp processing Porous structures Customizable binders	Weak mechanical strength Post-processing needed	Soil probes Air filters Biodegradable casings
Direct ink writing (DIW)	Straw pulp paste	Viscosity: 10^3–10^5 mPa s Extrusion pressure: 50–300 kPa	Rheology control Functional surfaces Multi-material potential	Ink formulation complexity Slow print speed	Smart irrigation Bio-sensors Wearables
Selective laser sintering (SLS)	Wood-derived powders	Laser power: 10–100 W Temp: >200 °C	High resolution No support structures Durable parts	Thermal degradation risk Energy-intensive	Structural panels Tooling components
Stereolithography (SLA)	Lignin-modified resin blends	UV wavelength: 405 nm Layer thickness: 25–100 μm	High detail Surface smoothness Functional photopolymers	Limited biomass compatibility Resin toxicity concerns	Decorative items Microfluidics Bio-optics
Liquid deposition modeling (LDM)	Straw slurry or gel	Flow rate: 0.1–1 mL/min Layer height: ~0.5 mm	Soft material printing Low energy Food-safe inks	Low mechanical strength Drying shrinkage	Edible packaging Biodegradable decor
Laminated object manufacturing (LOM)	Straw paper sheets	Sheet thickness: 0.1–1 mm Laser/cutter guided	Fast prototyping Low material waste Simple setup	Limited resolution Layer adhesion issues	Educational models Eco-artifacts

A comparative overview of the major 3D printing techniques used in bio-based additive manufacturing. It highlights the form of biomass required, technical parameters, advantages, limitations, and strategic applications, helping researchers and designers align material behavior with printing capabilities health without leaving behind persistent waste (Hoang et al. 2024). This regenerative quality becomes especially powerful in the realm of 3D printing, where biodegradable feedstocks offer a low-impact alternative for prototyping and fabrication.

Adding to this momentum, recent breakthroughs in biotechnology are unlocking new ways to boost the circularity of these biomaterials—making them even more adaptable, functional, and environmentally aligned. Microbial conversion techniques—leveraging bacteria, fungi, and algae—enable the breakdown of lignocellulosic structures into reusable monomers and bioactive compounds. Enzymatic hydrolysis, for instance, facilitates the depolymerization of cellulose and hemicellulose into fermentable sugars, which can be repurposed for bioplastic synthesis or energy generation (Vural Gursel et al. 2022; Malik 2022). These bio-based recycling pathways not only reduce dependency on fossil-derived inputs but also support the development of high-performance, low-impact materials suitable for additive manufacturing. Importantly, the integration of biotechnological strategies into material life cycles enables a shift from linear to adaptive systems—where waste is not merely minimized but actively valorized. For example, microbial enzyme technologies are being explored to depolymerize complex thermoset residues and convert them into feedstocks for new biocomposites (Orlando et al. 2023). Such innovations are critical for scaling sustainable 3D printing practices, especially in sectors like agriculture, architecture, and consumer goods, where material circularity can directly influence ecological resilience and economic viability. In sum, circularity in biomaterials is not a static attribute but a dynamic capability—one that is increasingly shaped by interdisciplinary advances in biotechnology, material science, and ecological design. By harnessing the renewable nature of agricultural residues and embedding them within bio-circular frameworks, we move closer to a future where fabrication is not only intelligent and efficient, but also regenerative and restorative.

4.1.2 Biodegradability Metrics

Biodegradability is a critical performance metric in evaluating the environmental compatibility of bio-based materials, particularly those intended for additive manufacturing and short-life applications. By definition, biodegradable materials are capable of decomposing into carbon dioxide (CO_2), water (H_2O), inorganic compounds, and biomass through the metabolic activity of microorganisms under aerobic or anaerobic conditions (Cakmak 2024). This decomposition process is governed by factors such as polymer structure, environmental parameters (temperature, humidity, microbial diversity), and the presence of additives or reinforcements. Among the most promising biodegradable polymers are polylactic acid (PLA) and polyhydroxyalkanoates (PHAs)—both derived from renewable feedstocks and widely used in sustainable 3D printing. When blended with lignocellulosic fillers such as straw or wood particles, these polymers exhibit accelerated biodegradation rates compared to their neat counterparts. The fibrous reinforcements act as microbial attractants, increase surface area, and disrupt polymer crystallinity, thereby enhancing enzymatic accessibility and hydrolytic breakdown (Cakmak 2024; Ferrari et al. 2022).

Compostability tests conducted under controlled aerobic conditions—typically at 58 ± 2 °C and regulated moisture levels—have demonstrated that straw-based PLA composites can degrade within 90 to 180 days, depending on the binder chemistry, particle dispersion, and matrix porosity (Wyman and Salmon 2024). For instance, composites with alkali-treated straw and biodegradable binders such as starch or PHA tend to exhibit faster degradation kinetics due to improved interfacial bonding and reduced hydrophobicity. Conversely, composites with synthetic or crosslinked binders may require extended composting durations or specialized microbial consortia to achieve complete mineralization. Importantly, biodegradability is not a binary attribute but a spectrum of behavior influenced by material formulation, processing history, and end-of-life conditions. Standards such as ASTM D5338 and ISO 14855 provide protocols for quantifying biodegradation through CO_2 evolution, mass loss, and molecular weight reduction, offering robust frameworks for validating compostability claims (Wyman and Salmon 2024). In the context of circular design, the ability of straw and wood-based composites to biodegrade efficiently not only reduces landfill burden but also enables nutrient cycling, soil enrichment, and carbon neutrality—making them ideal candidates for eco-functional applications in agriculture, packaging, and consumer goods.

4.1.3 Carbon Footprint Comparisons of Straw and Wood-Based Biocomposites

Straw and wood-based biocomposites are gaining attention not just for their sustainability claims, but for how they actually perform across their entire lifespan. Life cycle assessments (LCAs) are increasingly being used to evaluate these materials—providing a detailed, cradle-to-grave picture of their environmental impact, from raw material sourcing to end-of-life disposal. These assessments consistently demonstrate that biocomposites derived from agricultural residues such as straw and wood flour exhibit a 30–70% lower carbon footprint compared to conventional petroleum-based plastics, particularly when the biomass is locally sourced and processed using low-energy, decentralized methods (Ahmad et al. 2022; Kamau-Devers and Miller 2020). This reduction is attributed to the carbon sequestration potential of lignocellulosic feedstocks during their growth phase, as well as the minimal fossil energy inputs required for their transformation into printable materials. Achieving carbon neutrality with these biocomposites is feasible when biomass harvesting adheres to sustainable forestry and agricultural practices, and when processing is powered by renewable energy sources such as solar, wind, or bioenergy itself (Maurya et al. 2021). For instance, decentralized fabrication units using solar-powered extrusion systems have shown promising results in reducing scope 2 emissions during additive manufacturing. Moreover, the biogenic carbon embedded in straw and wood structures can offset emissions from fabrication and transportation, especially when materials are used in short-life-cycle applications like packaging or agricultural sensors.

Beyond production, carbon recycling strategies play a pivotal role in minimizing post-use emissions. Techniques such as pyrolysis, which thermochemically convert biocomposite waste into syngas, bio-oil, and biochar, enable partial recovery of embedded carbon while generating renewable energy (Kafle et al. 2021; Giwa et al. 2019). Similarly, anaerobic digestion of biodegradable biocomposites facilitates the production of methane-rich biogas and nutrient-rich digestate, contributing to circular bioenergy systems and soil enrichment (Yang et al. 2025). These end-of-life valorization pathways not only reduce landfill dependency but also close the carbon loop, reinforcing the regenerative ethos of bio-inspired design. Importantly, LCAs also highlight that transportation distance, binder composition, and feedstock variability significantly influence the net carbon footprint. For example, importing biomass from distant regions may negate the environmental benefits due to fuel consumption and emissions during transit (Fitzgerald et al. 2021). Therefore, regional biomass hubs, standardized feedstock specifications, and AI-guided process optimization are emerging as strategic enablers for scaling low-carbon biocomposite production. In essence, straw and wood-based biocomposites offer a viable pathway toward carbon neutrality, provided that their life cycle is managed holistically—from sustainable sourcing and energy-efficient processing to intelligent end-of-life recovery. As global climate targets become more urgent and breakthroughs in material science continue to unfold, biomaterials like straw and wood-based composites are emerging as game-changers—ready to reshape how we calculate and manage carbon in the manufacturing world.

4.1.4 Thermodynamic and Mechanical Challenges in Processing

Wood and straw-based biomaterials bring clear ecological benefits, but integrating them into additive manufacturing (AM) presents a complex set of challenges—especially when it comes to their thermochemical behavior and mechanical consistency.

From a thermal standpoint, the main hurdle lies in the sensitivity of their lignocellulosic components, particularly lignin and hemicellulose. These begin to degrade at relatively low temperatures (~180–220 °C), making them incompatible with high-heat AM techniques like Selective Laser Sintering (SLS) (Buettner and Weidner 2025). Straw fibers, in particular, retain significant moisture due to their porous vascular structure. During extrusion, this moisture can turn into steam, causing bubbling, warping, and poor interlayer adhesion—issues that compromise both structural integrity and print quality. To counter this, energy-intensive drying and conditioning steps are required, including pre-dehydration, thermal stabilization, and strict humidity control during filament preparation.

Mechanically, one of the biggest challenges is the variability in particle size. In Fused Deposition Modeling (FDM), inconsistent milling or dispersion of straw and wood fillers can lead to nozzle clogging, uneven extrusion, and erratic layering. Fiber agglomeration and poor distribution within the polymer matrix further contribute to uneven mechanical properties—often resulting in layer delamination, reduced tensile strength, and unpredictable failure modes (Tamayo-Vegas et al. 2022; Müller et al. 2024). Achieving strong interfacial bonding between natural fibers and polymer binders is critical, often requiring chemical or plasma-based surface treatments or coupling agents to ensure cohesive integration.

To address these issues, researchers are turning to advanced pre-treatment techniques like plasma activation, alkaline hydrolysis, and enzymatic conditioning. These methods improve fiber–matrix bonding and help stabilize thermal behavior (Singh et al. 2020). Hybrid composites—where straw is blended with other natural fibers such as flax, hemp, or kenaf—are also showing promise in enhancing printability, mechanical strength, and visual appeal. Perhaps most groundbreaking is the use of AI-guided thermodynamic modeling. By leveraging machine learning and simulation tools, researchers can now optimize extrusion parameters, predict heat flow, and anticipate structural defects before printing even begins (Lei et al. 2023).

In essence, successfully processing wood and straw for 3D printing requires more than just technical tweaks—it demands a strategic blend of material science, computational modeling, and adaptive design thinking. Eco-processing of agricultural residues is a multidimensional challenge, balancing sustainability, material behavior, and manufacturing feasibility. As the field advances, we can expect thermodynamic modeling, life cycle analysis, and mechanical optimization to converge—ushering in a new generation of biocomposites that are not only eco-friendly but also structurally sound, versatile, and commercially viable.

5 Smart Applications in Agriculture and Environmental Monitoring

5.1 Precision Agriculture: Biomaterials as Functional Components

The fusion of biomaterials with precision agriculture is transforming the way farming systems engage with their environment. By converting wood and straw residues into printable composites, we unlock a sustainable pathway for producing essential farming tools—sensor housings, irrigation nozzles, and soil probes. These components not only reduce reliance on conventional plastics but also align with the ecological ethos of modern agriculture, where efficiency meets environmental stewardship.

5.1.1 Smart Sensor Integration

Bringing straw and wood-based biomaterials into sensor systems marks a major breakthrough in sustainable agri-tech and environmental monitoring. One standout application is the use of biomass-derived casings for sensors that measure soil moisture, temperature, and nutrient levels. Made from lignocellulosic residues, these casings combine low thermal conductivity with biodegradability and mechanical strength—making them well-suited to shield sensitive electronics from harsh field conditions like UV radiation, temperature swings, and microbial activity. Their naturally porous structure also helps regulate temperature passively and reduces condensation, which is vital for keeping sensors functional over long periods in soil environments (Chavanne and Frangi 2024).

At the same time, researchers are pioneering flexible straw-based substrates for plant-wearable sensors—a new generation of devices that gently adhere to leaf surfaces and track plant health in real time. These sensors can monitor chlorophyll fluorescence, leaf temperature, and water stress without harming the plant, offering valuable insights into photosynthesis and drought response. Thanks to the biocompatibility and flexibility of straw composites, they integrate smoothly with plant tissues, minimizing abrasion or disruption. Even better, they're designed to work with Internet of Things (IoT) platforms, enabling wireless data transmission and cloud-based analytics to support smarter, more responsive farming systems (Chen et al. 2023).

Plant tissue culture plays a pivotal role in transforming wood and harvested crop straws into sustainable feedstocks for 3D printing, offering precision in propagating and optimizing plant material with desired structural and biochemical traits. Through somatic embryogenesis and organogenesis techniques, such as those demonstrated in *Oldenlandia umbellata* L., researchers can amplify biomass yield from dye- and fiber-rich plants, enhancing raw input for biocomposite printing (Siva et al. 2009). Efficient micropropagation protocols, as developed for *Ajuga parviflora*, further enable the mass production of small-flowered herbaceous species with potential EPS-like properties that improve adhesion and texture in bio-ink formulations (Kaur and Mudgal 2020). Studies on *Anagallis arvensis* underscore tissue culture's utility in manipulating cell wall properties and metabolite profiles that influence filament stiffness and printability (Kaur et al. 2023b, 2022; Kaur and Mudgal 2020). Similarly, rapid in vitro multiplication protocols like those established for *Sansevieria trifasciata* var.*Laurentii* (snake plant) highlights how hard-to-propagate ornamentals and fiber-rich tropical species can be scaled up for applications in biodegradable 3D structures and eco-friendly packaging (Kaur et al. 2023a; Kaur and Mudgal 2021). By integrating such methods with targeted molecular and biochemical modulation, tissue culture serves as a gateway to engineering plant-derived materials tailored for greener fabrication technologies.

Complementing these innovations are electrochemical sensors embedded in straw composites, which are being engineered to detect pesticide residues, soil pH, and salinity levels. These sensors utilize conductive pathways formed by carbonized straw fibers or lignin-derived inks, enabling sensitive detection of ionic and molecular analytes. Such systems support data-driven fertilization, irrigation scheduling, and contaminant monitoring, thereby enhancing resource efficiency and environmental stewardship. The use of straw as both structural and functional material reduces reliance on synthetic polymers and metals, aligning with circular design principles and reducing e-waste in sensor networks (Sajib and Sayem 2025).

Collectively, these smart sensor integrations exemplify how biomaterials can serve dual roles—as ecological substrates and as active components in sensing architectures. They pave the way for low-impact, high-resolution monitoring systems that are not only technically robust but also environmentally regenerative.

5.1.2 Field-Ready Applications

The deployment of straw and wood-based biomaterials in field-ready agricultural components exemplifies the convergence of ecological design and functional engineering. One of the most impactful innovations is the fabrication of 3D printed irrigation nozzles using straw–PLA biocomposite blends, which significantly reduce reliance on petroleum-derived plastics while offering customizable flow geometries tailored to crop-specific watering regimes. These nozzles, designed for precision irrigation, can be optimized for droplet size, pressure thresholds, and directional flow, thereby enhancing water-use efficiency and supporting climate-resilient farming practices (Playán et al. 2010; He et al. 2024).

Complementing irrigation systems are biodegradable soil probes fabricated from wood pulp composites, which serve as temporary diagnostic tools for monitoring soil moisture, temperature, and nutrient levels. These probes are engineered to decompose naturally within a single growing season, eliminating the need for post-harvest retrieval and minimizing labor and waste management costs. Studies have shown that wood pulp-based containers and probes degrade efficiently under standard cultivation conditions, especially when paired with organic soil amendments and irrigation protocols (Kumarasingha et al. 2023; Harris et al. 2020). In greenhouse and open-field environments, sensor enclosures printed from straw composites are gaining traction due to their UV resistance, thermal insulation, and biocompatibility. These enclosures protect sensitive electronics from solar radiation, temperature fluctuations, and particulate intrusion, while maintaining structural integrity over the crop cycle. Their porous microstructure also facilitates passive cooling and moisture regulation, making them ideal for housing sensors that monitor microclimatic variables such as humidity, CO_2 concentration, and light intensity. The integration of such enclosures into smart agriculture systems supports real-time data acquisition and automated environmental control, contributing to yield optimization and resource conservation (Bhujel et al. 2020).

Taken together, these field-ready innovations show how bio-based additive manufacturing is redefining agricultural tools—not just in form and function, but in purpose. By turning agricultural waste into high-performance components, we're creating solutions that are scalable, sustainable, and deeply aligned with ecosystem health—tools that serve both the farmer and the planet.

5.2 *Environmental Monitoring: Sustainable Sensing Platforms*

Beyond agriculture, wood and straw-based biomaterials are finding new life in environmental sensing—especially in monitoring air quality, water purity, and ecosystem health. This shift reflects a growing movement toward decentralized, biodegradable, and sensor-integrated systems designed for real-world diagnostics across natural and urban landscapes.

In air quality surveillance, porous straw composites are being engineered as passive sampling platforms. Their fibrous structure and high surface area make them ideal for capturing particulate matter (PM) and volatile organic compounds (VOCs), offering a low-cost, compostable alternative to synthetic filters. These platforms are well-suited for short-term deployment in industrial zones and city environments, where rapid, scalable monitoring is essential.

For water purity, researchers are turning to wood-derived hydrogels—natural polymers with crosslinked networks that exhibit strong hydrophilicity and selective binding to heavy metals like lead, cadmium, and arsenic. These hydrogels can be embedded with optical or electrochemical sensors, enabling real-time detection and simultaneous remediation (Cai et al. 2022). Meanwhile, straw-biocomposite sensor floats are being deployed in ponds, wetlands, and irrigation reservoirs to monitor pH, dissolved oxygen, and temperature. Fabricated through extrusion or low-temperature molding, these floats are designed to degrade harmlessly after use, aligning with circular design principles and reducing the environmental footprint of traditional monitoring tools (Pratama et al. 2019).

Complementing these innovations are printed straw-based surfaces embedded with microfluidic channels. These mimic natural vascular systems, guiding fluid flow and capturing analytes for soil and water pollutant tracking. They interface seamlessly with miniaturized sensor arrays, making them ideal for field deployment. In urban green infrastructure and reforestation zones, wood-infused smart panels are under investigation for thermal mapping and humidity sensing. By combining thermally sensitive lignin blends with piezoresistive cellulosic layers, these panels provide real-time feedback on microclimate conditions, supporting data-driven strategies for biodiversity restoration and climate adaptation (Yu et al. 2025). These technologies highlight a powerful synergy between biomaterial science, sensor innovation, and environmental stewardship, showing how agricultural waste can be transformed into intelligent tools for ecological resilience.

5.3 *Futuristic Horizons and Strategic Vision*

The future of smart agriculture and environmental monitoring is being reshaped by the rise of programmable biomaterials—engineered composites made from wood and straw that can respond to environmental cues, store data, and degrade in a controlled manner after use. These materials sit at the intersection of biofabrication, machine learning, and regenerative design, offering a compelling alternative to conventional plastics and metals. One of the most exciting developments in this space is the creation of AI-enhanced sensor networks built on straw-based substrates. These systems enable autonomous crop monitoring by combining real-time data collection with predictive analytics. Equipped with plant-wearable sensors, soil probes, and wireless communication modules, they support dynamic decision-making in irrigation, fertilization, and disease control (Gayathri et al. 2024). Alongside these innovations, researchers are exploring 4D printed straw composites for shape-shifting irrigation systems that respond to soil moisture levels. Using stimuli-responsive polymers and cellulose-based actuators, these systems morph over time based on hydration gradients—regulating water flow passively, without the need for external power. Inspired by natural phenomena like seed dispersal and leaf curling, they embody biomimicry and temporal design, offering scalable solutions for climate-resilient farming (Dixit and Pandey 2025). Complementing these advances are bio-electronic interfaces that use lignin-derived conductive inks for low-cost environmental sensing. Sourced from carbonized lignin—a byproduct of the paper industry—these inks deliver strong electrochemical performance, flexibility, and biocompatibility. They're ideal for screen-printed sensors, wearable biosensors, and remote monitoring platforms. Applications range from detecting soil pH and heavy metals to tracking airborne pollutants, with particular relevance for underserved and ecologically sensitive regions (Zappi et al. 2021; Kose et al. 2022).

Importantly, the integration of wood and straw biomaterials into smart systems is not merely a technical enhancement—it reflects a philosophical shift toward regenerative design, where materials are selected for their ecological intelligence and lifecycle harmony. These biomaterials align with global sustainability goals by offering low-impact, high-functionality alternatives that support climate resilience, resource efficiency, and ecosystem stewardship (see Table 3). As sensor miniaturization, AI-driven analytics, and biofabrication technologies continue to evolve, the role of biomaterials will expand—transforming fields, forests, and watersheds into intelligent ecosystems that learn, adapt, and regenerate.

6 Architectural and Structural Implementations

As we shift from conceptual frameworks to tangible applications, Sect. 6 explores how wood and straw-based biomaterials are being integrated into the built environment. In this phase, additive manufacturing moves beyond experimental prototypes to deliver sustainable building panels and interior fixtures that match—and often exceed—conventional materials in both performance and visual appeal. The section highlights how topology-optimized, low-load structural printing and modular assembly techniques open new doors for rapid construction, easy disassembly, and regenerative reuse. From honeycomb-style partitions to formwork-free shelters infused with straw composites, these innovations demonstrate how bio-based

Table 3 Application areas of wood and straw biomaterials with environmental benefits and lifecycle outcomes

Application area	Printed component	Environmental benefit	Lifecycle outcome
Precision agriculture	Straw-PLA sensor housings	Reduces petroleum plastic use; biodegradable in soil	Decomposes post-harvest; no retrieval needed
Smart irrigation systems	Moisture-responsive straw actuators	Conserves water; adapts to soil conditions	Compostable after seasonal use
Soil health monitoring	Wood pulp-based probes	Enables data-driven fertilization; low-impact materials	Biodegrades or reused as soil amendment
Greenhouse automation	Straw composite enclosures for sensors	UV-resistant; thermally insulating; low carbon footprint	Recyclable or compostable depending on binder
Air quality monitoring	Porous straw casings for VOC sampling	Passive sensing; avoids synthetic filters	Eco-disposed after data collection
Water purification	Wood hydrogel filter matrices	Filters heavy metals; renewable and non-toxic	Biodegradable or reusable in soil remediation
Wearable plant sensors	Flexible straw-based substrates	Tracks plant stress; integrates with IoT systems	Degrades naturally or recycled with electronics
Environmental sensing floats	Straw biocomposite buoys	Monitors pH, temperature, oxygen in wetlands	Biodegradable in aquatic environments
Urban green monitoring	Wood-infused smart panels	Tracks humidity and thermal shifts in reforestation zones	Reused or composted; minimal e-waste footprint
Pollutant detection surfaces	Straw microfluidic channels	Real-time soil/water pollutant tracking	Single-use; compostable or recyclable

This table presents key use cases for wood and straw-based 3D printed components across agriculture and environmental monitoring. It highlights the functional role of each application, its ecological contribution, and the expected lifecycle behavior—offering a strategic lens for evaluating sustainability and impact

materials can redefine architecture and infrastructure. It's a compelling fusion of ecological vision and engineering precision—where design meets sustainability in real-world construction.

6.1 Sustainable Building Panels and Interior Fixtures

The construction industry is undergoing a profound transformation, driven by the need for materials that minimize environmental impact while maximizing performance. At the forefront of this shift are biocomposites made from wood and straw residues—materials that offer sustainable, high-functionality alternatives for interior fixtures, partition panels, and lightweight structural elements. Through additive manufacturing and advanced composite formulation, these bio-based materials are stepping in to replace conventional options like gypsum boards, plastic laminates, and mineral panels. They bring a lower carbon footprint and greater design flexibility, aligning with the principles of eco-efficient construction. A standout example includes Oriented Strand Boards (OSB) and wood–cement panels, which combine the structural strength of wood strands with the fire-resistant properties of cementitious binders. These hybrids excel in thermal insulation, acoustic dampening, and mechanical durability—making them ideal for partition walls, prefabricated modules, and energy-efficient cladding systems.

In parallel, honeycomb panels crafted from recycled straw and paper cores offer lightweight construction, enhanced thermal resistance, and sound absorption. However, further innovation is needed to improve their fire retardancy and moisture resilience in humid environments (Alwis and Arooz 2023). Straw fibers are also being integrated into drywall alternatives, showing promise in terms of eco-performance and ease of installation. Yet, the environmental costs of gypsum mining and long-distance transport remain challenges to full lifecycle sustainability. Beyond paneling, wood and straw biocomposites are gaining traction in interior design—especially within the realm of green architecture. Their natural textures, renewable origins, and functional versatility make them well-suited for spaces that prioritize both aesthetics and environmental responsibility. Components such as furniture frames, door inserts, and decorative linings are being fabricated from straw–PLA blends, offering customizability through 3D printing, biodegradability, and aesthetic versatility suited to modern minimalist and biophilic interiors. These biocomposites are lightweight, thermally stable, and compatible with modular assembly techniques, making them attractive for circular furniture systems and adaptive interior layouts. Furthermore, printed ceiling tiles and wall claddings using wood pulp composites are supporting the rise of biophilic design, wherein natural textures and biodegradable surfaces foster sensory well-being, cognitive comfort, and ecological connectivity within built environments (Delgado 2020). Hence, these innovations signal a paradigm shift toward regenerative building systems, where agricultural residues not only displace high-impact materials but also encode cultural value, aesthetic resonance, and ecological logic into architectural practice.

6.2 Design Innovations in Low-Load Structural Printing

Low-load structural domains—such as non-load-bearing walls, roofing elements, and temporary shelters—present fertile ground for the application of biomaterial-based 3D printing, where ecological performance and material efficiency take precedence over high-load capacity. Recent advances in performance-based design have enabled the fabrication of structurally optimized components using topology optimization and non-horizontal printing techniques, which allow for the strategic deposition of material along stress trajectories. These methods have demonstrated the potential to reduce concrete and steel usage by up to 60%, particularly in beam and panel systems where traditional formwork and reinforcement are replaced by digitally guided, anisotropic geometries (Wu et al. 2023). By leveraging multi-material topology optimization, designers can distribute stiffness and density gradients across a component, tailoring mechanical performance to localized stress conditions while minimizing overall material consumption (Steck et al. 2024).

In parallel, material synergies are being explored to enhance the functional and environmental performance of printed structures. Wood–cement composites, when extruded via robotic arm-based systems, offer a compelling alternative for prefabricated wall assemblies, combining the thermal mass of cement with the acoustic dampening and fire resistance of lignocellulosic fillers. These composites are particularly suited for modular construction, where robotic precision enables the integration of insulation layers, conduits, and surface textures in a single pass. Meanwhile, straw-infused concrete is gaining traction as a formwork-free printing material for temporary and emergency structures, such as disaster relief shelters and transitional housing. Its low density, high porosity, and biodegradability reduce both labor intensity and carbon emissions, while enabling rapid on-site fabrication without the need for molds or reinforcement cages (Hassan 2025; Alami et al. 2023).

Together, these innovations signal a shift toward adaptive, low-impact construction paradigms, where material intelligence, digital fabrication, and ecological design converge to redefine how we build in resource-constrained and climate-sensitive contexts.

6.3 Circular Construction and Modular Design

The integration of wood and straw-based biomaterials into modular construction systems exemplifies a strategic alignment with the principles of the circular economy, emphasizing disassembly, reuse, and material traceability. When processed into prefabricated components, straw and wood-based biomaterials unlock architectural workflows that are both ecologically responsible and operationally agile. Imagine straw panels engineered via additive manufacturing to interlock like Lego blocks—allowing for rapid on-site assembly, easy dismantling, and reuse at the component level. These modular systems maintain structural integrity and thermal performance while supporting temporary shelters, transitional housing, and scalable infrastructure in disaster-prone or resource-limited regions (Gibbs and O'Neill 2015). At the same time, digital design platforms are reshaping how architects and engineers work with biomaterials. By embedding biomaterial databases into parametric modeling environments, these platforms enable eco-informed decision-making—evaluating materials based on embodied carbon, biodegradability, thermal conductivity, and end-of-life pathways. The integration of material passports—digital records that trace the origin, composition, and recyclability of each component—adds a layer of transparency and accountability, supporting circular lifecycle planning across diverse projects (Gershenfeld 2012). Ultimately, the architectural use of straw and wood biomaterials reflects a powerful convergence of sustainability, design innovation, and digital technology. Through low-load structural printing, modular assembly, and performance-driven optimization, these materials are redefining the built environment—from interior aesthetics to resilient infrastructure. As AI-driven design, robotic fabrication, and biomaterial science continue to evolve, the future of architecture will be shaped not merely by form, but by function, ecological intelligence, and adaptive resilience.

7 Consumer Products and Everyday Green Solutions

7.1 Biodegradable Packaging: From Waste to Wrap

Packaging remains one of the most visible and environmentally consequential domains for sustainable innovation, accounting for a significant share of global plastic waste. In response, researchers and industries are increasingly turning to biomaterials derived from wood and straw residues to develop biodegradable packaging solutions that align with circular economy principles, reduce reliance on fossil-based polymers, and promote end-of-life compostability.

7.1.1 Material Innovations

Recent advances in biopolymer engineering have enabled the transformation of agricultural waste into functional packaging materials:

- **Straw-based bioplastics**, when blended with starch or polylactic acid (PLA), are being used to fabricate food containers, wrappers, and takeaway boxes. These materials exhibit compostability within 60–180 days, depending on environmental conditions and blend ratios. Studies show that starch-based bioplastics reinforced with straw fibers offer improved flexibility and biodegradation rates, making them suitable for single-use food applications (Hamid et al. 2022; Navasingh et al. 2023).

- **Wood pulp composites** are being engineered into multi-layer films with enhanced barrier properties, particularly for moisture-sensitive goods such as snacks, powders, and pharmaceuticals. These films leverage the hydrophilic–hydrophobic layering of cellulose and PLA to achieve oxygen and water vapor resistance, while remaining biodegradable and recyclable (Kadea et al. 2024; Kristak et al. 2025).
- **Active packaging systems** incorporating straw-derived layers infused with antimicrobial agents—such as natamycin and nisin—are extending the shelf life of perishables like cheese and fish. These systems function by releasing antimicrobial compounds into the headspace or food surface, reducing microbial growth and spoilage without direct chemical additives (Berti 2025; Upadhyay et al. 2024).

7.1.2 Functional Enhancements

Beyond material substitution, functional enhancements are being integrated to improve performance, usability, and consumer engagement:

- **Nanocellulose additives** derived from wood—particularly microfibrillated cellulose (MFC) and cellulose nanocrystals (CNCs)—are being incorporated into bioplastic matrices to enhance mechanical strength, thermal stability, and oxygen resistance. These additives increase the crystallinity and interfacial bonding of the polymer network, resulting in stronger, more durable films suitable for high-performance packaging applications (Vineeth et al. 2019).
- **Smart labeling and eco-design principles** are being adopted to improve consumer trust, usability, and transparency. Labels now integrate QR codes, augmented reality, and interactive storytelling to communicate sustainability credentials, product origin, and disposal instructions. Studies show that visual complexity, green–blue color palettes, and minimalist typography enhance consumer perception of sustainability and increase adoption of eco-labeled products (Donato 2025; Pogačar and Gregor-Svetec 2025).

7.2 Eco-furniture and Lifestyle Goods: Design Meets Sustainability

The convergence of biomaterials, additive manufacturing, and sustainable design principles is reshaping the furniture and lifestyle sectors, offering a compelling alternative to conventional materials and production methods (Mudgal et al. 2025; Thakur and Kesari 2025; Kaur et al. 2023b, 2025a; Sharma et al. 2021; Jayaprakash and Kesari 2025; Das and Kesari 2025; Mudgal and Kesari 2025a; Rao and Kesari 2025; Seetharaman et al. 2025; Singh et al. 2024, 2025a). In particular, 3D printed furniture components fabricated from straw–PLA blends are gaining traction for their lightweight, modular, and biodegradable properties. These composites not only reduce dependence on petroleum-based plastics and virgin wood but also enable mass customization and on-demand fabrication, making them ideal for urban living and circular design ecosystems (Somsuk et al. 2025; Agrawal and Bhat 2025). Studies show that alkali-treated straw fibers improve interfacial bonding with PLA matrices, enhancing mechanical performance and print fidelity—critical for load-bearing furniture elements and ergonomic designs.

Simultaneously, wood-based filaments are being employed in the production of decorative panels, lampshades, and tableware, combining natural textures with digital fabrication techniques such as Fused Deposition Modeling (FDM) and Selective Laser Sintering (SLS). These filaments, often infused with lignin or cellulose derivatives, offer aesthetic warmth, tactile richness, and biodegradability, aligning with the principles of biophilic design and eco-functional aesthetics (Meraj and Jawaid 2024; Saad 2016).

From a strategic standpoint, Life Cycle Design (LCD) principles are guiding the development of furniture systems that minimize environmental impact across all stages—from raw material sourcing and manufacturing to use, disassembly, and reuse. LCD frameworks emphasize modular construction, material passports, and design for disassembly, enabling components to be tracked, repaired, and reintegrated into future products without loss of value (Li et al. 2023b). This approach supports closed-loop material flows, reduces landfill burden, and enhances product longevity—key metrics in sustainable furniture innovation. Moreover, the rise of eco-functional aesthetics—where sustainability is embedded in the visual and tactile experience—is resonating strongly with Gen Z and eco-conscious consumers. Today's conscious consumers value more than just functionality—they seek authenticity, transparency, and ethical sourcing. They gravitate toward brands that express environmental values through design language, material choices, and even interactive packaging (Theocharis and Tsekouropoulos 2025). In this context, furniture becomes more than décor; pieces that highlight natural grain, recycled textures, or biodegradable finishes act as narrative artifacts, reflecting the user's identity and commitment to sustainability. Together, these innovations mark a paradigm shift in furniture and lifestyle design. Biomaterials, digital fabrication, and life cycle thinking are converging to produce objects that are not only visually compelling but also environmentally restorative and culturally resonant.

7.3 *Branding Bio-prints as Eco-luxury*

The rise of eco-luxury is redefining what premium truly means—placing sustainability at the heart of high-end branding. For consumer goods crafted from wood and straw biomaterials, this shift positions them not just as functional products, but as ethical luxury artifacts. Attributes like artisanal craftsmanship, natural origin, and low carbon footprint are now central to brand identity. This evolution reflects a deeper cultural rethinking of luxury: exclusivity is no longer about rarity alone, but about responsibility, regeneration, and the stories materials carry. Brands that communicate their sustainability commitments through transparent narratives, eco-labels, and digital material passports are shown to significantly shape consumer perception, loyalty, and purchase behavior—especially among Gen Z and millennials, who value authenticity and environmental ethics (Theocharis and Tsekouropoulos 2025).

From a psychological standpoint, the appeal of eco-luxury is deeply rooted in brand warmth, competence, and morality—dimensions that shape consumer impressions and behavioral intentions. Studies reveal that consumers are more likely to engage with brands that exhibit compassion, fairness, and environmental stewardship, perceiving them as trustworthy and socially aligned with their values (Gidaković et al. 2022). Moreover, visual design elements—including color palettes, material tactility, and surface finishes—play a pivotal role in shaping perceptions of sustainability and quality. For instance, earth tones, natural textures, and minimalist packaging are associated with eco-consciousness and premium craftsmanship, while biodegradable finishes and organic contours reinforce the narrative of regenerative design (Donato 2025; Gidaković et al. 2022).

In this context, consumer products made from wood and straw biomaterials transcend their utilitarian function to become cultural statements—embodying a philosophy of planet-centered design and ethical consumption. By embedding sustainability into every layer of design, branding, and user experience, these products redefine everyday consumption as a regenerative act. As eco-consciousness, digital fabrication, and brand storytelling converge, the future of consumer goods will be shaped by biomaterials that speak not only to the planet, but also to the people—inviting users to participate in a narrative of restoration, responsibility, and refined simplicity.

8 Futuristic Horizons: Programmable Bio-composites and Smart Surfaces

8.1 Programmable Bio-composites: Materials That Think and Respond

The evolution of sustainable additive manufacturing is increasingly defined by the emergence of programmable bio-composites—materials that not only serve structural functions but also exhibit responsive behavior, adaptive morphologies, and environmental intelligence. These composites, often derived from straw, wood, and lignin-based polymers, are being engineered to sense and react to external stimuli such as humidity, UV light, and temperature fluctuations, enabling a new class of self-regulating systems. For instance, moisture-sensitive straw composites have demonstrated the ability to expand, contract, or morph in response to ambient humidity levels, making them ideal for self-adjusting irrigation components, climate-adaptive building skins, and passive ventilation systems (Kuzmin et al. 2024; Hassanpour and Karbhari 2024). These behaviors are driven by the hygroscopic swelling of lignocellulosic fibers and the anisotropic orientation of cellulose microfibrils, which channel deformation along programmed axes.

In parallel, lignin-infused smart polymers are being developed to exhibit UV-responsive functionalities, including self-healing, color modulation, and permeability control. These polymers leverage the aromatic structure of lignin, which absorbs UV radiation and initiates reversible photochemical reactions within the polymer matrix. Applications range from smart coatings and sensor surfaces to adaptive packaging films that respond to light exposure and environmental wear (Moreno and Sipponen 2020).

Perhaps most transformative is the integration of these materials into 4D printing platforms, where time-dependent shape transformations—such as twisting, bending, and folding—are programmed into the composite's microstructure. Inspired by natural seed dispersal mechanisms, these printed forms mimic biological systems like pinecones, orchid pods, and wheat awns, where cellulose microfibril orientation and moisture gradients drive dynamic shape change upon hydration or drying (Jiang et al. 2024; Li et al. 2022). These bioinspired morphing systems are being explored for deployable shelters, responsive facades, and soft robotics, where material behavior is encoded not just in chemistry but in microstructural geometry and stimulus-responsive kinetics.

Collectively, programmable bio-composites represent a paradigm shift—from passive materials to active agents in sustainable design. As AI-guided modeling, multi-material printing, and biomimetic engineering converge, these materials will increasingly shape the future of intelligent ecosystems, regenerative infrastructure, and adaptive manufacturing.

8.2 *Smart Surfaces: Interfaces That Interact*

The advancement of smart surfaces derived from wood and straw biomaterials marks a pivotal shift in sustainable interface design—where materials not only serve structural or aesthetic roles but actively sense, communicate, and regulate their interactions with the environment. These surfaces are being engineered to exhibit functional responsiveness and embedded intelligence, enabling dynamic performance across sectors such as biomedicine, agriculture, and adaptive infrastructure.

Among their functional capabilities, one of the most compelling is switchable wettability, where surfaces toggle between hydrophobic and hydrophilic states in response to external stimuli such as temperature, pH, or light. Inspired by biological systems like lotus leaves and rose petals, these surfaces leverage micro/nano-structuring and stimuli-responsive polymers to modulate liquid adhesion and mobility, with applications in microfluidics, anti-fog coatings, and programmable liquid transport (Li et al. 2021a). For instance, wood-based surfaces functionalized with photo-switchable molecules can reversibly alter their wetting behavior under UV or visible light, enabling smart irrigation membranes and self-cleaning panels.

In parallel, biofouling resistance is being achieved through straw-based coatings infused with antimicrobial agents such as silver nanoparticles, zwitterionic polymers, or plant-derived phenolics. These coatings inhibit bacterial adhesion, reduce biofilm formation, and extend the lifespan of surfaces exposed to humid or biologically active environments—ideal for medical devices, water filtration systems, and food packaging (Cai et al. 2022; Zhu et al. 2020; Asha et al. 2023; Wang et al. 2025). The coatings often combine superhydrophobicity with self-healing properties, ensuring long-term performance even under mechanical stress or abrasion.

Another breakthrough lies in thermo-responsive interfaces, where wood-derived polymers—such as poly(N-isopropylacrylamide) or poly(vinylcaprolactam)—alter their surface energy, adhesion, and mechanical properties near body temperature. These materials are being explored for wearable biosensors, skin-contact adhesives, and drug delivery patches, where temperature-triggered transitions enable on-demand activation, controlled release, or gentle detachment (Ward and Georgiou 2011; Doberenz et al. 2020; Manfredini et al. 2023).

Beyond passive responsiveness, smart surfaces are now embedding intelligence through AI-integrated panels and self-reporting materials. These systems incorporate embedded sensors, wireless communication, and machine learning algorithms to monitor environmental conditions, predict degradation, and adjust surface behavior in real time. For example, AI-enhanced agricultural panels can track humidity, UV exposure, and microbial load, dynamically modulating surface permeability or antimicrobial activity to optimize crop protection (Alotaibi and Nassif 2024; Miller et al. 2025). In infrastructure, self-reporting composites use strain gauges, moisture sensors, and thermal probes to detect stress accumulation, enabling predictive maintenance and adaptive performance in bridges, railways, and building facades (Ayeni 2025; Duan and Liu 2025).

Collectively, these smart surface technologies represent a convergence of biomaterials science, responsive chemistry, and digital intelligence, paving the way for interfaces that interact, learn, and evolve with their surroundings. As IoT integration, edge computing, and bioinspired design continue to mature, these surfaces will become central to next-generation ecosystems—from regenerative architecture to personalized healthcare.

8.3 Synergies with Smart Agriculture and Environmental Analytics

The convergence of programmable biomaterials, AI-driven analytics, and edge-enabled sensing systems is redefining the landscape of precision agriculture and environmental monitoring, transforming passive sustainability into dynamic regeneration. In agricultural contexts, smart irrigation systems fabricated from straw-based composites are being engineered to autonomously respond to soil moisture fluctuations, adjusting water flow in real time without external inputs. These systems utilize the hygroscopic expansion of straw fibers and embedded microfluidic channels to regulate irrigation based on localized hydration gradients, reducing water waste and enhancing crop resilience in drought-prone regions (García et al. 2020; Simić et al.

2016). Complementing this are bio-printed plant wearables, which conform to leaf and stem surfaces to monitor growth metrics, nutrient uptake, and stress indicators such as chlorophyll fluorescence and stomatal conductance. These wearables feed high-resolution physiological data into machine learning models, enabling predictive insights into crop performance, disease onset, and optimal fertilization schedules (Chen et al. 2023).

In parallel, environmental monitoring platforms are integrating smart surfaces derived from wood and straw biomaterials to detect airborne pollutants, temperature shifts, and microbial activity across terrestrial and aquatic ecosystems. These surfaces are functionalized with stimuli-responsive coatings, nanocellulose additives, and biofouling-resistant layers, enabling real-time sensing of VOCs, heavy metals, and microbial biofilms in urban, industrial, and agricultural zones (Shahid et al. 2025). The integration of these platforms with IoT frameworks and edge computing architectures allows for real-time data processing, remote diagnostics, and automated environmental responses, such as activating filtration systems or adjusting greenhouse ventilation based on sensor feedback (Quy et al. 2023).

Collectively, these innovations signal a profound shift from static sustainability to eco-intelligence, where materials and systems are designed to learn, adapt, and co-evolve with their surroundings. As AI, material science, and biomimicry converge, the future of additive manufacturing will be shaped by materials that sense, structures that respond, and systems that regenerate—all rooted in the humble brilliance of wood and straw, now elevated to the forefront of environmental stewardship and agricultural innovation.

9 Policy, Commercialization, and Scaling Challenges

9.1 Market Readiness and Commercialization Pathways

The commercialization of bio-based 3D printing technologies—particularly those utilizing straw and wood-derived composites—remains uneven across global markets, despite a surge in pilot projects and niche applications that have demonstrated technical feasibility. While sectors such as packaging, agriculture, and construction have begun integrating bio-composite printing into select workflows, mainstream adoption is constrained by fragmented supply chains, limited consumer awareness, and the absence of uniform performance standards. In regions with supportive policy frameworks, government incentives—including subsidies for bio-based materials, accelerated depreciation for green equipment, and tax breaks for sustainable manufacturing—have catalyzed early adoption and industrial experimentation (Ullah et al. 2022). These fiscal instruments not only reduce upfront costs but also signal long-term policy commitment to circular manufacturing practices.

Equally pivotal are public–private partnerships, which serve as translational bridges between academic research, industrial prototyping, and market deployment.

These collaborations—often structured around interdisciplinary consortia or innovation hubs—enable the co-development of scalable technologies, facilitate access to pilot infrastructure, and foster regulatory alignment across sectors (Meissner 2019). In parallel, open innovation platforms and technology incubators are nurturing a new generation of startups focused on bio-composite printing, offering mentorship, prototyping facilities, and investor matchmaking. Initiatives such as Terrarium India, C-CAMP BioNEST, and KIIT-TBI Bioincubator exemplify how regional ecosystems are supporting early-stage ventures in sustainable materials and climate biotechnology (Vaz 2024; Reddanna et al. 2024; Conicella 2014).

From a technical standpoint, most straw and wood-based printing technologies currently reside within Technology Readiness Levels (TRLs) 4–6, indicating successful lab validation, component integration, and early prototyping, but limited field deployment and commercial scalability (Bisogni 2023). Advancing these technologies to TRL 7–9—which entails full system demonstration, operational qualification, and market deployment—requires robust standardization, certification, and performance benchmarking protocols. These frameworks are still under development for bio-composites, with efforts underway to define biocircularity metrics, material passports, and sustainability certification schemes that align with circular economy principles (Sierra et al. 2021).

Together, these commercialization pathways underscore the need for multi-stakeholder coordination, policy alignment, and technical maturation to unlock the full market potential of bio-based 3D printing—transforming agricultural residues into regenerative infrastructure and consumer goods.

9.2 Scaling Challenges: From Lab to Landscape

Scaling bio-based 3D printing technologies—particularly those utilizing straw and wood composites—from laboratory success to industrial deployment remains a formidable challenge, shaped by intertwined technical, economic, and regulatory factors. Technically, one of the most pressing barriers lies in material inconsistency due to seasonal and geographic variability in agricultural feedstocks. Variations in lignin content, moisture levels, and fiber morphology not only influence print fidelity but also undermine mechanical reproducibility across batches. Additionally, these biomaterials often exhibit moisture sensitivity and low thermal thresholds, complicating extrusion processes and limiting compatibility with conventional high-temperature printing platforms. Since most commercial 3D printers are optimized for synthetic polymers like PLA and ABS, printing with bio-composites often requires retrofitting extruders, adjusting nozzle temperatures, and modifying flow dynamics to accommodate non-homogeneous feedstocks (Kristak et al. 2025; Jiang et al. 2024).

Economically, the scalability of bio-based printing is hampered by high upfront costs for feedstock pre-treatment, binder formulation, and quality assurance. Fragmented biomass supply chains and decentralized sourcing further limit economies of scale, increasing transportation costs and reducing batch uniformity. As highlighted in the (Possidónio et al. 2025), investor confidence in bio-composite startups remains low due to uncertain return on investment and the lack of proven industrial traction—particularly in applications beyond packaging or decor.

Regulatory frameworks present another critical bottleneck. Currently, there is an absence of unified standards governing the safety, durability, and biodegradability of bio-composite materials, impeding certification and delaying market entry. Novel materials often face protracted evaluation timelines, compounded by insufficient testing protocols and a lack of performance benchmarks tailored to bio-based products. Moreover, circular design policies—such as material passports and end-of-life traceability—are either poorly defined or inconsistently enforced across jurisdictions. Robust policy frameworks are essential to support the commercialization of sustainable technologies and ensure regulatory parity with incumbent materials. In sum, overcoming these scalability challenges will require coordinated action across research institutions, industry stakeholders, and policymakers, anchored in standardization, education, and infrastructural investment.

9.3 Strategic Recommendations

To unlock the full potential of wood and straw-based 3D printing, overcoming existing scalability and adoption barriers requires a coordinated multi-stakeholder strategy that spans policy, research, industry, and international cooperation. These strategic recommendations serve as a blueprint for enabling the transition from innovation to widespread impact—driven by systemic integration, digital infrastructure, and circular design.

From a policy standpoint, governments must institute green procurement mandates that prioritize the use of bio-based materials in public infrastructure, packaging, and construction—thereby creating stable demand and market legitimacy. Such mandates are already gaining traction within the EU and India through initiatives aligned with Sustainable Public Procurement (SPP) frameworks (Holden et al. 2023). In parallel, targeted funding for translational research—bridging the gap between laboratory success and industrial deployment—is essential. Programs like *SERB-TETRA* and *DBT-BIRAC ATGC* are already enabling this transition by supporting prototype validation, pilot scaling, and academia–industry linkage. Furthermore, policy innovation should extend to material passports and traceability systems, which digitize information on material origin, composition, and recyclability—thus supporting circularity, compliance, and trade. Platforms like *Matrak* and *Circuland* exemplify how blockchain-enabled traceability can enhance trust and accountability across supply chains.

Equally vital is the deepening of industry–academia collaboration. The joint development of standardized feedstock specifications, printability indices, and mechanical benchmarks will help stabilize performance variability and create market confidence in bio-composites (Li et al. 2021b). Establishing regional biomass hubs—centralized nodes for straw and wood preprocessing—can reduce logistical inefficiencies, aggregate biomass flow, and promote uniformity in feedstock quality. This strategy is supported by frameworks developed under International Energy Agency (IEA) (Valipour et al. 2024). Additionally, the integration of AI and digital twin systems into processing workflows holds promise for predictive modeling, parameter optimization, and failure detection. Digital twin platforms such as *Valere* and *Tonex* offer real-time insights that can adaptively guide print resolution, binder dispersion, and energy efficiency across manufacturing units.

On the global front, participation in international consortia focused on sustainable additive manufacturing is imperative. Networks like *SIRAMM* and *America Makes* enable cross-border knowledge sharing, standard harmonization, and collective investment in materials innovation. To assess and report impact, organizations should adopt UN SDG-aligned metrics, which allow for multi-dimensional evaluation of environmental, social, and governance performance. Resources such as the *Global SDG Indicators Database* support transparent benchmarking and reporting. Finally, harmonization of certification protocols—especially across ASTM, ISO, and regional bodies—will be critical to facilitate global trade, interoperability, and scalability for bio-composite products (Sierra et al. 2021; Ladu and Blind 2017).

Ultimately, the journey from biomass to bio-print represents not only a technical progression, but a broader transformation in how industries, governments, and communities engage with sustainability. By aligning policy mechanisms, commercialization levers, and global standards, stakeholders can foster an ecosystem in which wood and straw waste becomes the very engine of green innovation, economic resilience, and planetary stewardship.

10 Reimagining Waste: A Manifesto for Bio-inspired Design Culture

10.1 Waste as a Beginning, Not an End

In conventional industrial paradigms, waste is viewed as an inevitable byproduct of production and consumption, destined for landfills or incineration. In stark contrast, nature operates on a closed-loop principle in which every fallen leaf, shed skin, or decayed organism becomes nutrient, structure, or energy for another system. This regenerative logic—where "waste" simply doesn't exist—is the cornerstone of bio-inspired design, which seeks to emulate nature's efficiency, adaptability, and capacity for continuous renewal. Agricultural residues such as wood and straw, often dismissed as detritus, can be reimagined through additive manufacturing as functional artifacts, smart systems, and aesthetic expressions—transforming what was once discarded into sources of value, narrative, and solution.

10.2 Bio-inspiration as a Design Ethos

Bio-inspired design transcends mere mimicry of natural forms; it embodies a philosophy of co-evolution that invites designers, engineers, and researchers to study biological systems for their ethics, resilience, and interconnectedness rather than form alone (Vincent et al. 2006). At its core lie three guiding principles:

- Function follows ecology: every design decision is evaluated against its environmental impact and regenerative potential.
- Adaptation over optimization: materials and systems are conceived to evolve with context, embracing flexibility rather than rigid perfection.
- Collaboration over competition: interdisciplinary and cross-sector partnerships mirror nature's symbiotic relationships, enabling richer, more resilient solutions.
- Together, these principles replace the linear "take–make–dispose" model with a circular narrative of "observe–emulate–regenerate," embedding ecological intelligence at every stage of the design process.

10.3 Designing for Regeneration

The future of design isn't about producing more—it's about creating smarter, longer-lasting, and self-renewing systems. Wood and straw-based 3D printing embodies this shift, offering a blueprint for regenerative innovation across sectors.

- **Material circularity**: Products made from these biomaterials can return safely to the earth at the end of their life cycle, leaving no toxic residue behind.
- **Functional evolution**: Stimuli-responsive composites allow printed objects to adapt, respond, and even improve over time—ushering in a new era of dynamic functionality.
- **Cultural resonance**: Locally sourced biomaterials root products in place, connecting them to regional traditions, ecological knowledge, and community narratives.

Case studies in adaptive architecture, smart agriculture, and eco-luxury branding illustrate how these bio-inspired strategies are not just reshaping industries—they're redefining what value means in a world that demands both innovation and responsibility (Roggema 2022).

10.4 A Call to Designers, Makers, and Thinkers

This manifesto is a clarion call for everyone who designs, teaches, researches, and builds: to see waste as a resource rather than a burden; to design with nature rather than against it; and to create systems that heal rather than harm. As climate change intensifies, resources dwindle, and ecological fragility grows, the role of design must evolve from problem-solving to planet-healing. Wood and straw-based 3D printing stands as both symbol and strategy for this transition—a way to "print the future" with materials that remember the past and regenerate the present. Reimagining waste through a bio-inspired design culture is not merely a creative exercise but a moral imperative—one that demands our collective commitment to ecological harmony and regenerative innovation (Fig. 2).

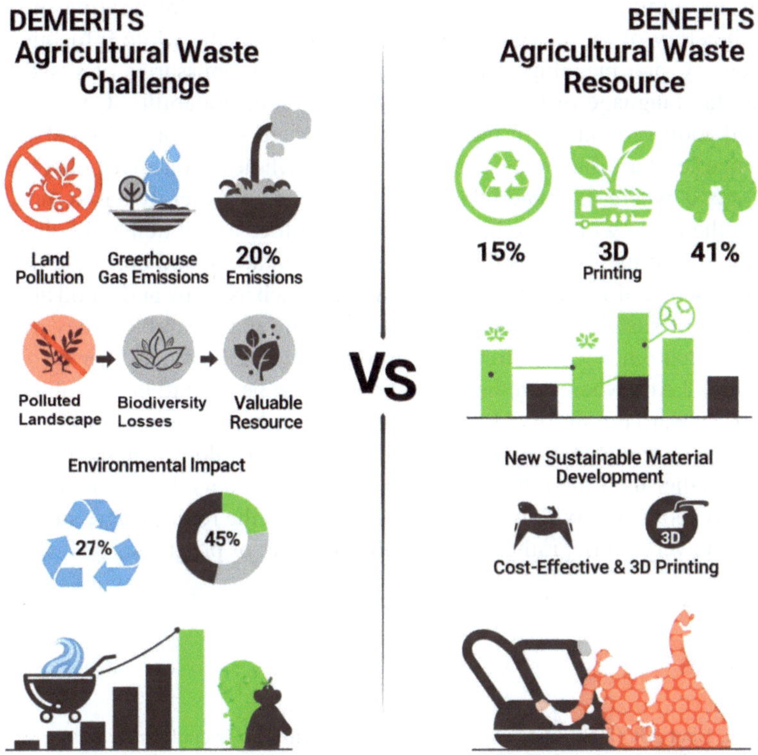

Fig. 2 From burden to benefit: agricultural waste in 3D printing. This infographic provides a compelling visual contrast between the Demerits of Agricultural Waste and the Benefits of its use in 3D Printing. On the left, the "Demerits" section illustrates negative impacts such as land pollution, greenhouse gas emissions, and water contamination through relevant icons and a laptop depicting a polluted environment. Conversely, the "Benefits" section on the right highlights advantages like reduced environmental impact, sustainable development, and the transformation of waste into a valuable resource, further exemplified by symbols for cost-effectiveness and a laptop showcasing a 3D-printed object. This clearly emphasizes the transformative potential of utilizing agricultural waste for greener 3D printing solutions

11 Conclusion: Cultivating a Future with Every Layer

The passage through wood- and straw-based 3D printing uncovers more than a change in the constituent material—it exposes a road map to ecological intelligence, technological synchrony, and design-driven regeneration. From the molecular details of the lignocellulosics to the macro-scale potential of smart agriculture and shapeshifting urban infrastructure, biomaterials are rewriting the meaning of build, package, sense, and interact. Challenging the organic matter of the agriculture industry, we're working not only to eliminate waste, but to reclaim narratives. We're printing pieces that embue the story of the field, the forest, the farmer, all the while infusing innovation, memory, and meaning. These materials become emissaries of sustainability, speaking the language of texture, response, and biodegradability. On the future of bio-based additive manufacturing won't be measured solely on technical specifications or market share, but on the cultural sense as a whole to conceive holistically—to build systems where the materials are being selected not merely based on cost, but based on their role within the broader tapestry of life. It will demand collaboration across the discipline, policies that foster experimentation, and curricula that foster eco-design sensibilities. Ultimately, biomaterials such as straw and wood are less the culmination of sustainable innovation than they are the preamble. They ask us the question: What other waste streams wait to be unlocked? What other resources do we overlook? And where can the wonders of design be used to both problem-solve, as well as steward possibility? Let this page serve as both reference point, as well as rallying cry—a guidepost, if you will, to those who would rather see us print with intention, rather than mere haste, who would rather see us design both empathetically, and responsibly. Who would rather see us build the future where sustainability becomes the very fabric, rather than the afterthought, of progress.

References

Agrawal K, Bhat AR (2025) Advances in 3D printing with eco-friendly materials: a sustainable approach to manufacturing. RSC Sustain 3(6):2582–2604

Ahmad H et al (2022) A comprehensive review on construction applications and life cycle sustainability of natural fiber biocomposites. Sustainability 14(23):15905

Alami AH et al (2023) 3D concrete printing: recent progress, applications, challenges, and role in achieving sustainable development goals. Buildings 13(4):924

Alotaibi E, Nassif N (2024) Artificial intelligence in environmental monitoring: in-depth analysis. Discov Artif Intell 4(1):84

Alwis M, Arooz F (2023) A preliminary experimental study to reuse fabric waste as an alternative partitioning material for building interiors. In: 2023 Moratuwa engineering research conference (MERCon). IEEE

Asha AB et al (2023) Bioinspired antifouling and antibacterial polymer coating with intrinsic self-healing property. Biomater Sci 11(1):128–139

Ayeni O (2025) Integration of Artificial Intelligence in predictive maintenance for mechanical and industrial engineering. Int Res J Mod Eng Technol Sci 7(3):1–23

Bahrami M, Abenojar J, Martínez MÁ (2020) Recent progress in hybrid biocomposites: mechanical properties, water absorption, and flame retardancy. Materials 13(22):5145
Balu UR et al (2024) Linking EDC-laden food consumption and modern lifestyle habits with preeclampsia: a non-animal approach to identifying early diagnostic biomarkers through biochemical alterations. Food Chem Toxicol 194:115073
Berti S et al (2025) A sustainable and edible solution for cheese packaging: active films with Natamycin and Nisin. Food Bioprocess Technol 1–13
Bhujel A et al (2020) Sensor systems for greenhouse microclimate monitoring and control: a review. J Biosyst Eng 45(4):341–361
Bisogni A (2023) Assessing the innovation readiness level of technologies: a new framework. Politecnico Milano, Italy, p 128
Buettner A, Weidner E (2025) Springer handbook of circular plastics economy. Springer Nature
Cai Y et al (2022) A wood sponge sensor for heavy metal ion detection and adsorption. Wood Sci Technol 56(4):1175–1190
Cakmak OK (2024) Biodegradable polymers—a review on properties, processing, and degradation mechanism. Circul Econ Sustain 4(1):339–362
Chavanne X, Frangi J-P (2024) A sensor to monitor soil moisture, salinity, and temperature profiles for wireless networks. J Sens Actuat Netw 13(3):32
Chen C et al (2020) Structure–property–function relationships of natural and engineered wood. Nat Rev Mater 5(9):642–666
Chen R et al (2023) Recent advances and prospects in wearable plant sensors. Rev Environ Sci Bio/Technol 22(4):933–968
Conicella F, Buffa I (2014) Lifesciences incubators
Das D et al (2025) Biocompatible cellulose derivatives: green chemistry and its sustainable applications. In: Kesari KK et al (eds) Functionalized cellulose materials: sustainable manufacturing and applications. Springer Nature Switzerland, Cham, pp 159–190
Delgado JM (2020) Sustainable materials in building construction, vol 11. Springer Nature
Dixit G, Pandey PM (2025) Multi-objective optimization for shrinkage, mechanical, and shape-memory behavior in 4D printed polymer composites. Int J Adv Manuf Technol 136(7):3345–3366
Doberenz F et al (2020) Thermoresponsive polymers and their biomedical application in tissue engineering—a review. J Mater Chem B 8(4):607–628
Donato C (2025) Eco-label visual design and sustainability: the impact on consumer perceptions and market trends. Springer
Duan L, Liu J (2025) Smart composite materials and IoT: revolutionizing real-time railway health monitoring. MRS Commun 15(1):64–80
Ferrari F et al (2022) An overview on wood waste valorization as biopolymers and biocomposites: definition, classification, production, properties and applications. Polymers 14(24):5519
Fitzgerald A et al (2021) A life cycle engineering perspective on biocomposites as a solution for a sustainable recovery. Sustainability 13(3):1160
Fotina NV et al (2024) Enhancement of wheat growth by plant growth-stimulating bacteria during phytopathogenic inhibition. Biocatal Agric Biotechnol 60:103294
García L et al (2020) IoT-based smart irrigation systems: an overview on the recent trends on sensors and IoT systems for irrigation in precision agriculture. Sensors 20(4):1042
Garima, Mishra DK, Gupta D (2024) Introduction to additive manufacturing technologies for bioengineering. In: Additive manufacturing for biomedical applications: recent trends and challenges. Springer, pp 1–17
Gayathri R et al (2024) AI-enhanced farming: a real-time monitoring app powered by IoT. In: 2024 5th International conference on smart electronics and communication (ICOSEC). IEEE
Gershenfeld N (2012) How to make almost anything: the digital fabrication revolution. Foreign Aff 91:43
Gibbs D, O'Neill K (2015) Building a green economy? Sustainability transitions in the UK building sector. Geoforum 59:133–141

Gidaković P et al (2022) The influence of brand sustainability on purchase intentions: the mediating role of brand impressions and brand attitudes. J Brand Manag 29(6):556–568
Giwa AS et al (2019) Pyrolysis coupled anaerobic digestion process for food waste and recalcitrant residues: fundamentals, challenges, and considerations. Energy Sci Eng 7(6):2250–2264
Hamid L et al (2022) Fabricating starch-based bioplastic reinforced with bagasse for food packaging. Circul Econ Sustain 2(3):1065–1076
Harika R et al (2025) Sweetener synapse: exploring non-nutritive sweeteners' role in zebrafish glucose homeostasis. Environ Technol Innov 39:104322
Harris BA, Habteselassie MY, Pennisi SV (2020) Mineral nutrients, organic amendment, and water impact decomposition of biodegradable containers under controlled conditions. Water Air Soil Pollut 231(10):493
Hassabo AG et al (2022) Review on utilizing plasma pretreatment on some fabrics to enhance their printability. Mater Int
Hassan A et al (2025) 3D printed concrete for sustainable construction: A review of mechanical properties and environmental impact. Arch Comput Methods Eng 1–31
Hassanpour B, Karbhari VM (2024) Characteristics and models of moisture uptake in fiber-reinforced composites: a topical review. Polymers 16(16):2265
He L et al (2024) A method to integrate hydraulic structure models into 3D terrain models for irrigation infrastructure visualization. Sci Rep 14(1):21255
Hoang TD et al (2024) Valorization of agriculture residues into value-added products: a comprehensive review of recent studies. Chem Rec 24(8):e202300333
Holden NM et al (2023) Biocircularity: a framework to define sustainable, circular bioeconomy. Circul Econ Sustain 3(1):77–91
Iqbal H et al (2024) Status of polymer fused deposition modeling (FDM)-based three-dimensional printing (3DP) in the pharmaceutical industry. Polymers 16(3):386
Jayaprakash M et al (2025) Collagen fusion with natural biopolymers in food packaging: an in-depth review from conceptualization to consumer application. In: Kesari KK et al (eds) Functionalized cellulose materials: sustainable manufacturing and applications. Springer Nature Switzerland, Cham, pp 65–87
Jiang H et al (2024) 4D printing of liquid crystal elastomer composites with continuous fiber reinforcement. Nat Commun 15(1):8491
Kadea S et al (2024) Utilization of microfibrillated cellulose as a filler in adhesive for improvement of mechanical–thermal performance and biodegradation of wood pulp/PLA laminate biocomposite. J Polym Environ 32(4):1741–1751
Kafle A et al (2021) 3D/4D printing of polymers: fused deposition modelling (FDM), selective laser sintering (SLS), and stereolithography (SLA). Polymers 13(18):3101
Kamau-Devers K, Miller SA (2020) The environmental attributes of wood fiber composites with bio-based or petroleum-based plastics. Int J Life Cycle Assess 25(6):1145–1159
Kantaros A et al (2025) Biomimetic additive manufacturing: engineering complexity inspired by nature's simplicity. Biomimetics 10(7):453
Kaur J, Mudgal G (2021) An efficient and quick protocol for in vitro multiplication of snake plant, Sansevieria trifasciata var. Laurentii [Prain]. Plant Cell Tissue Organ Culture (PCTOC) 147(2):405–411
Kaur J, Mudgal G (2020) An efficient protocol for micro propagation of small-flowered bugleweed (Ajuga parviflora). Plant Cell Biotechnol Mol Biol 16–22
Kaur J et al (2022) An exopolysaccharide-producing novel Agrobacterium pusense strain JAS1 isolated from snake plant enhances plant growth and soil water retention. Sci Rep 12(1):21330
Kaur J et al (2023a) GC-MS validated phytochemical up-leveling with in vitro-raised Sansevieria trifasciata [Prain]: the mother in law's tongue gets more antibacterial. Curr Plant Biol 35–36:100308
Kaur J et al (2023b) Reactive Black-5, Congo Red and Methyl Orange: chemical degradation of azo-dyes by agrobacterium. 15(9):1664

Kaur J, Mudgal G, Dhar SK (2025a) Decoration plants and their role in remediation, in sustainable remediation for pollution and climate resilience. In: Abdel Latef AAH, Zayed EM, Omar AA (eds) Springer Nature Singapore, Singapore, pp 695–724

Kaur S et al (2025b) Poisoned seas: chemical threats to marine life and human health. In: Prakash C, Kesari KK, Negi A (eds) Sustainable development goals towards environmental toxicity and green chemistry: environment and sustainability. Springer Nature Switzerland, Cham, pp 167–200

Khazaee S et al (2025) Development of an environmentally friendly and low-cost binder for 17-4PH metal part printing via fused deposition modeling. In: TMS annual meeting & exhibition. Springer

Konan D et al (2025) Extrusion-biodelignification approach for biomass pretreatment. Waste. MDPI

Kose U et al (2022) Artificial intelligence and smart agriculture technology. CRC Press

Kottaridis S et al (2025) Life cycle assessment on osmotically dehydrated cut potatoes: effects of shelf-life extension on cultivation, waste, and environmental impact reduction. Waste. MDPI

Kristak L et al (2025) Advances in wood-based composites. MDPI 1104

Kumarasingha S, Abeywickrama C, Edirisinghe U (2023) Biodegradable nursery containers: a sustainable solution for plant growth. In: International research symposium-2023 (IRS2023-UoVT) "emerging technologies and skills for resilient industries". Ratmalana, Sri Lanka

Kuzmin A et al (2024) Mechanical, thermal, and water absorption properties of HDPE/barley straw composites incorporating waste rubber. Sci Rep 14(1):25232

Ladu L, Blind K (2017) Overview of policies, standards and certifications supporting the European bio-based economy. Curr Opin Green Sustain Chem 8:30–35

Lei J et al (2023) Insights into pyrolysis kinetics, thermodynamics, and the reaction mechanism of wheat straw for its resource utilization. Sustainability 15(16):12536

Li S et al (2021a) Smart bionic surfaces with switchable wettability and applications. J Bionic Eng 18(3):473–500

Li Y-F et al (2021b) Static and dynamic performances of chopped carbon-fiber-reinforced mortar and concrete incorporated with disparate lengths. Mater Int 14(4):972

Li Y, Zhu N, Chen J (2023a) Straw characteristics and mechanical straw building materials: a review. J Mater Sci 58(6):2361–2380

Li Y, Xiong X, Qu M (2023b) Research on the whole life cycle of a furniture design and development system based on sustainable design theory. Sustainability 15(18):13928

Li Q et al (2022) Hybrid shape-memory, natural fibres bio-based composites for shape morphing and adaptive structures under environmental moisture and temperature actuation. In: Bioinspiration, biomimetics, and bioreplication XII. SPIE

Mahajan A, Devgan S, Zitoune R (2024) Additive manufacturing of bio-implants, 1st edn. In: Amit Mahajan SD, Zitoune R (eds) Biomedical materials for multi-functional applications. Springer Singapore

Malik JA (2022) Microbial and biotechnological interventions in bioremediation and phytoremediation. Springer

Manfredini N et al (2023) Thermo-responsive polymers as surface active compounds: a review. Eur Polym J 198:112421

Maurya PK et al (2021) Roadmap to sustainable carbon-neutral energy and environment: can we cross the barrier of biomass productivity? Environ Sci Pollut Res 28(36):49327–49342

Meissner D (2019) Public-private partnership models for science, technology, and innovation cooperation. J Knowl Econ 10(4):1341–1361

Meraj A, Jawaid M (2024) Plant biomass-based materials: an overview. Plant Biomass Appl 1–24

Miller T et al (2025) Integrating artificial intelligence agents with the internet of things for enhanced environmental monitoring: applications in water quality and climate data. Electronics 14(4):696

Mohanty AK et al (2018) Composites from renewable and sustainable resources: challenges and innovations. Science 362(6414):536–542

Moreno A, Sipponen MH (2020) Lignin-based smart materials: a roadmap to processing and synthesis for current and future applications. Mater Horiz 7(9):2237–2257

Mudgal G et al (2025a) Game changer: how cellulose-based bioleather is transforming the market. In: Kesari KK et al (eds) Functionalized cellulose materials: sustainable manufacturing and applications. Springer Nature Switzerland, Cham, pp 277–295
Mudgal G et al (2025b) Strategic conservation plan. In: Abdel Latef AAH, Zayed EM, Omar AA (eds) Sustainable remediation for pollution and climate resilience. Springer Nature Singapore, Singapore, pp 139–206
Müller T et al (2024) Mechanical recycling: processing challenges. Springer handbook of circular plastics economy. Springer, pp 407–417
Narra M-M et al (2025) Maximizing energy recovery from waste tires through cement production optimization in Togo—a case study. Waste MDPI
Nature EJ (2022) To get serious on the circular economy, upend how global business works. 612(7939):190
Navasingh RJH et al (2023) Sustainable bioplastics for food packaging produced from renewable natural sources. Polymers 15(18):3760
Olariu MA et al (2023) Improving printability of polytetrafluoroethylene (PTFE) with the help of plasma pre-treatment. Polymers 15(16):3348
Orlando M et al (2023) Microbial enzyme biotechnology to reach plastic waste circularity: current status, problems and perspectives. Int J Mol Sci 24(4):3877
Playán E et al (2010) Application of a topographic 3D scanner to irrigation research. Irrig Sci 28(3):245–256
Pogačar A, Gregor-Svetec D (2025) Eco-friendly design for sustainable gift packaging. Appl Sci 15(6):2973
Possidónio C et al (2025) Exploring supply-side barriers for commercialization of new biopolymer production technologies: a systematic review. Sustainability 17(3):820
Pratama RP, Rusdinar A, Wibawa IPD (2019) Floating robot control system for monitoring water quality levels in citarum river. In: 2019 IEEE international conference on internet of things and intelligence system (IoTaIS). IEEE
Quy NM et al (2023) Edge computing for real-time Internet of Things applications: future internet revolution. Wirel Pers Commun 132(2):1423–1452
Rao PM et al (2025) Cellulose-based packaging: an emerging sustainable packaging option for future use. In: Kesari KK et al (eds) Functionalized cellulose materials: sustainable manufacturing and applications. Springer Nature Switzerland, Cham, pp 89–115
Reddanna P, Shaik J, Gandla J (2024) Innovations in the pharma and biopharma industries: role of academia-industry interactions. Biotechnology in India-reworking a strategy. Springer, pp 253–279
Roggema R (2022) Designing for regeneration. Design for regenerative cities and landscapes: rebalancing human impact and natural environment. Springer, pp 1–21
Saad RM (2016) The revolution of materials used in 3D Printing applications in furniture and interior design. Int Des J 6(3):143–163
Sacchi F et al (2025) Bio-based photopolymers for additive manufacturing. J Mater Sci 1–30
Sajib MMH, Sayem ASM (2025) Innovations in sensor-based systems and sustainable energy solutions for smart agriculture: a review. Encyclopedia 5(2):67
Scheers Q et al (2024) Assessing the potential of biodegradable plastics in a circular economy: a methodological outlook. Circul Econ Sustain 1–15
Seetharaman B et al (2025) The piteous price of progress: the environmental and health costs of modern society. In: Prakash C, Kesari KK, Negi A (eds) Sustainable development goals towards environmental toxicity and green chemistry: environment and sustainability. Springer Nature Switzerland, Cham, pp 291–313
Shahid S et al (2025) Innovations in air quality monitoring: sensors, IoT and future research. Sensors 25(7):2070
Sharma A et al (2021) Toxicity with waste-generated ionizing radiations: blunders behind the scenes. In: Kesari KK, Jha NK (eds) Free radical biology and environmental toxicity. Springer International Publishing, Cham, pp 305–325

Sierra AR et al (2021) The bio-based industries joint undertaking: a high impact initiative that is transforming the bio-based industries in Europe. New Biotechnol 60:105–112
Simić M et al (2016) Multi-sensor system for remote environmental (air and water) quality monitoring. In: 2016 24th telecommunications forum (TELFOR). IEEE
Singh DK et al (2020) Finite element modeling of the fiber-matrix interface in polymer composites. J Compos Sci 4(2):58
Singh GB et al (2024) Azo dye bioremediation: an interdisciplinary path to sustainable fashion. Environ Technol Innov 36:103832
Singh A et al (2025a) Unveiling the transformative power of smart cellulosic nanomaterials: revisiting potential promises to sustainable future. In: Kesari KK et al (eds) Functionalized cellulose materials: sustainable manufacturing and applications. Springer Nature Switzerland, Cham, pp 1–42
Singh GB, Dey S, Mudgal G (2025b) Microbial contributions to a circular economy. In: Kaur S, Dhiman S, Tripathi M (eds) Microbial metabolomics: recent developments, challenges and future opportunities. Springer Nature Singapore, Singapore, pp 419–439
Siva R, Rajasekaran C, Mudgal G (2009) Induction of somatic embryogenesis and organogenesis in Oldenlandia umbellata L., a dye-yielding medicinal plant. Plant Cell Tissue Organ Cult (PCTOC) 98(2):205–211
Siyoum YB, Kindie FG, Gebeyehu MA (2025) A review of current research and prospects of fused deposition modelling: application, materials, performance, process variables, parameter optimization, and numerical study. Int J Adv Manuf Technol 1–37
Somsuk N et al (2025) Enhancing mechanical properties of 3D-printed PLA composites reinforced with natural fibers: a comparative study. J Compos Sci 9(4):180
Song J et al (2024) Comparative analysis of macro/microstructures and constituents of sorghum and reed straw. Biomimetics 9(2):107
Steck P et al (2024) A new slicer-based method to generate infill inspired by sandwich-patterns for reduced material consumption. Mater Horiz 17(22):5596
Tamayo-Vegas S et al (2022) The effect of agglomeration on the electrical and mechanical properties of polymer matrix nanocomposites reinforced with carbon nanotubes. Polymers 14(9):1842
Thakur H et al (2025) Smart sustainable materials: a blueprint for a better tomorrow. In: Kesari KK et al (eds) Functionalized cellulose materials: sustainable manufacturing and applications. Springer Nature Switzerland, Cham, pp 253–276
Theocharis D, Tsekouropoulos G (2025) Sustainable consumption and branding for Gen Z: how brand dimensions influence consumer behavior and adoption of newly launched technological products. Sustainability 17(9):4124
Ullah S, Khan FU, Ahmad N (2022) Promoting sustainability through green innovation adoption: a case of manufacturing industry. Environ Sci Pollut Res 29(14):21119–21139
Upadhyay P et al (2024) An overview of advanced antimicrobial food packaging: emphasizing antimicrobial agents and polymer-based films. Polymers 16(14):2007
Valipour M et al (2024) Integrating bio-hubs in biomass supply chains: insights from a systematic literature review. J Clean Prod 467:142930
Vaz E (2024) Artificial terrariums as urban habitats innovative solutions for environmental, economic, and social development. Habitat Int 144:103017
Vinayak A, Mudgal G, Singh GB (2021) Environment persistent free radicals: long-lived particles. In: Kesari KK, Jha NK (eds) Free radical biology and environmental toxicity. Springer International Publishing, Cham, pp 1–19
Vincent JF et al (2006) Biomimetics: its practice and theory. J R Soc Interface 3(9):471–482
Vineeth S, Gadhave RV, Gadekar PT (2019) Nanocellulose applications in wood adhesives. Open J Polym Chem 9(04):63
Vural Gursel I et al (2022) Defining circular economy principles for biobased products. Sustainability 14(19):12780
Wang J et al (2025) Design and application of antifouling bio-coatings. Polymers 17(6):793

Ward MA, Georgiou TK (2011) Thermoresponsive polymers for biomedical applications. Polymers 3(3):1215–1242
Wu H et al (2023) Structural performance-based 3D concrete printing for an efficient concrete beam. In: World congress of architects. Springer
Wyman DA, Salmon S (2024) Critical factors in lab-scale compostability testing. J Polym Environ 32(12):6182–6210
Yang Q et al (2011) Structural characterization of lignin from wheat straw. Wood Sci Technol 45(3):419–431
Yang Z et al (2025) Review of anaerobic digestion models for organic solid waste treatment with a focus on the fates of C, N, and P. Energy Ecol Environ 10(1):1–14
Yildiz HY, Pulatsü S (2022) Towards zero waste: sustainable waste management in aquaculture. J Fisher Aquat Sci (Su Ürünleri Dergisi)
Yu X et al (2025) Urban green spaces enhanced human thermal comfort through dual pathways of cooling and humidifying. Sustain Cities Soc 118:106032
Zappi D et al (2021) Innovative eco-friendly conductive ink based on carbonized lignin for the production of flexible and stretchable bio-sensors. Nanomaterials 11(12):3428
Zhu Y, Ke J, Zhang L (2020) Anti-biofouling and antimicrobial biomaterials for tissue engineering. Racing for the Surface: antimicrobial and interface tissue engineering. Springer, pp 333–354

Sustainable Cellulose Films as Packaging Materials to Reduce Plastic Waste

M. Kilburt and R. Emilin Renitta

Abstract Plastics are non-biodegradable and take a long time to decompose, their extensive use, especially in food packaging, has raised major environmental and health issues. Cellulose and its derivatives are numerous, biodegradable, renewable, and exhibit higher mechanical and barrier properties, they have become known as potential alternatives. This paper highlights cellulose's structure, origins, and physicochemical properties, focusing particular consideration to the material's potential for application in eco-friendly packaging. Cellulose, one of the most versatile natural polymers, is mostly made from plant-based biomass, which includes industrial leftovers, microbial fermentation, and agricultural waste. Its use in active food packaging has been further increased by modern processing methods and functional changes such the addition of antioxidants and antimicrobials. Due to its high tensile strength, crystallinity, and reactive surface area, nano-cellulose provides improved reinforcement in coatings and films, extending the shelf life and preserving food. Recent advances in cellulose functionalization and solvent systems show that cellulose has a great deal of promise to replace traditional plastics, considering issues with cost, moisture sensitivity, and large-scale production. In overall, cellulose-based coatings and films enable waste reduction and promote a circular bioeconomy, making them an essential step toward sustainable packaging solutions.

1 Introduction

Plastics are essential in modern life due to their versatility and many useful properties. Their common use, especially in packaging, poses serious risks to human health, ecosystems, and the environment. Most plastics do not break down, leading to huge amounts of waste worldwide. Current disposal methods, such as landfilling and ocean dumping, are not effective. Some plastics can take hundreds of years to decompose (Xu et al. 2024; Romão et al.2022). As the global population grows

M. Kilburt · R. Emilin Renitta (✉)
Division of Food Processing Technology, Karunya Institute of Technology and Sciences, Coimbatore, India
e-mail: emilinrenitta@karunya.edu

K. K. Kesari et al. (eds.), *Smart Sustainable Materials Advancing Re-manufacturing and Technology*, World Sustainability Series, https://doi.org/10.1007/978-3-032-17907-4_6

and non-renewable resources are used unsustainably, plastic pollution has become a major concern. Governments have created policies to limit single-use plastics like bags and straws to address these environmental problems and encourage sustainable practices (FAO 2011). The COVID-19 pandemic also increased the demand for food packaging materials due to disruptions in global manufacturing and transportation (Hussain et al. 2024). Biodegradable options like bio-bags have emerged, they initially offered only slight improvements over traditional plastics. Modern biodegradable plastics show better properties, making them promising alternatives to petroleum-based products (Romão et al. 2022).

Natural polymers like cellulose, a carbohydrate polymer found in plants, agricultural by-products, marine organisms, and microorganisms, provide an eco-friendly solution for food packaging. The cellulose fibers in plant cell walls are arranged as microfibrils embedded in a matrix of hemicellulose, lignin, pectin, and other substances (Yu et al. 2020). In various industries there is a growing demand for developing a cellulose-based food packaging films to replace the traditional plastics. To ensure the food safety and prolonging shelf life food packaging plays a important role during transport and storage. Antimicrobial and antioxidant packaging systems are the recent innovations that aim to prevent food contamination and spoilage (Xu et al. 2024). Mechanical strength, recyclability, and environmental sustainability are the major factors to design a new packaging material (Hussain et al. 2024). Recycling agricultural by-products like citrus and melon peels also helps in environmental benefits. These peels, usually seen as waste, which is rich in bioactive compounds such as pectin, flavonoids, and phenolic acids which contains more antimicrobial properties (FAO 2011). Incorporating these compounds into biodegradable matrices like carboxymethyl cellulose (CMC) has shown promise for creating active packaging films (Romão et al. 2022). Techniques like X-ray diffraction (XRD), Scanning Electron Microscopy (SEM), and Fourier Transform Infrared (FTIR) spectroscopy are used to examine the physical and chemical properties of these materials.

Nano-cellulose is a form of cellulose that measures less than 100 nm in diameter. It is suitable for food packaging due to its mechanical and barrier properties. Due to its high tensile strength, crystallinity, and surface area it carry antioxidants and antimicrobial agents that can extend the shelf life of food (Hussain et al. 2024). Cellulose derivatives such as methyl Cellulose (MC) and carboxymethyl cellulose (CMC) are combined to improved the barrier properties and mechanical strength of biodegradable films. This makes them good alternatives to petroleum-based plastics. Replacing stiff and flexible packaging materials made from petroleum still presents challenges. Because of their layered structures and food contamination, these materials cannot be recycled. Every year, they contribute millions of tons of plastic garbage, the majority of which leaks into the environment or ends up in landfills (Yu et al. 2020). Because they are readily available, renewable, and biodegradable, biomass-derived polymers such as chitin and cellulose are attractive substitutes. Enhancing their practical qualities through research could revolutionise sustainable packaging and lessen the negative effects of plastic waste on the environment.

2 Structure and Properties of Cellulose

The most prevalent organic polymer on Earth is cellulose. It is essential for the primary cell walls of oomycetes, algae, and green plants. Cellulose is a polysaccharide composed of a linear chain of D-glucose units connected by $\beta(1 \rightarrow 4)$. A few hundred to several thousand units may make up this chain. Because of intermolecular hydrogen bonds, the linear shape enables densely packed chains. The polymer has excellent compressive and tensile strength because of this bonding (Gupta et al. 2019). A naturally occurring linear polymer composed of β-1,4-linked D-glucose units is cellulose. Because of its high mechanical strength due to its semi-crystalline and fibrous character, it is used in a variety of industries, such as construction, textiles, and packaging. To make the micro fibrillar aggregates stiff and extremely resistant to strain, the polymer chain forms intra- and intermolecular hydrogen bonds (Madhushree et al. 2024). Cellulose's biocompatibility, which makes it safe for use in pharmaceutical and medical applications, is one of its main advantages. It is useful for structural applications since it is both strong and lightweight. Its fibrous structure makes it possible to convert it into composites, films, or sheets. Its value as a green material is increased by its low manufacturing cost, recyclability potential, and renewable nature. Cellulose's biodegradable qualities make it a popular substitute for synthetic polymers in packaging. It is also used as a starting material for nano composites in food applications, where it improves product quality and shelf life. The chemical modifiability of cellulose also enhances its range of application, allowing tailoring of surface functionality, solubility, and mechanical properties for particular industry requirements. Consequently, cellulose and its nano structured variants are progressively used in sophisticated applications in electronics, health care, printing, and other areas (Madhushree et al. 2024).

Cellulose is a long polymer made of repeating β-1,4-D-glucopyranose units. These units arrange themselves into highly ordered microfibrils through extensive hydrogen bonding. This structure gives cellulose its high crystallinity, strength, and thermal stability. The same hydrogen-bonded network also makes native cellulose insoluble in all but a few usual solvents, presenting a principal difficulty in its processing for application in regenerated fibres and films. The developments in solvent systems, especially those without chemical derivatization, have allowed considerable advancement in cellulose dissolution and regeneration (Jadhav et al. 2024). One of these developments is the utilization of zwitterionic liquids (ZILs), more specifically aqueous (6-(1-(2-methoxyethyl)imidazol-3-yl)-hexanoate), which contain covalently tethered oxyethylene-substituted imidazolium cations and carboxylate anions. The latter have exhibited the ability to dissolve cellulose as well as enable fibre formation by dry-jet wet spinning (Fig. 1).

The cellulose fibers made from zwitterionic liquids (ZIL) solutions have good properties, like a denier of 1.2–1.6, a tenacity of 3.5–4.0 g/d, and a diameter at break of 8–10%. These are similar to the properties of commercial Lyocell fibers. The regenerated fibre also has a crystallinity of about 60%, reflecting retention of ordered domains necessary for mechanical stability. These features highlight the versatility

Fig. 1 Shows the structure of cellulose

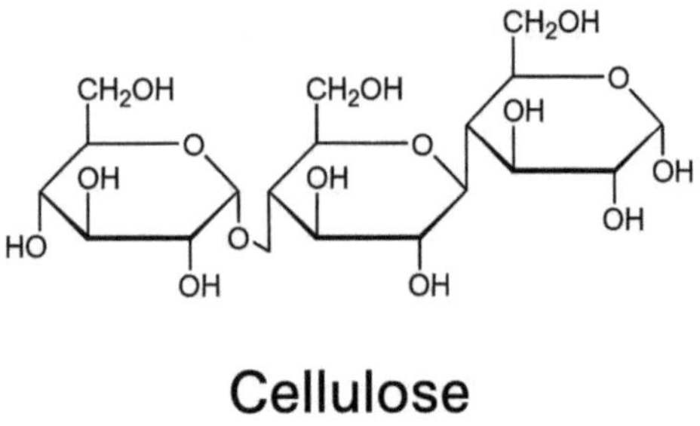

of cellulose during processing with new solvent systems and reiterate its appropriateness for high-performance biodegradable fibre end-uses (Jadhav et al. 2024) because of its non-toxic, biodegradable, and biocompatible properties. Cellulose finds a wide variety of applications. Its inherent mechanical robustness and chemical modifiability make it a preferable component in nanotechnology, pharmaceuticals, food and cosmetic products, textiles, and the paper industry. Cellulose's security features and functional flexibility make it a crucial part of drug delivery systems, especially in cancer treatment. Natural cellulose is purified and turned into cellulose derivatives through different extraction and modification methods, including oxidation, etherification, and esterification. Derivatives like microcrystalline cellulose are commonly applied as excipients in food and pharmaceutical applications owing to their superior binding, tableting, and plasticizing capabilities (Gupta et al. 2019). Specifically, bacterial cellulose is characterized by its good dispersibility, lack of taste, and odorlessness, which render it appropriate for a wide range of industrial uses. Its purity and nanofibrous nature increase its value in biomedical and food uses. Crucially, a large percentage of India's municipal waste—up to 50% cellulose—represents an invaluable, yet underutilized, resource for sustainable material production (Gupta et al. 2019).

2.1 Sources of Cellulose

Cellulose is the most common organic polymer in the world. Lignocellulosic biomass, including wood, non-woody plant fibers, agricultural residues, and by-products resulting from urban or industrial processes, is its major source (Pennells et al. 2021). The percentage of cellulose, hemicellulose, and lignin differs in biomass depending on its origin. Hardwood trees such as oak, maple, and birch and softwood trees such as pine, spruce, and fir are some lignocellulosic materials. Agricultural residues such as coconut coir, wheat straw, maize stalks, rice straw, and palm tree empty fruit bunches are also sources of cellulose. Industrial and municipal waste is another source of cellulose by the examples of sugarcane bagasse, spent brewing grains, and papermaking sludge. These sources are major contributors from the paper and agro-food industries (García et al. 2016).

2.1.1 Woody Biomass

Woody biomass is still a ubiquitous global source of cellulose, harvested from forest and sawmill residues and representing 90–95% of the world's cellulosic pulp (Jin et al. 2021). The wood structure, species and tissue type specific, is composed of cellulose, hemicellulose, lignin, and minor constituents like pectin, xyloglucan, stilbenes, flavonoids, waxes, tannins, and ash-forming minerals. Typical woody cellulose sources are needles, branches, treetops, and forest litter, harvested either directly or as residues of processing. Wood species fall into two broad groups: Softwoods (Conifers) e.g., pine, fir, spruce Hardwoods (Deciduous trees) e.g., birch, eucalyptus, acacia, oak. Softwoods tend to have longer, thicker fibres with more lignin and resin content. Hardwoods, by contrast, have shorter fibres with proportionally more cellulose and hemicellulose (Suota et al. 2021). The two groups are promising raw materials for the manufacture of biopolymers, textile fibres, and biofuels due to their renewable carbohydrate content.

2.1.2 Non-Woody Biomass

Crop residues, municipal solid waste, and energy crops like switchgrass and miscanthus are some of the examples of non-woody biomass. Non-woody biomass has lower cellulose with lignin content and requires less processing than wood (Owonubi et al. 2021). Non-woody biomass can be used to reduce waste and environmental sustainability due to their abundance and affordability (Bajpai 2018; Alila et al. 2013). They can be used to generate biofuels, nanocellulose, and biodegradable plastics.

Crop residues, both primary (straw, stalks) and secondary (husks, bagasse), are useful lignocellulosic biomass. Rice straw and wheat straw, corn stover, sugarcane bagasse are good sources of structural polysaccharides and have wide applications in bioethanol and value-added bioproducts (Mehmood et al. 2018). Annual global production of crop residues ranges from 2,802 to 3,758 million metric tons, or about more than one billion barrels of diesel fuel.

Industrial and municipal waste is also being used for the recovery of cellulose to a greater extent. Products like food waste, garden waste, cardboard, and textiles have high cellulose content. The food processing industry, for instance, produces cellulose-bearing residues like apple pomace, grape skins, citrus peel, potato and carrot peels (Dilucia et al. 2020). Other industrial residues like spent grains, coffee, and dairy effluents (e.g., whey) also have double advantage in waste valorization and cellulose availability, although their suitability is with waste composition, processing technology, and purity demands for cellulose (Munawar et al. 2023).

2.1.3 Microbial Sources

Certain bacteria, such as Acetobacter, Sarcina ventriculi, and Agrobacterium, are capable of synthesising cellulose extracellularly (Fernandes et al. 2020). Plant-derived cellulose are differ from bacterial cellulose; It exhibits high in crystallinity, superior mechanical properties, a nanoscale fibrous network, and high water-holding capacity. Produced as a biofilm, bacterial cellulose is being optimised in laboratory conditions to improve yields for biomedical and food-grade applications (Abeer et al. 2014; Wang et al. 2019).

The primary differences between plant and bacterial cellulose are that plant cellulose is typically associated with hemicellulose and lignin, whereas bacterial cellulose is synthesised in a pure form. Whereas plant cellulose is synthesised through photosynthesis and is found within the walls of plant cells, bacterial cellulose is synthesised through microbial fermentation and is secreted as a protective casing. Bacterial cellulose is more homogenous, more biodegradable and more biocompatible, and is therefore particularly suited to certain applications such as edible hydrogels, wound dressings and implants. Micro algae Spirogyra and Chlorella are other microbial producers of cellulose, which have high photosynthetic efficiency and cellulose productivity. Algal cellulose is also being explored for producing biofuel. Fungal cellulose from mushroom species is also of industrial and biotechnological interest (Bi et al. 2014) (Table 1).

3 Physicochemical Properties of Cellulose and Antioxidant Functionalization

3.1 Solubility

Cellulose is insoluble in water and other conventional organic solvents and hence difficult to utilize in biomedical devices. Intra- and intermolecular hydrogen bonds, hydrophobic, and electrostatic interactions stabilize the cellulose chains and cause this insolubility. Bergenstråhle et al. (2010) and Lindman et al. (2017) noted that these interactions form densely packed, crystalline fibrils with regularly spaced hydroxyl groups along the polymer backbone. Although hydrogen bonding has been most studied, recent evidence is that electrostatic and hydrophobic interactions also contribute to imparting structure (Lindman et al. 2017). Special solvent systems are therefore needed to disrupt these stabilizing interactions.

Low-volatile, thermal-stable, recyclable, and low-toxicity cellulose, ionic liquids (ILs) are a potential green solvent for dissolving cellulose (Li et al. 2016). Various ILs have dissolved cellulose, including 1-butyl-3-methylimidazolium formate (BMIMFmO) (Xu et al. 2010), 1-butyl-3-methylimidazolium chloride (BmimCl) (Erdmenger et al. 2007), and mixed organic–inorganic systems like

Table 1 Different sources of cellulose and its advantages

Sources	Examples	Features	References
Woody biomass—vegetable fibres	• Softwoods: spruce, pine, and fir • Hardwoods: acacia, birch, oak, and eucalyptus	• A significant contributor worldwide (90–95% of cellulosic pulp) • Softwoods have thick, lengthy fibers and a greater lignin/resin content • Hardwoods have higher cellulose and hemicellulose and shorter fibers	Jin et al. (2021), Suota et al. (2021)
Non-woody biomass—agro waste	• The primary remnants are maize stalks, corn stover, and rice and wheat straw • Secondary residues include coir, bagasse, and husks	• Reduction in lignin content facilitates processing; residual production ranges from 2,800 to 3,700 million MT annually; abundant, affordable, and sustainable	Owonubi et al., (2021), Mehmood et al. (2018), Munawar et al. (2023)
Industrial wastes	• Food waste: apple pomace, grape skins, citrus peels, potato and carrot peels • Industrial by-products: sugarcane bagasse, spent grains, coffee grounds, dairy effluents	• Food and packaging waste has a high cellulose content. It serves two purposes: waste valorization and cellulose recovery • Purity and processing techniques determine utility	Dilucia et al. (2020), Munawar et al. (2023)
Microbial sources	• Bacteria: acetobacter, sarcina ventriculi, Agrobacterium • Microalgae: *spirogyra, chlorella* • Fungi: certain mushroom species	• Bacterial cellulose: high crystallinity, nanofibrillar, pure, strong water-holding capacity • Plant cellulose: bound with lignin/hemicellulose • Algal cellulose: efficient photosynthesis, promising for fuels • Fungal cellulose: industrial potential	Fernandes et al. (2020), Wang et al. (2019)

N,N-dimethylacetamide/lithium chloride (DMAc/LiCl) (Potthast et al. 2002). Low-temperature aqueous alkali/urea systems like NaOH/thiourea (Jiang et al. 2017), LiOH/urea, and NaOH/urea (Cai and Zhang 2005) are also effective.

One of the most effective ionic liquids is 1-ethyl-3-methylimidazolium diethyl phosphate ([EMIM] DEP), which has a low melting point and a viscosity of 19–21 °C. It creates stable solutions when dissolved in pyridine (Vitz et al. 2009). There are some drawbacks, including high energy costs, high processing costs, and difficult solvent recovery (de Oliveira Ribeiro et al. 2018; Zhu et al. 2018). Organic solvents like N-methylmorpholine-N-oxide monohydrate (NMMO·H_2O) are excellent alternatives. The solvents are recycled with 99% efficiency due to its non-toxicity. we must produce more cost- and environment-friendly solvents to improve the application of cellulose in the biomedical field (Zhang et al. 2017).

3.2 Mechanical Properties

Cellulose and cellulose derivatives are increasingly engineered into high-performance products by blending them with inorganic materials and synthetic or biopolymers. This is primarily due to their exceptional inherent stiffness as well as tensile strength, particularly in the shape of cellulose I crystals (Benitez and Walther 2017). These characteristics render the cellulose products appropriate for biomedical products that need mechanical properties.

Ordered crystalline domains and strong hydrogen bonds among polymer chains, cellulose nanofibers (CNFs) have high modulus and specific strength (Yu et al. 2012). Several research papers (Šturcová et al. 2005; Tanaka and Iwata 2006; Tashiro and Kobayashi 1991) suggest that crystalline cellulose I has an axial elastic modulus of 124–155 GPa. This has been compared with highly advanced materials like steel and Kevlar (Eichhorn et al. 2010; Moon et al. 2011; Xiong et al. 2016).

Bacterial cellulose (BC) also exhibits superior mechanical properties. Though its elastic modulus ranges from 60 to 115 GPa under dry conditions (Benitez and Walther 2017; Guhados et al. 2005; Hsieh et al. 2008), BC is similar to soft tissues under wet conditions (Hu and Catchmark 2011; Svensson et al. 2005). Its microstructure is comparable to collagen matrices, which are already well characterized in tissue engineering. Its dual mechanical role has generated interest in applying it as a scaffold material (Bäckdahl et al. 2006; Berglund et al. 2003; L'Heureux et al. 1993; Seliktar et al. 2000). The potential of cellulose as a strengthener in drug delivery, wound healing, and tissue regeneration follows from evidence in the comparison of the mechanical properties of different cellulosic materials.

3.3 *Hygroscopicity*

Cellulose materials have good mechanical properties in a dry state, but their water-uptaking characteristics are a constraint in their application. Surface hydroxyl groups in cellulose are liable to hydrogen bond formation with water molecules (Mertaniemi et al. 2016). Water is absorbed mainly in the disordered areas of the cellulose matrix, which are approximately 30% the fiber saturation point (Kolin and Janezic 1996).

Cellulose structures expand when they take up water. This is beneficial in water treatment but may also result in undesirable swelling and distortion in other uses. Nanocellulose materials possess more surface area than conventional fiber-based materials, and their swelling is therefore more noticeable (Benítez et al. 2013). Osmotic pressure results in water absorption in chemically modified cellulose materials with ionizable functional groups (Walther et al. 2020). Such hygroscopicity essentially limits the degree to which cellulose is viable for biomedical uses, especially in conditions where resistance to moisture and size stability are essential.

3.4 *Insertion of Natural Antioxidants in Cellulose-Based Films for Active Food Packaging*

Lipid oxidation is one of the significant reasons for food spoilage, enabling loss of nutritional and energy value and concomitant undesirable change in flavor and color—factors playing significant roles in consumer acceptability (Li et al. 2021). In an effort to avoid such issues, antioxidants have been incorporated into food and food packaging materials. Antioxidants can be either natural or synthetic; however, natural antioxidants such as tocopherols, essential oils and plant extracts are of natural origin such as herbs and spices that have attracted significant interest due to concerns regarding the potential toxicity of synthetic antioxidants, e.g., polyphenols, organophosphates, and thioesters.

Carotenoids, a group of natural pigments with high antioxidant activity, have been researched for application in active biodegradable films. For instance, the incorporation of carotenoids such as norbixin, lycopene, and zeaxanthin into cellulose acetate films was shown to improve mechanical and physical barrier properties of the films according to Assis et al. (2020). Addition of fruit extracts such as the murta fruit and pomegranate seed improves the antioxidant activity of the films. This is due to the fact that they possess a range of phenolic compounds. Studies on cellulose derivatives such as carboxymethyl cellulose (CMC), hydroxypropyl methylcellulose (HPMC), hydroxyethyl cellulose (HEC), and methylcellulose (MC) support this fact.

CMC derivatives are also of interest due to their high film-forming property. He et al. (2021) synthesized CMC-gelatin composite films with functionalized bamboo leaf antioxidants, which showed excellent light barrier property and high DPPH radical scavenging activity with dosage-dependent antioxidant concentration. Some other antioxidants, including curcumin, epigallocatechin gallate, Chinese chives and α-tocopherol, have been incorporated into CMC films successfully, which reflects their possible use as active antioxidant agents for food packaging (Zhou et al. 2022).

Berry extracts have also been comprehensively studied as functional cellulose films. The leaves and fruits of Maquis (Aristotelia chilensis), traditionally used to treat sore throats, ulcers, migraine, and other diseases, exhibit antioxidant, cardioprotective, digestive, anti-migraine and anti-inflammatory activities based on higher phenolic content (Fernandez et al. 2021). Ethyl cellulose derivatives blended with polydimethylsiloxane (PDMS) and clove essential oil have been shown to have greater flexibility, antioxidant activity, and antimicrobial activity, effectively inhibiting biofilm formation on the film surface (Wang and Li 2019). MC-based films incorporating Lippia alba extract and silver nanoparticles showed greater antioxidant and antimicrobial activity due to phenolic compounds catalyzing the biosynthesis of metal nanoparticles (Ahmed et al. 2021).The films with strong antioxidant activity, improved UV–Vis light blocking, and increased mechanical strength—all essential qualities for food packaging materials—have been created using MC combined with chitosan ascorbate. All things considered, the findings show that natural antioxidants embedded in cellulose-based films have the potential to be used to create functional packaging materials that provide antibacterial and antioxidant protection, protecting food preservation and consumer health.

4 Nanocellulose: Properties, Applications, and Potential in Sustainable Materials

The elasticity, stability, and optical characteristics of aqueous media are all strongly affected by the size and form of the nanocellulose particles (Tiwari and Sanjog 2025). The use of nanocellulose in various applications is determined by its structural properties. It has been discovered that square or spherical nanocellulose particles have significant potential as drug delivery agents for encapsulation and stabilisers in Pickering emulsions (Chen et al. 2021).

Many recent research have shown that nano-cellulose is a promising reinforcement material in composite materials (Patel et al. 2023). Researchers can use a variety of chemo-mechanical treatments to modify the nano-cellulose fibers' morphological and structural properties. They are therefore ideal for the creation of green nano composites. The composites have been made into sheets, films, and papers. They have better mechanical qualities, such as increased toughness, strength, and Young's modulus. The materials are being used more and more in visible filtration, wound healing, and water purification because of their enhanced optical transmittance. Research on cellulose-based materials has advanced rapidly across a range of industries due to the increasing need for ecologically benign cellulose fibers, both chemically extracted and produced by bacteria (Singh et al. 2020); Wang et al. 2019).

The many applications of nanocellulose stem from a number of significant characteristics. This material is inexpensive, readily accessible, eco-friendly, and biodegradable (Gupta and Sharma 2023). Nano-cellulose may be converted into a variety of practical goods, such as paper and cellophane, in addition to being utilized as a feedstock for biofuels like ethanol, which is a cleaner fuel than fuels derived from petroleum (Zhou et al. 2018; Fernandez et al. 2017). Nano-cellulose may be effectively recovered from waste materials with comparatively high yields via chemo-mechanical methods. Its applications include the production of food thickeners (Ahmed et al. 2022), adhesives, and fillers for composites due to its non-abrasive properties (Huang and Li 2019). The high quality of the material can be used in advanced technical applications because of its high specific strength, low density, and Young's modulus. Different functional groups can connect to nanocellulose's relatively reactive surface, making it easy to modify it for a particular purpose (Tiwari and Sanjog 2025).

5 Conclusion

The need for sustainable packaging alternatives has been brought to light by the growing environmental problems associated with typical plastic trash. Because cellulose films are biodegradable, numerous, and functionally diverse, they have become a highly future research solution, particularly those derived from renewable plant and microbial sources. Progress in processing methods and material alteration, addition of antioxidants and antimicrobial compounds, is further increasing their applicability for food packaging through improvement of barrier, mechanical, and bioactive properties. Nano-cellulose and cellulose derivatives opened up greater possibilities of developing high-performance, sustainable packaging materials that can displace petroleum-based plastics. Despite challenges such as cost, scalability, and moisture sensitivity, research and innovations are being done to pave the way for cellulose-based goods to play a significant role in reducing plastic pollution. Cellulose films and coatings are excellent investments for creating a circular and sustainable packaging economy as businesses and governments move toward more environmentally friendly practices.

References

Abeer MM, Mohd Amin MCI, Martin C (2014) A review of bacterial cellulose-based drug delivery systems: their biochemistry, current approaches and future prospects. J Pharm Pharmacol 66(8):1047–1061. https://doi.org/10.1111/jphp.12253

Ahmed S, Ikram S, Akhtar MJ (2021) Development of methylcellulose-based bionanocomposite films with *Lippia alba* extract and silver nanoparticles: enhanced antimicrobial and antioxidant properties. Food Packag Shelf Life 30:100739. https://doi.org/10.1016/j.fpsl.2021.100739

Ahmed S, Ikram S, Naz S (2022) Extraction, characterization and food application of cellulose nanocrystals from fruit peels. Carbohyd Polym 276:118738. https://doi.org/10.1016/j.carbpol.2021.118738

Ajala EO, Olatunji OA, Akinlabi ET (2021) Sustainable use of sugarcane bagasse for the production of cellulose nanocrystals. Heliyon 7(1):e06023. https://doi.org/10.1016/j.heliyon.2021.e06023

Alila S, Besbes I, Vilar MR, Vilaseca F, Boufi S, Mutjé P (2013) Non-wood plant fibers: extraction and characterization. In: Gandini A, Belgacem MN (eds) Monomers, polymers and composites from renewable resources. Elsevier, pp 275–294

Assis OBG, Bernardes Filho R, da Silva FF, Tavares MIB (2020) Incorporation of carotenoids into cellulose acetate films for packaging applications. Food Hydrocolloids 101:105509. https://doi.org/10.1016/j.foodhyd.2019.105509

Bäckdahl H, Helenius G, Bodin A, Nannmark U, Johansson BR, Risberg B, Gatenholm P (2006) Mechanical properties of bacterial cellulose and interactions with smooth muscle cells. Biomaterials 27(9):2141–2149. https://doi.org/10.1016/j.biomaterials.2005.11.012

Bajpai P (2018) Biotechnology for pulp and paper processing. Springer. https://doi.org/10.1007/978-981-10-8619-6

Benitez AJ, Walther A (2017) Cellulose nanomaterials: science and technology of the bio-based nanostructure. Angew Chem Int Ed 56(26):6742–6769. https://doi.org/10.1002/anie.201607820

Benítez AJ, Siqueira G, Daisuke H, Walther A (2013) Humidity-controlled water uptake and release in nanopaper films derived from cellulose nanofibrils. ACS Appl Mater Interfaces 5(13):6650–6657. https://doi.org/10.1021/am401643e

Berglund JD, Mohseni MM, Starly B (2003) In vivo remodeling of a tissue-engineered vascular graft. Biomaterials 24(25):4273–4278. https://doi.org/10.1016/S0142-9612(03)00347-2

Bergenstråhle M, Wohlert J, Himmel ME, Brady JW, Henriksson A (2010) Hydrogen bonding in cellulose: the roles of van der Waals forces in cellulose structure and reactivity. Cellulose 17(3):523–530. https://doi.org/10.1007/s10570-009-9405-4

Bi Y, Zhang Z, Guo J, Zhang Y (2014) Isolation and characterization of cellulose from fungal biomass. Cellulose 21:4571–4580. https://doi.org/10.1007/s10570-014-0435-7

Cai J, Zhang L (2005) Rapid dissolution of cellulose in LiOH/urea and NaOH/urea aqueous solutions. Macromol Biosci 5(6):539–548. https://doi.org/10.1002/mabi.200400222

Chen H, Huang C, Zhang C, He Y (2021) Applications of nanocellulose as pickering emulsifiers in food and pharmaceuticals: A review. Food Hydrocolloids 112:106287. https://doi.org/10.1016/j.foodhyd.2020.106287

de Oliveira Ribeiro AP, Silvério HA, Teixeira EM (2018) Ionic liquids and deep eutectic solvents for cellulose dissolution: recent advances and perspectives. Carbohyd Polym 196:140–150. https://doi.org/10.1016/j.carbpol.2018.05.048

Dilucia F, Lacivita V, Conte A, Del Nobile MA (2020) Sustainable use of fruit and vegetable by-products to enhance food packaging performance. Food Packag Shelf Life 26:100569. https://doi.org/10.1016/j.fpsl.2020.100569

Eichhorn SJ, Dufresne A, Aranguren M, Marcovich NE, Capadona JR, Rowan SJ, Weder C (2010) Review: Current international research into cellulose nanofibres and nanocomposites. J Mater Sci 45:1–33. https://doi.org/10.1007/s10853-009-3874-0

Eom J, Park S, Jin HJ, Kwak HW (2020) Multiscale hybridization of natural silk–nanocellulose fibrous composites with exceptional mechanical properties. Front Mater 7:98p.https://doi.org/10.3389/fmats.2020.00098

Erdmenger T, Vitz J, Wiesbrock F, Schubert US (2007) Recent developments in the use of ionic liquids for cellulose chemistry: a short review. Macromol Rapid Commun 28(9):1000–1018. https://doi.org/10.1002/marc.200700059

Fernandes M, Salgado AJ, Gomes ME (2020) Bacterial cellulose: production, properties and applications in regenerative medicine. Appl Microbiol Biotechnol 104(16):6867–6881. https://doi.org/10.1007/s00253-020-10729-3

Fernandez J, Etxabide A, Uranga J, Guerrero P, de la Caba K (2017) Influence of gelatin concentration on properties of gelatin-based biodegradable films. Food Hydrocolloids 66:355–364. https://doi.org/10.1016/j.foodhyd.2016.12.020

Fernandez K, Rojas MJ, Salgado N (2021) Functional characterization of *Aristotelia chilensis* extract and its application in food packaging films. Food Res Int 141:110151. https://doi.org/10.1016/j.foodres.2021.110151

Food and Agriculture Organization of the United Nations (2011) Global food losses and food waste—extent, causes and prevention. FAO. https://www.fao.org/3/i2697e/i2697e.pdf

García A, Egüés I, Labidi J (2016) Lignocellulosic biomass as a source of valuable products. In: Bioenergy research: advances and applications, pp 267–289. Elsevier

Guhados G, Wan W, Hutter JL (2005) Measurement of the elastic modulus of single bacterial cellulose fibers using atomic force microscopy. Langmuir 21(14):6642–6646. https://doi.org/10.1021/la0504311

Gupta PK, Singh A, Katiyar V (2019) Cellulose: a review on its structure, properties, sources, and applications. Carbohyd Polym 210:1–12. https://doi.org/10.1016/j.carbpol.2019.01.069

Gupta R, Sharma V (2023) Nanocellulose-based composites: recent advances and emerging trends. Mater Today: Proc 69(2):507–514. https://doi.org/10.1016/j.matpr.2022.07.332

He Q, Ren Y, Wang L (2021) Functional properties of antioxidant films developed from gelatin and carboxymethyl cellulose incorporated with bamboo leaf extract. Food Packag Shelf Life 28:100643. https://doi.org/10.1016/j.fpsl.2021.100643

Hsieh YL, Yano H, Nakagaito AN, Nogi M (2008) Strength of transparent nanocomposites composed of cellulose nanofibers and acrylic resin. Cellulose 15(4):619–624. https://doi.org/10.1007/s10570-008-9215-2

Hu W, Catchmark JM (2011) Mechanical properties of single bacterial cellulose fibers. Biomacromol 12(9):3434–3440. https://doi.org/10.1021/bm200885g

Huang X, Li M (2019) Green nanocomposite adhesives from nanocellulose and soy protein isolate for wood composites. Compos B Eng 177:107385. https://doi.org/10.1016/j.compositesb.2019.107385

Hussain A, Ahmad T, Niazi MBK, Ahmad Z (2024) Impact of COVID-19 pandemic on global food packaging and sustainable alternatives. J Environ Manage 345:119862. https://doi.org/10.1016/j.jenvman.2024.119862

Jin K, Tang Y, Liu J, Wang J, Ye C (2021) Nanofibrillated cellulose as coating agent for food packaging paper. Int J Biol Macromolecules 168:331–338. https://doi.org/10.1016/j.ijbiomac.2020.12.066

Jadhav AG, Mhaske ST, Gite VV (2024) Advances in cellulose dissolution using novel zwitterionic solvents for fibre regeneration. Carbohyd Polym 322:121093. https://doi.org/10.1016/j.carbpol.2024.121093

Jiang Z, Liu X, Chen Y (2017) Dissolution of cellulose in NaOH/thiourea aqueous solution: role of urea and thiourea. Cellul Chem Technol 51(9–10):897–904

Kolin A, Janezic T (1996) Water uptake and swelling in cellulose materials. Cellul Chem Technol 30(3–4):353–364

Li H, Zhang D, Zhou J (2016) Ionic liquid-based solvent systems for cellulose dissolution and functional materials preparation: a review. Prog Mater Sci 78–79:1–38. https://doi.org/10.1016/j.pmatsci.2016.03.001

Li T, Chen C, Brozena AH, Zhu JY, Xu L, Driemeier C, Dai J, Rojas OJ, Isogai A, Wågberg L, Hu L (2021) Developing fibrillated cellulose as a sustainable technological material. Nature 590(7844):47–56. https://doi.org/10.1038/s41586-020-03167-7

L'Heureux N, Paquet S, Labbe R, Germain L, Auger FA (1993) A completely biological tissue-engineered human blood vessel. FASEB J 7(9):1243–1249. https://doi.org/10.1096/fasebj.7.9.8370485

Lindman B, Karlström G, Stigsson L (2017) On the mechanism of dissolution of cellulose. J Mol Liq 228:927–933. https://doi.org/10.1016/j.molliq.2016.10.056

Madhushree M, Vairavel P, Mahesha GT, Bhat KS (2024) A comprehensive review of cellulose and cellulose-based materials: Extraction modification and sustainable applications. J Nat Fibers 21(1):p 2418357. https://doi.org/10.1080/15440478.2024.2418357

Mertaniemi H, Escobedo-Lucea C, Sanz-Garcia A, Gandía C, Mäkitie A, Partanen J, Ylipert-tula M (2016) Human stem cell decorated nanocellulose threads for biomedical applications. Biomaterials 82:208–220. https://doi.org/10.1016/j.biomaterials.2015.12.046

Munawar M, Khan AH, Rehman ZU, Rahim A, Aziz M, Almuaythir S, El Kheir BS, Haider F (2023) Micro to nanolevel stabilization of expansive clay using agro-wastes. Adv Civ Eng (1):2753641. https://doi.org/10.1155/2023/2753641

Moon RJ, Martini A, Nairn J, Simonsen J, Youngblood J (2011) Cellulose nanomaterials review: structure, properties and nanocomposites. Chem Soc Rev 40(7):3941–3994. https://doi.org/10.1039/C0CS00108B

Owonubi SJ, Agwuncha SC, Malima NM, Shombe GB, Makhatha EM, Revaprasadu N (2021) Non-woody biomass as sources of nanocellulose particles. Rev Extr Proced Frontiers Energy Res 9:608825. https://doi.org/10.3389/fenrg.2021.608825M

Potthast A, Rosenau T, Kosma P (2002) Analysis of oxidized functionalities in cellulose. Cellulose 9:41–53. https://doi.org/10.1023/A:1015874121347

Patel DK, Ganguly K, Dutta SD, Patil TV, Lim KT (2023) Cellulose nanocrystals vs. cellulose nanospheres: A comparative study of cytotoxicity and macrophage polarization potential. Carbohydr Polym 303:120464. https://doi.org/10.1016/j.carbpol.2022.120464

Seliktar D, Black RA, Vito RP, Nerem RM (2000) Dynamic mechanical conditioning of collagen-gel blood vessel constructs induces remodeling in vitro. Ann Biomed Eng 28(4):351–362. https://doi.org/10.1114/1.284

Singh S, Dhakar GL, Kapgate BP, Maji PK, Verma C, Chhajed M, Rajkumar K, Das C (2020) Synthesis and chemical modification of crystalline nanocellulose to reinforce natural rubber composites. Polym Adv Technol 31(12):3059–3069. https://doi.org/10.1002/pat.v31.1210.1002/pat.5030

Šturcová A, Davies GR, Eichhorn SJ (2005) Elastic modulus and stress-transfer properties of tunicate cellulose whiskers. Biomacromol 6(2):1055–1061. https://doi.org/10.1021/bm049291k

Svensson A, Nicklasson E, Harrah T, Panilaitis B, Kaplan DL, Brittberg M, Gatenholm P (2005) Bacterial cellulose as a potential scaffold for tissue engineering of cartilage. Biomaterials 26(4):419–431. https://doi.org/10.1016/j.biomaterials.2004.02.049

Sofia, Romão Ana, Bettencourt Isabel AC, Ribeiro (2022) Novel features of cellulose-based films as sustainable alternatives for food packaging. Polym 14(22):4968. https://doi.org/10.3390/polym14224968

Suota MJ, Kochepka DM, Moura MG, Pirich CL, Matos M, Magalhães WE, Ramos LP (2021) Lignin functionalization strategies and the potential applications of its derivatives–A Review BioResources 16(3):6471–6511. https://doi.org/10.15376/biores.16.3.Suota

Tanaka R, Iwata T (2006) Elastic modulus of single cellulose microfibrils from tunicate measured by atomic force microscopy. Cellulose 13(4):509–517. https://doi.org/10.1007/s10570-006-9064-6

Tashiro K, Kobayashi M (1991) Theoretical evaluation of three-dimensional elastic constants of native and regenerated celluloses: role of hydrogen bonds. Polymer 32(8):1516–1526. https://doi.org/10.1016/0032-3861(91)90068-B

Tiwari A, Sanjog J (2025) Morphological, structural, and thermal properties of cellulose nanocrystals extracted from Indian water chestnut shells (agricultural waste). Next Materials 8:100653. https://doi.org/10.1016/j.nxmate.2025.100653

Vitz J, Erdmenger T, Haensch C, Schubert US (2009) Extended dissolution studies of cellulose in imidazolium-based ionic liquids. Green Chem 11(3):417–424. https://doi.org/10.1039/B818376B

Walther A, Bjurhager I, Malho J, Pere J, Ruokolainen J, Berglund LA, Ikkala O (2020) Large-area, lightweight, and flame-resistant nanocomposites based on clay and cellulose nanofibrils assembled in a paperlike structure. ACS Nano 14(2):1058–1071. https://doi.org/10.1021/acsnano.9b06930

Wang J, Tavakoli J, Tang Y (2019) Bacterial cellulose production, properties and applications with different culture methods – Rev Carbohydr Polym 219:63–76. https://doi.org/10.1016/j.carbpol.2019.05.008

Wang Y, Li Y, Zhang Y, Zhang Z, Li Y, Li W (2021) Nanocellulose aerogel for highly efficient adsorption of uranium (VI) from aqueous solution. Carbohydr Polym 267:118233. https://doi.org/10.1016/j.carbpol.2021.118233

Wang H, Li S, Wu T, Wang X, Cheng X, Li D (2019) A comparative study on the characterization of nanofibers with cellulose I, I/II, and II polymorphs from wood. Polym 11(1):153. https://doi.org/10.3390/polym11010153

Xu A, Zhang Y, Zhang H (2010) Cellulose dissolution in ionic liquids: a review. Carbohyd Polym 82(4):893–896. https://doi.org/10.1016/j.carbpol.2010.06.043

Xu T, Wang X, Shi Q, Liu H, Chen Y, Liu J (2024) Review of soil microplastic degradation pathways and remediation techniques. Int J Environ Res 18(5):77. https://doi.org/10.1007/s41742-024-00615-4

Xiong R, Zhang Z, Chai W, Huang Y, Chrisey DB (2016) Freeform 3D printing of vascularized tissues: where we are and what's next. Appl Phys Rev 3(4):041101. https://doi.org/10.1063/1.4962867

Yu H, Qin Z, Liang B, Liu N, Zhou Z, Chen L (2012) Facile extraction of thermally stable cellulose nanocrystals with a high yield of 93% through H_2SO_4–methanol–H_2O hydrolysis. Carbohyd Polym 89(2):727–733. https://doi.org/10.1016/j.carbpol.2012.03.062

Yu W, Wang C, Yi Y, Wang H, Zeng L, Li M, Yang Y, Tan Z (2020) Comparison of deep eutectic solvents on pretreatment of raw ramie fibers for cellulose nanofibril production. ACS omega 5(10):5580–5588. https://doi.org/10.1021/acsomega.0c00506

Zhou X, Liu X, Wang Q, Lin G, Yang H, Yu D, Cui SW, Xia W (2022) Antimicrobial and antioxidant films formed by bacterial cellulose, chitosan and tea polyphenol–Shelf life extension of grass carp. Food Packag Shelf Life 33:100866. https://doi.org/10.1016/j.fpsl.2022.100866

Zhang H, Wu J, Zhang J, He J (2017) 1-Allyl-3-methylimidazolium chloride room temperature ionic liquid: a suitable medium for cellulose dissolution. Cellulose 24(8):3811–3818. https://doi.org/10.1007/s10570-017-1326-5

Zhou Y, Huang X, Huang J, Zhang L, Zhou Z (2018) Predicting the dielectric properties of nanocellulose-modified presspaper based on the multivariate analysis method. Mol 23(7):1507. https://doi.org/10.3390/molecules23071507

Zhu S, Wang P, Li Y (2018) Challenges and opportunities of ionic liquids in the dissolution and processing of cellulose for sustainable materials. Chem Rev 118(2):2305–2318. https://doi.org/10.1021/acs.chemrev.7b00539

Replacing Plastic with Cellulose Film-Based Plastic Towards Greener Solutions

A. Philomena Joy Lindsey, Murugan Sevanan, and R. Emilin Renitta

Abstract Global dependence on petroleum-based plastics has elevated environmental impact and health issues owing to their accumulation in various ecosystems, non-biodegradability, and the release of toxic by-products. This chapter explores the feasibility of utilizing cellulose and its derivatives as sustainable and eco-friendly alternatives for packaging and other applications that align with the principles of circular economy and sustainability goals. The environmental hazards of conventional plastics and their impact on marine ecosystems, agriculture, food safety, and human health are explored to necessitate the need for sustainable alternatives. Investigations on cellulose-based film, highlighting their manufacturing techniques—including solvent casting, extrusion blow molding, layer-by-layer assembly, and electrospinning have been documented along with their tailored properties, limitations, and functional advantages. Life-cycle assessment of cellulose-based films emphasizes the need for process optimization and the utilization of compatible renewable resources that are cost-effective. It evaluates the ecological impact of each step of the production process to attain sustainability. The limitations in employing cellulose-based film encounters technological constraints in retaining hydrophobicity, mechanical stability, production costs, and scalability. These issues are addressed by considering the future perspective of replacing petroleum-based plastics through chemical modification and nano structuring for greener solutions. This chapter concludes that cellulose-based films present a promising pathway to reduce plastic pollution and its

A. P. J. Lindsey · R. E. Renitta (✉)
Division of Food Processing Technology, School of Engineering and Technology, Karunya Institute of Technology and Sciences, Coimbatore, Tamil Nadu, India
e-mail: emilinrenitta@karunya.edu

A. P. J. Lindsey
e-mail: apjlindseygct@gmail.com

A. P. J. Lindsey
Department of Food Technology, Karpagam Academy of Higher Education, Coimbatore, Tamilnadu, India

S. Murugan
Division of Biotechnology, School of Engineering and Technology, Karunya Institute of Technology and Sciences, Coimbatore, Tamil Nadu, India
e-mail: murugan@karunya.edu

K. K. Kesari et al. (eds.), *Smart Sustainable Materials Advancing Re-manufacturing and Technology*, World Sustainability Series, https://doi.org/10.1007/978-3-032-17907-4_7

jeopardizing impact on the environment by promoting the utilization of sustainable materials that support global well-being.

Keywords Biodegradable film · Cellulose and its derivatives · Environmental sustainability · Life cycle assessment · Petroleum-based plastic

1 Introduction

Synthetic plastics have been a cornerstone of packaging industry for many years because of their long-lasting performance, versatile nature and and cost-effectiveness. Plastics are widely used in a range of environments, from domestic to industrial to produce items such as shopping bags, storage bins, water bottles, pitchers, glassware, plumbing pipes, furniture, and agricultural goods. However, the overuse of plastic has resulted in significant environmental challenges, primarily because it is not biodegradable and the accumulation of plastic waste is increasing. More than a million tons of plastic waste are generated from various sources every year, and packaging plastics contribute to half of the global plastic waste. This poses a challenge in maintaining environmental sustainability and has fostered interest towards the development of biodegradable plastics (Choi et al. 2024; Kumari et al. 2023; Ncube et al. 2021). Biodegradable polymers have emerged as significant alternatives to petroleum-based plastics owing to their abundant availability and eco-friendly properties. They have been identified as potent sources in replacing conventional plastics despite their limited market acceptability owing to their long term performance and cost (Goncalves et al. 2024; Ncube et al. 2020).

Petroleum based plastics continue to dominate the packaging market because of their efficacy, cost, and long-term utilization. The lack of proper disposal and utilization of single-use plastics increase the need for extensive research in developing bio-based plastics from renewable resources including starch, cellulose, protein and polymers produced by bacteria (Kumari et al. 2023). However, significant hurdles such as large scale production for real-world application, integration of innovative technologies in the production process for feasible production exists in achieving a complete transition from synthetic to sustainable packaging and bridging the gap in supply and demand (Gonzalez-Lopez et al. 2023). To address this, cellulose film-based materials have been identified as promising alternatives to petroleum-based plastics as their scope complies with global sustainability goals in promoting circular economy. Biodegradability is the primary characteristic of cellulose-based materials and has significantly influenced their feasibility for various applications in the healthcare, construction, textile and packaging sector (Madhushree et al. 2024). Advancements in material science and packaging studies have also enabled the development of active polymer films incorporating antimicrobial agents and antioxidants (Al-Khalili et al. 2025). Recent investigations have also focused on scaling up the production of cellulose-based films through process optimization methods including solvent casting (Cindradewi et al. 2021), extrusion (Hosseini et al. 2023), and multilayer film development (Xie 2023). Life cycle assessment (LCA), study of biodegradability and carbon emissions from cellulose-based films are crucial in ensuring greener solutions that eventually offer functionality and renewability (Chen et al. 2023). This chapter provides a comprehensive summary of the technical exploration of cellulose-based films and highlights their eco-friendly role in replacing petroleum-based plastics, particularly in packaging applications. It offers a balanced perspective on the potential constraints of cellulose films, addressing practical insights and scalable strategies to confront the growing challenge of plastic pollution.

2 Types of Petroleum Based Plastics

Petroleum based plastics can be categorized as biodegradable or non-biodegradable. Polybutylene succinate (PBS), poly (butylene adipate-co-terephthalate) (PBAT), polyglycolic acid (PGA), polycaprolactone (PCL) and polyvinyl alcohol (PVA) are non-biodegradable plastics whereas polypropylene (PP), polyethylene (PE), polyvinyl chloride (PVC), polyethylene terephthalate (PET) and polystyrene (PS) are biodegradable plastics (Mori 2023). They can also be categorized based on their thermal properties as thermoplastics and thermosetting plastics. Polyvinyl chloride, polystyrene, polypropylene and polyethylene-Low Density Polyethylene (LDPE), High Density Polyethylene (HDPE) and Medium Density Polyethylene (MDPE) are categorized as thermoplastics and phenol formaldehyde, polyurethane, epoxy resins and urea formaldehyde are grouped under thermosetting plastics (Kumari et al. 2023). There also exist single use plastics and reusable plastics (Pilapitiya and Ratnayake 2024). These plastics are widely utilized for various applications in the agricultural, pharmaceutical and packaging sectors. Table 1 summarizes the applications of petroleum-based plastics in various sectors.

3 Challenges in Plastic Recycling

Plastic recycling faces numerous challenges that hinder its effectiveness and sustainability. One of the primary issues is low global recycling rates, which remain at approximately 9% despite increasing plastic production, indicating inefficiencies in the recycling process (Singh and Walker 2024). Although mechanical recycling of plastics is favoured, successive cycles result in the deterioration of material characteristics ultimately leading to complications in processing heterogeneous and contaminated plastic materials (Sambyal et al. 2025). Alternatively, chemical recycling converts plastics into monomers that can potentially be utilized for different applications. One major constraint is the consumption of high energy involved in converting complex plastics and the associated environmental concerns linked to the release of undesirable by-products. The potential application of microorganisms in enzymatic degradation is a bio-based approach that requires further investigation in terms of

Table 1 Summary of various applications of petroleum-based plastics in day-to-day life

Plastic name	Applications
Polypropylene (PP)	General-use plastic products, packaging containers, packaging electronics, household appliances, battery cases, filaments, bottle caps, ketchup bottles, yogurt and margarine containers, syrup bottles, drinking straws, baby bottles, plastic pressure pipe system
Polyethylene (PE)	Plastic bags, containers, bottles, electric wires and cables, pipes, medical devices, automotive and space uses
Low-Density Polyethylene (LDPE)	Containers, plastic films and wraps, plastic bags, frozen food bags, squeezable bottles, outdoor furniture, floor tiles, shower curtains
High-Density Polyethylene (HDPE)	Water pipes, liquid food containers, detergent and shampoo bottles, garbage bags, food cups for yogurt and margarine, molded plastic cases
Medium-Density Polyethylene (MDPE)	Carry bags, gas pipes and fittings
Polyvinyl chloride (PVC)	Cables, flooring, plumbing materials, toys, food packages and wraps, squeeze bottles, shampoo bottles, cooking oil and butter jars, detergent and window cleaner bottles, shower curtains, medical tubing, construction products, electrical cable/wire insulation
Polyethylene terephthalate (PET)	Beverage bottles, textiles (polyester), food packaging film, containers/bottles for beverages, detergents, butter jars, microwavable packaging
Polystyrene (PS)	Disposable cutlery, foam containers, furniture, glasses, utensils, styrofoam products, egg cartons, plastic tableware, food containers, plastic cutlery, compact disk cases
Polybutylene succinate (PBS)	Agricultural films, cosmetic containers, packaging, bottles, compostable bags
Poly (butylene adipate-co-terephthalate) (PBAT)	Agricultural films, flexible packaging, fishing gear, coatings
Polyvinyl alcohol (PVA)	Films, textiles, medical applications, adhesives, emulsion stabilizers
Polyglycolic acid (PGA)	Medical sutures, industrial parts, shale gas drilling applications
Polycaprolactone (PCL)	Agricultural films, compost bags, medical applications, adhesives, inks
Polytetrafluoroethylene (PTFE) (Teflon)	Cookware (non-stick), waterproof coatings
Polyamide PA	Fibers, fishing line, toothbrush bristles, tubing
Polycarbonate PC	Compact disks, eyeglasses, security windows, traffic lights and lenses

scalability and practical implementation (Kaushal et al. 2021; Mohanan et al. 2020). The sorting of plastics is limited by the presence of diversed plastics. Although, AI-based sorting is feasible, high implementation costs remain a barrier (Bernat 2023).

Internationally, recycling the insufficient capacity of plastics in importing countries poses challenges to governance and sustainability eventually leading to social and ecological implications (Thapa et al. 2024). Advanced technologies such as blockchain development aid in improving the recycling process through better waste management systems that require substantial investment and collaboration between stakeholders (Bulkowska et al. 2024). Despite these challenges, the development of methods for recycling plastics remains critical. Recent research on catalytic and thermo-chemical recycling techniques holds promise for converting plastic waste into valuable products, thus contributing to a circular economy (Achilias 2025). Enhanced recycling strategies, supported by robust policies and international cooperation, are essential for mitigating plastic pollution and advancing global sustainability efforts (Castro-Amoedo et al. 2024).

4 Environmental Hazards of Petroleum-Based Plastics

The extensive utilization of petroleum-based plastics has resulted in substantial carbon dioxide emissions during their production process since twentieth century. Despite the introduction of bioplastics, 97% of petroleum-based plastics are preferred owing to their lower capital investment in production, market price and long-lasting characteristics for packaging applications. Most plastic items introduced into the global market consist of single-use plastics from materials, particularly polypropylene, polyethylene, and polyvinylchloride (Mori 2023). In recent decades, the worldwide demand for plastics has surged four times, with projections indicating that it will double by 2050. Since 1950, annual production has increased from 2 million tons to 400 million tons in 2022 with an annual growth rate of 8.4%. The World Health Organization (WHO) predicts that global plastic consumption could reach 516 million tons by 2025, and if current patterns continue, it might surpass 1.2 billion tonnes annually by 2060. Regarding consumption, China stands as the top polymer consumer with 20% contribution towards global plastic usage while United States follows closely with 18%. The EU28 contributes 16%, and other Asian regions account for 12%. India and Japan contribute 6% and 4% respectively, whereas Middle East countries and Africa account for 7 and 5% of the total global plastic consumption (Houssini et al. 2025; WHO 2025).

Understanding the supply chain management of plastics is crucial to address the environmental impacts of global production and consumption. Annually, 500 million tons of plastic products are produced worldwide, and these products take 800 years to break down in soil and 400 years in oceanic environments. Plastic containers add 120 million tons of waste annually. Alarmingly, only 20% of globally produced plastics are recycled, while 90% of plastics make up 90% of oceanic waste. Each year, 150 tons of plastic waste are dumped into the oceans, leading to the creation of large waste islands. There are currently five such islands: two in the Pacific Ocean, two in the Atlantic Ocean, and one in the Indian Ocean. Plastic waste poses a significant threat to wildlife, including birds and fish, resulting in the deaths of millions of marine creatures annually (Abdullayeva and Yamil 2022). Owing to its durability and water-repellent properties, plastic debris fosters the growth of diverse microbial communities, forming an ecosystem known as the "plastisphere" (Zettler et al. 2013).

Plastics are highly stable and durable, making them resistant to natural decomposition in the environment. The degradation of plastic materials into microplastics involves photodegradation, mechanical, chemical, thermal and biological degradation through the influence of factors such as sunlight, heat, wind, waves, oxidation and characteristics such as polymer size, density, molecular weight and microbes present in the marine environment and soil ecosystem (Andrady 2011; Bajt 2021). Microplastic particles (MPs) are tiny plastic fragments less than 5 mm in size that have endured for hundreds of years and accumulate in organisms, sediments, and food chains. Ingesting MPs can cause physical damage, introduce toxic chemicals and disrupt cellular function, nutrient uptake, metabolism, and behavior of living organisms (Issac and Kandasubramanian 2021; Pilapitiya and Ratnayake 2024). To tackle the issue of microplastic pollution, it is recommended to improve monitoring and evaluation processes, implement policies and regulations aimed at reducing the production and escape of plastic waste, encourage innovation and cooperation in managing and recycling plastic waste, and increase public awareness and education about microplastic pollution and its possible solutions (Matavos-aramyan 2024). The environmental hazards of petroleum-based plastics are multifaceted, impacting different ecosystems and human health. Plastic pollution, accumulation of microplastics, and toxic chemical leaching are the primary hazards that impose jeopardy on human health, wildlife and marine ecosystem with the deposition of trillions of microplastics that persist for hundreds of years (Eriksen et al. 2014). These small particles can be ingested by marine organisms, leading to harmful effects on both marine life and human health (Yu and Singh 2023). They invariably affect terrestrial and marine ecosystem, food chains, and agricultural soil and pollute the atmosphere through the emission of greenhouse gases and toxic chemicals (Han et al. 2024; Kehinde et al. 2020; Lakhiar et al. 2024).

5 Impact of Petroleum-Based Plastics on Marine Ecosystem

The effects of petroleum-based plastics on marine ecosystems are significant and complex, influencing biodiversity, habitat stability, and anthropogenic activities. As plastic pollution continues to increase, its harmful impacts on marine life and ecosystems have become more apparent. Large amounts of microplastics from floating debris and pollutants are transported and are widespread by ocean currents which is highly challenging to monitor and eliminate (Sulaiman et al. 2023). Although, millions of tons of such microplastics enter the oceans, less than 1% are visible on the surface as they either settle as sediments or are suspended within water bodies (Safahieh et al. 2025). Microplastics, made of substances such as PCBs, fibers, PET fragments, and polyesters, are reported to be toxic and widespread across most seafood (Lebreton et al. 2017). They are present in all forms of marine ecosystem including coral reefs, beaches, deep-sea floors and backwaters with differing levels of contamination that significantly affect aquatic life (Nchimbi et al. 2022).

Microplastics have been documented to have sublethal effects on aquatic organisms leading to physical damage, inflammation, hormonal imbalance and blockages in organs such as the gills and digestive systems, which can influence feeding, growth, and survival. They also cause biomagnification, which possibly enters predators through the food chain, affecting biodiversity (Agbekpornu and Kevudo 2023). They have also been detected on the surface of aquatic plants, hindering photosynthesis and nutrient uptake (Katare et al. 2022; Madiraju et al. 2024; Saha and Saha 2024). Filter feeders have been reported to consume microplastics among which bivalve mollusks tend to have a higher tendency to consume microplastics because of their feeding habits. These microplastics reportedly originate from fishing nets, made of nylon, polyethylene and polypropylene (Magesh and Ajith 2024). Tracking and eliminating microplastics from the oceans presents a formidable task, considering their minute size and vastness (Matavos-aramyan 2024).

6 Impact of Petroleum-Based Plastics on Soil and Plant Growth

Plasticulture applications are among the most beneficial indirect inputs in agriculture, offering the potential to revolutionize Indian farming and possibly spark a "Second Green Revolution." This approach involves the utilization of plastics in agriculture, horticulture, water management, and food grain storage. They are specifically used for plant or soil coverage in the form of mulch films, row covers, low tunnels, and greenhouses (Akhter et al. 2021; Lakhiar et al. 2024). This widespread usage has resulted in negative impacts on soil fertility and crop growth. They contribute to soil pollution in the form of microplastics, which alter the physico-chemical properties of soil, potentially affecting soil microbiota, eventually leading to heightened ecological impact. This in turn enters the food chain and is an indirect threat to human health and other life forms (Yu et al. 2022). Plastic-based materials are also utilized to manage soil fumigation, maintain irrigation systems and pack agricultural goods. They also shield and protect crops from rainfall. Plants play a crucial role in maintaining healthy soil ecosystems and aid in the detoxification of contaminants. However, microplastics (MPs) disrupt water and nutrient uptake, potentially inducing the secretion of degradative enzymes. Rhizosphere microbial and faunal communities may aid in the co-degradation of MPs and their by-products. MPs can alter plant biomass, tissue structure, root morphology, and soil microbial dynamics. Despite ongoing studies, the interactions between MPs and soil biota remain poorly understood, with inconsistent findings. The effects of MPs vary depending on their polymer type, size, concentration, and environmental context. Many laboratory studies have used unrealistically high MP levels, whereas natural soils may mitigate MP effects through dilution and competitive sorption. Thus, long-term field-relevant studies are crucial to comprehensively assess the impact of MPs on soil ecosystems (Wang et al. 2019).

A comprehensive three-year study was undertaken to evaluate the impact of plastic residues from landfills on maize crop productivity utilizing three different plastic sizes wherein medium-sized plastic residues notably reduced crop yield in comparison to the control group. The study also indicated that all types and sizes of plastic would eventually degrade, potentially leading to a gradual decline in soil health (Zhao et al. 2024). Recent research has indicated that microplastics, which differ in shape and polymer type, can significantly affect soil properties and plant growth. Employing plastics in the form of fibers, foams, films and fragments to improve the shoot biomass of *Daucus carota* resulted in 27–60% increase due to improved moisture retention and aeration in the cultivable land whereas microbial activity was hindered as there was a reduction in soil aggregation. This study highlights the complex and shape-dependent effects of microplastics on crop growth (Lozano et al. 2021). Plastic particles also impart changes in the organic carbon content in the soil, which is overestimated as better soil quality. Furthermore, they influence soil nutrient properties by altering the composition of organic nitrogen, ammonium nitrogen, dissolved organic phosphorus, and phosphate levels (Yang et al. 2020). Although best management practices reduce pollution in the agricultural sector, the prevalence of microplastics influences soil characteristics, function, microbiota and nutrient cycling, which has driven the need for alternatives that can offer environmental sustainability and mitigate the adverse effect of petroleum-based plastics (Tiwari and Sistla 2024). The influence of petroleum-based plastics on agriculture is multifaceted, and bioplastics have been identified to mitigate this effect. The adoption of bioplastics to achieve sustainable agriculture by enhancing their barrier properties is a critical challenge for attaining ecological balance.

7 Impact of Petroleum-Based Plastics in Food Packaging

Plastic food packaging materials comprises a wide range of chemicals and synthetic materials derived from various chemical combinations that are specifically identified and designed to impart food safety during packaging and transport. Plastic materials are preferred in the food packaging sector because of their cost and barrier properties. However, their non-biodegradable properties negatively influence human health and the environment, highlighting the urgent need for alternative solutions (Simoes et al. 2024; Zhan et al. 2024). Municipal solid waste primarily consists of food packaging waste that leaches out harmful chemicals into the environment (Zhao et al. 2020). Low-molecular-weight polymers such as LDPE lack covalent attachment, and hence adopt branched configuration that prohibits the compact packing of monomer units. Consequently, these monomers diffuse from the polymer matrix (Bhunia et al. 2013). Research has demonstrated that these substances are intentionally and non-intentionally added substances (NIAS) as they migrate from packaging materials. The composition of intentionally added substances is monitored by regulatory authorities by setting limits to ensure safety. However, most migrating compounds are NIAS, which include impurities produced through degradation products (Kato and Conte-Junior 2021; Wang et al. 2023). The most notable instance of leaching plastic components into water involves 4,4-dioxydiphenyl-2,2-propane, commonly known as bisphenol A (BPA), which is a recognized endocrine disruptor. It is found in food packages that eventually enter through contact with the packaging material resulting in developmental and reproductive problems, cancer, and obesity in countries where plastics are extensively manufactured (Datta et al. 2024). Although regulatory limits exist, factors such as temperature, contact time, pH, food type, and packaging condition influence BPA migration. Temperature and packaging material are the most critical contributors to BPA leaching into food (Agarwal et al. 2025). Moreover, BPA contamination in food occurs through contact with epoxy resin coated polycarbonate plastic baby bottles, cans, milk jugs, bags and water pipes that are specifically used to store food. In packaged meat products, approximately 3% migration was observed due to contact with the surface area of packaging films (Wang et al. 2023). Bisphenol A (BPA) also infiltrates into food at elevated temperatures especially when used to contain boiling water wherein the rate of migration is 55 times higher than that at lower temperatures (Corre et al. 2015). At high temperatures, BPA monomer residues leach into food that is are packed in cans owing to incomplete polymerization (Ungureanu et al. 2019).

Phthalate diesters, commonly referred to as phthalates (PEs) have been reported to migrate into water stored in various containers. The type of polymer used to prepare the container, storage temperature, and bottling process have a potential impact on the leaching of phthalates into water (Jeddi et al. 2015; Massahi et al. 2025). Similar results were documented for formaldehyde and acetaldehyde, in which carbonation and the influence of sunlight also played a major role. The increased mobility of the polymer chain at the glass transition temperature (~80 °C) increased the rate of migration. Additionally, the type of water is crucial in the process of migration, as the pH and pressure exerted by the liquid (carbonated bottled water) influence the process. This type of migration has also been observed in polystyrene food containers (Manoli and Voutsa 2019). Degradation of these polymers has a potential impact on environmental sustainability. Brown et al. (2023) studied and compared the environmental degradation pattern of petroleum-based and plant-based plastic teabags for 12 months. Teabags containing petroleum-based plastic exhibited partial degradation through the persistent release of microplastic fragments whereas 3.5-month complete degradation was observed in plant-based teabags made of cellulose or cellulose-PLA. This study requires a comprehensive life cycle assessment to confirm its standard to validate its long-term application. Although petroleum-based plastics meet the necessary standards for effective food packaging, their impact on health and the environment necessitates a re-evaluation of their use.

8 Impact of Petroleum-Based Plastics on Human Health

The health hazards of petroleum-based plastics are currently in the limelight because each step of the production process significantly influences various life forms. Children and infants residing near plastic manufacturing sites were identified as vulnerable populations in low-income and minority "fenceline" communities and they also encounter exposures at home, school, and playgrounds located near plastic disposal sites. Workers in the plastic industry are regularly exposed to harmful substances, including toxic monomers, chemical additives, airborne particles such as asbestos, and hazardous vapors. These exposures can lead to elevated toxin levels in the body, even when workplace air quality meets legal standards, and can persist after time away from work. Health effects linked to such exposure include cancers (e.g., liver angiosarcoma, lung, ovarian, brain, and breast), soft tissue cancers, mesothelioma, infertility, heart disease, and nervous system damage (Fayshal 2024; Landrigan et al. 2023). Recent investigations in human cellular biology have demonstrated that exposure to nano- and microplastics can elicit a range of cellular responses, including cytotoxicity, genotoxicity, inflammation, apoptosis, and oxidative stress (Winiarska et al. 2024). Phthalate esters, which are employed as plasticizers exhibit low acute toxicity and pose significant risks to reproductive health resulting in abnormal sexual development and birth defects in infants (Cheng et al. 2013). Microplastics and nanoplastics are absorbed into the body primarily through food consumption and inhalation of polluted air, where they cause oxidative stress by producing reactive oxygen species (ROS) that abruptly damage the DNA, proteins, and lipids. This results in genetic mutations that lead to cancer, development of inflammation, genetic toxicity, immune reactions and increases protein expression in the liver and kidneys. They also interfere with metabolic activity by disrupting the uptake of amino acids, bile acids, and lipids which potentially increases the activity of neurotransmitters (Wang et al. 2020).

Micro- and nanoplastics (MNPs) have been found in the skin, and respiratory, digestive, circulatory, and reproductive systems, suggesting that they can bypass barriers and escape the immune system. The primary route of entry is inhalation followed by accumulation in various tissues. Determination in the human body is crucial and hence require advanced imaging and analytical tools and more investigations using in vitro models to understand its pathway and effect on various organs of the human body (Feng et al. 2023). The management of plastic waste through collection, sorting and recycling can reduce contamination and potential health risks. However, the cost of recycling and the lack of advanced recycling technologies pose significant challenge especially for low-income countries. These limitations in existing waste management practices have led to a growing interest in replacing plastics with conventional or alternative materials, such as natural polymers

(Pilapitiya and Ratnayake 2024). Bioplastics offer environmental advantages over conventional plastics by reducing reliance on fossil fuels and lowering the emission of greenhouse gases. They support feedstock recovery and can be converted with municipal waste into syngas for energy and chemical production. Unlike traditional plastics, bioplastics enable renewable energy generation from biogenic carbon and help reduce the use of landfills. Their biodegradation results in lower net CO_2 emissions, as the plants used for production absorb equivalent CO_2. Additionally, bioplastics positively influence soil carbon dynamics, making them suitable for sustainable agriculture and waste management (Islam et al. 2024).

9 Cellulose as a Renewable Raw Material for Greener Solution

Cellulose, the most abundant macromolecule on Earth, is primarily extracted from plant cell walls but is also found in algae, bacteria, fungi, protozoa, and agricultural residues. Plant biomass exists alongside hemicellulose and lignin, typically in a 4:3:3 ratio, along with minor components such as pectin and ash. Structurally, cellulose is a linear homopolymer of glucose units linked by β-(1–4)-glycosidic bonds, with strong hydrogen bonding, resulting in a stable crystalline structure (Kaya and Yucel 2023; Gamage et al. 2024; Romao et al. 2022). Cellulose is obtained from a variety of sources, including hardwoods, softwoods, agricultural residues, and certain bacteria and marine organisms. Despite its numerous advantages, cellulose exhibits limitations such as poor solubility, lack of thermoplasticity, and absence of antimicrobial properties. Chemical modifications, such as etherification are employed to enhance its solubility and functionality. Common cellulose derivatives include cellulose (CMC), hydroxypropyl methylcellulose (HPMC), methylcellulose (MC), hydroxyethyl cellulose (HEC), ethyl cellulose (EC), and cellulose acetate (CA). The hydrophobicity of cellulose can be improved by incorporating acrylate, methacrylate, or siloxane groups into its structure (Magalhaes et al. 2023). Important derivatives of cellulose that can act as alternatives to petroleum-based plastics include methyl cellulose, hydroxyethyl cellulose, cellulose acetate, cellulose butyrate, and cellulose acetate-butyrate. Despite their potential, these materials have received limited attention for the development of biodegradable plastics. These materials are attributable to slow degradation rates with high production costs, and the generation of noxious emissions during processing which further require process optimization for greener solution (Tajeddin 2014).

10 Manufacturing Techniques of Cellulose Film

The desirable properties of the polymer films developed using cellulose determines the method to be adopted for suitable applications. Cellulose films are a significant alternative to plastic-based packaging films because of their potential properties including biodegradability, renewability and environment sustainability. Recent investigations have focused on improving the properties of cellulose films to render them suitable for packaging applications. Hydrophilicity has been identified as the major hindrance that adversely affects the moisture absorption and tensile strength of films. Chemical modifications were introduced to enhance the barrier properties of the film. However, a few chemical modifications can influence the film characteristics (Jiang and Ngai 2022). Therefore, the method involved in developing such films is of prime importance for creating eco-friendly films for a wide range of applications. The techniques employed offer unique benefits by utilizing cellulose and its composites at various combinations with or without the incorporation of functional materials. Solvent casting is a conventional process that involves the dissolution of the polymer in an appropriate solvent followed by homogenization and casting onto a uniform surface. Extrusion blow molding is another film development technique that melts the polymer and shapes it either into films or into a required structure within a mold. Layer-by-layer assembly is an innovative technique for constructing multilayer films that are oppositely charged and that eventually retain film thickness and functionality. Electrospinning was utilized in film development to obtain a polymer film with a uniform surface area through the application of a high-voltage electric field that eventually draws ultrafine fibers from a filmogenic solution. This method is suitable for developing nanofibrous mats with uniform surface area and porosities (Mu et al. 2025). Figure 1 depicts the methods used to prepare the cellulose-based plastic films.

10.1 Solvent Casting

Solvent-casting is a conventional method of film development in which a volatile solvent is used to dissolve the polymer. The addition of a plasticizer and a cross-linking agent is advantageous for rendering films with good barrier, mechanical and optical properties. The obtained filmogenic solution was cast on a smooth surface, followed by drying at a particular temperature in an oven, convection chamber, or desiccator to evaporate the solvent. After drying, the films were cut according to the desired application (Dyartanti et al. 2021). Solvent selection is a crucial step as it defines toxicity, cost, recyclability and environmental impact. Advancement in solvent selection and the utilization of ionic liquids N-methylmorpholine-N-oxide (NMMO), N-dimethylacetamide/lithium chloride (DMAc/LiCl), molten inorganic salt solutions, metal complex solutions, and tetrabutylammonium fluoride/dimethyl

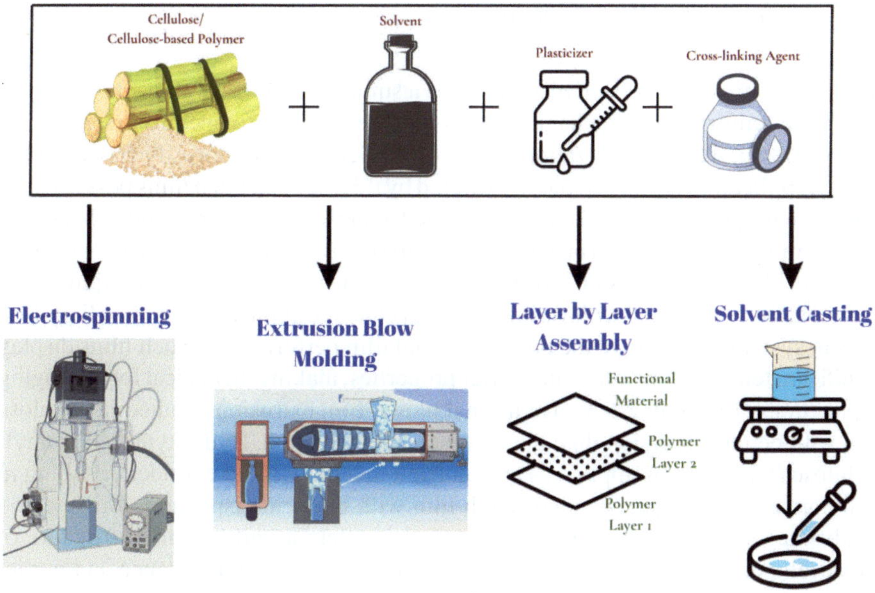

Fig. 1 Techniques involved in the preparation of cellulose-based film

sulfoxide (DMSO), effectively produce homogenized cellulose and cellulose-based filmogenic solutions to develop films with improvised quality (Liu et al. 2023).

Incorporation of fatty acid esters such as laurate, myristate, palmitate, and stearate into thermoplastic cellulose for the development of packaging film was studied by Kallakas et al. (2023). Improved hydrophobicity due to the increase in aliphatic side chain length of fatty acids and water barrier properties alongside identical tensile and mechanical properties to those of LDPE and HDPE films were observed. Significantly, rosemary oil imparted plasticizing properties to the film, which also imparted antimicrobial and antioxidant properties to the film potentiating the possibility for packaging applications (Bouftou et al. 2024). Moreover, the incorporation of cellulose nanofibrils into the cellulose acetate matrix by solvent casting enhanced the mechanical properties of the developed film (Cindradewi et al. 2021). The composite film developed with cellulose acetate and chitosan demonstrated excellent tensile strength, hydrophobicity and barrier properties through solvent casting technique, highlighting their synergistic characteristics (Zhou et al. 2021). The practicality of this method makes it a preferred option for the development and fabrication of cellulose-based film. With the aid of suitable solvents and functional additives such as essential oil and fatty acids, the performance of packaging films can be enhanced to meet the criteria for petroleum-based plastics for sustainable applications.

10.2 Extrusion Blow Molding

Blow film extrusion has been recognized as a suitable method for thin-film development with uniform surface characteristics that are resistant to chemical degradation and oxidation. The customization of the film thickness and porosity based on the desired application is a key advantage offered by this method. Such films possess electrokinetic properties that make them suitable for biosensors and biomedical applications (Kramar et al. 2023). The absence of natural plasticity in cellulose is crucial for the direct blow-film extrusion. To tackle this challenge, researchers have explored the combination of cellulose with various polymers and the application of ionic liquids as solvent systems to improve the processing and film performance. Such films display significant tensile, elongation and barrier properties, making them ideal for packaging applications (Liu et al. 2023). Traditional blown-film extrusion resulted in cellulose film with better dispersion than the solvent casting method. Hydroxypropyl methylcellulose (HPMC) was applied as an emulsion in a blow-mold extruder, resulting in a strengthened hydrogen bonded starch film with uniform dispersion. This method is advantageous over other methods because it offers a uniform microstructure of the developed film with enhanced mechanical strength, water resistance and barrier characteristics (Jiang et al. 2022). Cellulose nanocrystals (CNCs) were integrated into biodegradable thermoplastic Poly Lactic Acid (PLA) which was extracted from maize starch. Direct dry-blending and melt-blending methods were employed to develop composite films using a single-screw extruder. Inflation through the development of internal air pressure results in a blow-up effect, potentiating its application in blow-molded biopolymer packaging (Karkhanis et al. 2017). The addition of an extrusion die resulted in a thin film layer with a uniform surface. Furthermore, oxidation and spoilage of foods can be prevented by incorporating functional agents, such as antimicrobials and antioxidants to extend the shelf life of packaged foods (Lopes et al. 2020). Overall, this method fabricates thin film with sustainable eco-friendly material such as cellulose, to replace plastic materials in the food packaging and biomedical sectors offering resource conservation (Dhar et al. 2017).

10.3 Layer-By-Layer Technique

The layer-by-layer (LBL) technique is different from other methods of film development, as it functions based on the charge of polymers. Oppositely charged materials were assembled as a layer on a substrate to develop multilayered films. Charged biopolymers such as cationic guar gum (CGg), carboxylated cellulose nanocrystals (cCNCs), and hydroxypropyl methylcellulose (HPMC) are predominantly utilized in this method. A five-layer structure (CGg/cCNCs/HPMC/cCNCs/CGg) composite film with homogeneous surface characteristics was developed by electrostatic charge and hydrogen bonding. This multilayered film exhibited excellent mechanical and barrier properties, and was flexible, transparent and durable, offering resistance to harsh solvents including acetone, Tetrahydrofuran (THF) and N,N-Dimethylacetamide (DMAc) (Saini et al. 2021). Highly charged nanocelluloses (NCs) can be incorporated into layer-by-layer assembled films composed of polyelectrolytes, rendering colloidal stability and ductility. The ability of nanocellulose to form wet-stable aerogels and foams with reduced flammability has resulted in the development of phosphorylated cellulose nanofibril (CNFs) films for fire safety applications (Wagberg and Erlandsson 2021).

The practical applications of bacterial cellulose (BC) and chitosan (CT) are limited by their high moisture absorption. A bioplastic composite made of bacterial cellulose in combination with chitosan grafted using enzymes and lauryl gallate via green synthesis was developed through layer-by-layer assembly. Films with enhanced tensile strength and hydrophobicity that significantly blocked water droplets for 15 min were produced, which could potentially find applications in developing sustainable drinking straws and packaging bags (Cheung et al. 2023). A UV-shielding cellulose film was developed by layer-by-layer assembly wherein L-phenylalanine was grafted onto cellulose nanofibrils. Octadecyltrichlorosilane and 4ARM-PEG-NH_2 (a multiarm PEG derivative with amine groups at each terminal) were incorporated to improve the hydrophobicity and mechanical reinforcement of the film. Complete UV blocking (200–320 nm) with excellent mechanical properties and antibacterial activity enabled effective oil–water separation (up to 93.87%) offering itself as a green alternative to plastic-based materials (Ren et al. 2023). LBL assembly has been proven to enhance film characteristics through the formation of dense multilayer films that effectively find application in packaging where moisture retention is required with reduced permeability. LBL films can also be fabricated in accordance with the desired application in terms of thickness and surface properties by modifying the pH and ionic strength of the filmogenic solution. Moreover, the electrostatic properties of the LBL multifunctional films render them suitable for applications that enable water repellence, flame retardancy, oil resistance, and UV protection (Ganjeh et al. 2024). Reduction in flammability, UV protection and oil/water resistance are advantageous properties of multilayered cellulose-based films developed by this method. This highlights the suitability of this method and the widespread applicability of the developed films as alternatives to plastic films employed in various sector.

10.4 Electrospinning

Electrospinning is a cost-effective, electrohydrodynamic technique that is involved in the production of micro and nanofiber films through the application of a high-voltage electric field. Polymers, metals and composites that require precise control over the film characteristics and surface properties can be employed in this method. The application of polymer films with sustained release of active agents and encapsulated sensitive compounds that significantly promote gas exchange along with uniform dispersion within the film matrix is an advantageous attribute of this method

(Mohammadi et al. 2020; Topuz and Uyar 2024; Zhang et al. 2024). Fresh produce stored in containers was sealed with a tricomponent smooth nanofiber film consisting of carboxymethyl cellulose (CMC) combined with polyvinyl alcohol (PVA) and polyvinylpyrrolidone (PVP). This film exhibited a reduction in moisture accumulation for 40 days in containers with fruits and 10 days in vegetable containers compared to those sealed with plastic lids (Hashmi et al. 2021). A smart double-layer film made from natural materials for seafood packaging was developed by Ahangari et al. (2025). Carboxymethyl cellulose film impregnated with pomegranate peel anthocyanins served as the bottom layer that eventually changed colour upon spoilage. This layer acts as an indicator of freshness while the top layer was made of cellulose acetate nanofibers loaded with lipid carriers containing essential oil through electrospinning. The film exhibited enhanced mechanical strength, antibacterial activity, and visible colour changes during the storage of fish and shrimp, indicating changes in pH and oxidation.

This method aided in the preparation of antioxidant and antimicrobial films prepared with biodegradable polymers, such as ethyl cellulose and soy protein isolate enriched with bitter orange peel extract. The film exhibited high porosity with an efficient water vapor transmission rate, good thermal stability and durability for packaging use (Rashidi et al. 2021). Electrospun polymer films were developed to replace conventional plastics in the textile industry, which demonstrated high porosity and thermal stability, wherein the film was prepared with nanocrystalline cellulose-based composite fibers (Ge et al. 2021). This film can be applied to the development of thin film-like fabrics that require specific filtration, absorption and structural properties. Table 2 shows the various applications of electrospun cellulose and cellulose-based nanofibers, highlighting the composition of the developed film (Rezaei et al. 2015; Zhang et al. 2021).

Despite the advantages offered, challenges such as heat sensitivity, low bioavailability of the compounds, and electrospinning scalability limit the large-scale application of this method. Extensive research on novel blends with nano composites fabricated through electrospinning could aid in the development of sustainable and greener solutions to replace petroleum-based plastic films (Cui et al. 2024; Topuz and Uyar 2024).

11 Life-Cycle Assessment of Cellulose-Based Film

Life cycle assessment (LCA) is the process of evaluating the factors responsible for the ecological impairment caused by the production of a product through each stage until disposal. It aids in identifying potential areas that require improvement to reduce their environmental implications and provides a framework for cellulose production and assessment. In the production of nanocrystalline cellulose from cotton fibers, lower CO_2 emission was observed, whreas the utilization of solvents and electricity were identified as significant contributors to environmental impacts. Therefore, process optimization and utilization of renewable resources are recommended

Table 2 Application of electrospun cellulose and cellulose-based nanofibers

S. No	Type of cellulose and cellulose-based film	Application
I	*Water treatment*	
1	Electrospun chitosan/cellulose nanofibers	Adsorption of heavy metal ions (arsenic (V), lead (II), and copper (II))
2	Hydroxyapatite nanoparticles (a mineral fixator) impregnated into Cellulose Acetate (CA) matrix	Tackle the aggregation issue of hydroxyapatite nanoparticles
3	Addition of thiol groups, citric acid, and polyvinylamine to CA matrix	Removal of Cu (II), Cr (VI), and organo arsenic contaminants
4	Electrospun cellulose nanofibers incorporated with photocatalysts such as ZnO and metal–organic frameworks	Degradation of organic contaminants like effluent dye molecules
5	Electrospun fiber mats of cellulose acetate butyrate (CAB) and ZnO nanostructures	Adsorption and photocatalytic property
6	Incorporation of Cellulose Nanocrystals (CNCs) into a PVDF nanofibrous membrane	Oil/water separation
II	*Biomaterials*	
1	Electrospun CA/collagen nanofibers incorporated with latex from *C. procera*	Wound healing
2	Electrospun CA nanofibers loaded with gelatin and *Zataria multiflora*	Absorption of accumulated wound exudates, wound healing capacity and antibacterial property
3	Electrospun ethyl cellulose and cellulose diacetate nanofibers doped with silver sulfadiazine and protoporphyrin	Antibacterial activity
4	Encapsulation of flufenamic acid in electrospun carboxymethyl cellulose/PVA nanofibers	Prolonged drug release
III	*Active packaging*	
1	Electrospun ethylcellulose/poly (ethylene oxide) hybrid nanofibers encapsulated with ethyl formate	Preservation of strawberries
2	Electrospun zein/ethyl cellulose nanofibers incorporated with cinnamaldehyde essential oil	Improved water resistance of food
3	Electrospun anthraquinone-2-carboxylic acid grafted silk fibroin/CA nanofibers	Antibacterial activity
IV	*Other applications*	
1	Electrospun CA/polystyrene nanofibers	Collection of water from fog
2	Electrospun CA and deacetylated CA Nanofibers were used as electrolyte membranes	Energy conservation

(continued)

Table 2 (continued)

S. No	Type of cellulose and cellulose-based film	Application
3	Electrospun cellulose acetate nanofibers	Controlled release of immobilized Vitamin A and E
4	Immobilization of *Candida rugosa* lipase on electrospun cellulose acetate nanofibers	Retention of enzyme activity

through life cycle assessment (Rebec et al. 2024). In comparison with man-made cellulose fibers from non-renewable resources, cellulose from cotton has a high potential to curb the emission of green-house gases. From extraction techniques to end-product uses, ongoing innovations in cellulose technology aim to balance performance with ecological responsibility, demonstrating the value of LCA in guiding sustainable development (Shen and Patel 2008). Production of cellulose through bacterial fermentation involves water usage where the resource-intensive stages are raw material production and downstream processes (Forte et al. 2021). Process optimization can significantly reduce the ecological impact across the stages of BC production.

Numerous studies have identified the environmental impact associated with the production of cellulose and its derivatives. Rebec et al. (2024) identified biomass depolymerization and bleaching as the stages of consideration throughout the multistep production of nanocrystalline cellulose. The extraction of cellulose nanofibers from vegetable waste involves enzymes and spinning processes for film orientation, wherein electrospinning is better than wet spinning considering their potential environmental impact (Piccinno et al. 2015). A study by Righi et al. (2011) identified the possibility of reducing the ecological impact through process enhancement by optimizing energy consumption during the extraction of cellulose. It is also essential to include the life cycle assessment (LCA) of ionic liquids used in cellulose dissolution as they are advantageous in terms of human toxicity (Zhang et al. 2017). This highlights the complexity of cellulose film development with regard to the influential factors in production and outcomes to attain sustainability. These evaluations suggest the advantages of switching to cellulose-based materials as potent substitutes for petroleum-based plastics, supporting ecological sustainability.

12 Challenges and Technological Limitations

Cellulose and its derivatives are widely used as sustainable and eco-friendly resources for plastics. To balance sustainability, the use of eco-friendly, biodegradable, and compostable bioplastics is crucial, offering a green alternative to conventional plastic films. These materials align well with the principles of Closed-Loop Economy (Masek and Kosmalska 2022). However, long term durability and production costs significantly affect the application of cellulose and its derivatives. Cellulose-based functional materials encounter challenges in facilitating solute transport through their microporous network which hinders the mass transfer rate and eventually reduces the processing efficiency. In developing mechanically stable cellulose gels and films in terms of stretchability and flexibility, long term usage results in the breaking and collapse of cellulose fibers, eventually damaging the product quality (Li et al. 2024).

Despite possessing outstanding properties to function as films, fibers and membranes, cellulose nanomaterials are less effective at lower concentrations, which is influential in determining the product quality. Dispersion and interfacial computability are other constraints encountered during commercialization. Furthermore, inhalation of cellulose nanoparticles and migration risks must be managed to ensure consumer acceptance. Despite these limitations, cellulose and its nano composites can be used to create continuous fibers suitable for the textile industry, dense films for packaging, porous membranes for filtration, and hydrogels or aerogels for absorption and insulation purposes. Widespread availability with excellent quality and relatively low cost dictates the ultimate success of their application (Motaung 2021; Wang et al. 2021). Current methods of cellulose utilization involve energy and chemical intensive steps that eventually impact ecological sustainability requiring the need for greener alternatives. Therefore, utilization of recyclable chemicals and enzymes is recommended. Balancing biodegradability with durability is necessary as improving material hydrophobicity reduces the accessibility of natural enzymes and microbes to degrade the material in the soil environment.

Flammability is another constrain in the utilization of cellulose-based materials in structural engineering (Li et al. 2021). For "Good Health and Well-being," functional materials of cellulose can be utilized in medical devices to improve healthcare. It also finds application in renewable energy in developing advanced technologies, aiding in "Affordable and Clean Energy," "Industry, Innovation, and Infrastructure," and "Sustainable Cities and Communities." They can replace traditional materials and reduce environmental impact, supporting "Climate Action." In the future, cellulose-based materials will face challenges but offer a promising path to a greener future (Li et al. 2024).

13 Conclusion

Utilization of cellulose and its derivatives explore the potential engagement of greener alternatives to petroleum-based plastics embarking on the need for sustainability. As summarized in this chapter, the materialistic and functional properties of cellulose enable its transformative application in packaging, textiles, electronics and biomedical domains. Challenges in retaining mechanical stability, structural integrity during long-term usage, mass transfer efficiency and environmental concerns encountered in the utilization of chemicals through the multistep production of cellulose-based materials must be rectified to widen their application. Overcoming these limitations paves the way for a synergistic approach with nano-structuring, interfacial engineering, and application of green chemistry principles. By integrating advanced characterization techniques, process optimization and predictive modelling, future research on cellulose and its derivatives can be tailored to meet the needs of the industry. Overall, cellulose stands as a greener platform to achieve global sustainability objectives and aids in replacing non-biodegradable, petroleum-based plastics in multiple sectors.

References

Abdullayeva MY, Yamil GA (2022) Psychology the psychological role of ethnic identity. World Sci 5(September):77. https://doi.org/10.31435/rsglobal_ws/30092022/7867

Achilias DS (2025) Thermo-chemical recycling of plastics as a sustainable approach to the plastic waste issue. Euro-Mediterr J Environ Integr 2605–2618. https://doi.org/10.1007/s41207-025-00800-7

Agarwal A, Gandhi S, Tripathi AD, Gupta A, Iammarino M, Sidhu JK (2025) Food contamination from packaging material with special focus on the Bisphenol-A. Crit Rev Biotechnol 45(1):69–79. https://doi.org/10.1080/07388551.2024.2344571

Agbekpornu P, Kevudo I (2023). The risks of microplastic pollution in the aquatic ecosystem. https://doi.org/10.5772/intechopen.108717

Ahangari H, Ebrahimi A, Ehsani A, Amjadi S (2025) Multipurpose packaging system based on intelligent carboxymethyl cellulose film and activated cellulose acetate electrospun nanofibers for seafoods. Int J Biol Macromol 298(January). https://doi.org/10.1016/j.ijbiomac.2025.140115

Akhter A, Malik AA, Sultan A, Nabi A, Azrah Indrabi S, Javeed I, Bhat TA (2021) Plasticulture—a key step to second green revolution. Int J Curr Microbiol App Sci Spec Issue (11):2299–2315. http://www.ijcmas.com

Al-Khalili M, Rahman S, Al-Habsi N (2025) Date seed–added biodegradable films and coatings for active food packaging applications: a review. Packag Technol Sci 38(6):445–472. https://doi.org/10.1002/pts.2893

Andrady AL (2011) Microplastics in the marine environment. Mar Pollut Bull 62(8):1596–1605. https://doi.org/10.1016/j.marpolbul.2011.05.030

Bajt O (2021) From plastics to microplastics and organisms. FEBS Open Bio 11(4):954–966. https://doi.org/10.1002/2211-5463.13120

Bernat K (2023) Post-consumer plastic waste management: from collection and sortation to mechanical recycling. Energies 16(8). https://doi.org/10.3390/en16083504

Bhunia K, Sablani SS, Tang J, Rasco B (2013) Migration of chemical compounds from packaging polymers during microwave. Conventional Heat Treat Storage 12(5):63–76

Bouftou A, Aghmih K, Lakhdar F, Abidi N, Gmouh S, Majid S (2024) Enhancing cellulose acetate film with green plasticizers for improved performance, biodegradability, and migration study into a food simulant. Meas Food 15(January):100180. https://doi.org/10.1016/j.meafoo.2024.100180

Brown RW, Chadwick DR, Zang H, Graf M, Liu X, Wang K, Greenfield LM, Jones DL (2023) Bioplastic (PHBV) addition to soil alters microbial community structure and negatively affects plant-microbial metabolic functioning in maize. J Hazard Mater 441(September 2022):129959. https://doi.org/10.1016/j.jhazmat.2022.129959

Bulkowska K, Zielinska M, Bulkowski M (2024) Blockchain-based management of recyclable plastic waste. Energies 17(12). https://doi.org/10.3390/en17122937

Castro-Amoedo R, Granacher J, Kantor I, Dahmen A, Barbosa-Povoa A, Maréchal F (2024) On the role of system integration in plastic waste management. Resour Conserv Recycl 201(October 2023):107295. https://doi.org/10.1016/j.resconrec.2023.107295

Chen Z, Aziz T, Sun H, Ullah A, Ali A, Cheng L, Ullah R, Khan FU (2023) Advances and applications of cellulose bio-composites in biodegradable materials. J Polym Environ 31(6):2273–2284. https://doi.org/10.1007/s10924-022-02561-8

Cheng Z, Nie XP, Wang HS, Wong MH (2013) Risk assessments of human exposure to bioaccessible phthalate esters through market fish consumption. Environ Int 57–58:75–80. https://doi.org/10.1016/j.envint.2013.04.005

Cheung KM, Chong HL, Jiang Z, Ngai T (2023) Water-resistance chitosan film through enzymatic treatment and layer-by-layer assembly with bacterial cellulose for food packaging materials. Soft Matter 20(40):7344–7356

Choi HJ, Hwang D, Yoon YS, Jeon TW, Rhee SW (2024) Applying material flow analysis for sustainable waste management of single-use plastics and packaging materials in the Republic of Korea. Sustainability (Switzerland), 16(16). https://doi.org/10.3390/su16166926

Cindradewi AW, Bandi R, Park CW, Park JS, Lee EA, Kim JK, Kwon GJ, Han SY, Lee SH (2021) Preparation and characterization of cellulose acetate film reinforced with cellulose nanofibril. Polymers 13(17). https://doi.org/10.3390/polym13172990

Corre LL, Besnard P, Chagnon MC (2015) BPA, an energy balance disruptor. Crit Rev Food Sci Nutr 55(6):769–777. https://doi.org/10.1080/10408398.2012.678421

Cui X, You Y, Ding Y, Sun C, Liu B, Wang X, Guo F, Liu Q, Fan X, Li X (2024) Improving the function of electrospun film by natural substance for active packaging application of fruits and vegetables. Lwt 191(December 2023):115683. https://doi.org/10.1016/j.lwt.2023.115683

Datta S, Chauhan A, Ranjan A, Sardar AH, Tuli HS, Ramniwas S, Shahwan M, Sharma U, Jindal T (2024) A comparative review on bisphenol a sources, environmental levels, migration, and health impacts in India and global context. Nat Environ Pollut Technol 23(2):1095–1104. https://doi.org/10.46488/NEPT.2024.v23i02.043

Dhar P, Gaur SS, Soundararajan N, Gupta A, Bhasney SM, Milli M, Kumar A, Katiyar V (2017) Reactive extrusion of polylactic acid/cellulose nanocrystal films for food packaging applications: influence of filler type on thermomechanical, rheological, and barrier properties. Ind Eng Chem Res 56(16):4718–4735. https://doi.org/10.1021/acs.iecr.6b04699

Dyartanti ER, Sunaryati N, Alhakim RR, Putri LN, Putri AYM, Nurohmah AR, Nisa SS (2021) Recent development of polyvinylidene fluoride/cellulose membranes electrolyte separator for lithium ion batteries. IOP Conf Ser Mater Sci Eng 1096(1):012144. https://doi.org/10.1088/1757-899x/1096/1/012144

Eriksen M, Lebreton LCM, Carson HS, Thiel M, Moore CJ, Borerro JC, Galgani F, Ryan PG, Reisser J (2014) Plastic pollution in the world's oceans: more than 5 trillion plastic pieces weighing over 250,000 tons afloat at sea. PLoS ONE 9(12):1–15. https://doi.org/10.1371/journal.pone.0111913

Fayshal MA (2024) Current practices of plastic waste management, environmental impacts, and potential alternatives for reducing pollution and improving management. Heliyon 10(23):e40838. https://doi.org/10.1016/j.heliyon.2024.e40838

Feng Y, Tu C, Li R, Wu D, Yang J, Xia Y, Peijnenburg WJGM, Luo Y (2023) A systematic review of the impacts of exposure to micro- and nano-plastics on human tissue accumulation and health. Eco-Environ Health 2(4):195–207. https://doi.org/10.1016/j.eehl.2023.08.002

Forte A, Dourado F, Mota A, Neto B, Gama M, Ferreira EC (2021) Life cycle assessment of bacterial cellulose production. Int J Life Cycle Assess 26(5):864–878. https://doi.org/10.1007/s11367-021-01904-2

Gamage A, Thiviya P, Liyanapathiranage A, Wasana MLD, Jayakodi Y, Bandara A, Manamperi A, Dassanayake RS, Evon P, Merah O, Madhujith T (2024) Polysaccharide-based bioplastics: eco-friendly and sustainable solutions for packaging. J Compos Sci 8(10). https://doi.org/10.3390/jcs8100413

Ganjeh AM, Saraiva JA, Pinto CA, Casal S, Gonçalves I (2024) Recent advances in nano-reinforced food packaging based on biodegradable polymers using layer-by-layer assembly: a review. Carbohydr Polym Technol Appl 7(November 2023):100395. https://doi.org/10.1016/j.carpta.2023.100395

Ge L, Yin J, Yan D, Hong W, Jiao T (2021) Construction of nanocrystalline cellulose-based composite fiber films with excellent porosity performances via an electrospinning strategy. ACS Omega 6(7):4958–4967. https://doi.org/10.1021/acsomega.0c06002

Goncalves EM, Silva M, Andrade L, Pinheiro J (2024) From fields to films: exploring starch from agriculture raw materials for biopolymers in sustainable food packaging. Agriculture (Switzerland) 14(3). https://doi.org/10.3390/agriculture14030453
Gonzalez-Lopez ME, Calva-Estrada SJ, Gradilla-Hernández MS, Barajas-Álvarez P (2023) Current trends in biopolymers for food packaging: a review. Front Sustain Food Syst 7(August):1–20. https://doi.org/10.3389/fsufs.2023.1225371
Han M, Liu H, Zhu T, Tang S, Li Y, Zhu C, Zhou Z, Jiang Q (2024) Toxic effects of micro(nano)-plastics on terrestrial ecosystems and human health. TrAC Trends Anal Chem 172(November 2023). https://doi.org/10.1016/j.trac.2023.117517
Hashmi M, Ullah S, Ullah A, Saito Y, Haider MK, Bie X, Wada K, Kim IS (2021) Carboxymethyl cellulose (Cmc) based electrospun composite nanofiber mats for food packaging. Polymers 13(2):1–12. https://doi.org/10.3390/polym13020302
Hosseini S, Kadivar M, Shekarchizadeh H, Alsharif MA (2023) Preparation and characterisation of biodegradable polylactic acid/ethyl cellulose film produced using the extruder and roller mixer machine. Int J Food Sci Technol 58(2):777–784. https://doi.org/10.1111/ijfs.16232
Houssini K, Li J, Tan Q (2025) Complexities of the global plastics supply chain revealed in a trade-linked material flow analysis. Commun Earth Environ 6(1):1–11. https://doi.org/10.1038/s43247-025-02169-5
Islam M, Xayachak T, Haque N, Lau D, Bhuiyan M, Pramanik BK (2024) Impact of bioplastics on environment from its production to end-of-life. Process Saf Environ Prot 188(April):151–166. https://doi.org/10.1016/j.psep.2024.05.113
Issac MN, Kandasubramanian B (2021) Effect of microplastics in water and aquatic systems 19544–19562
Jeddi MZ, Ahmadkhaniha R, Yunesian M, Rastkari N (2015) Magnetic solid-phase extraction based on modified magnetic nanoparticles for the determination of phthalate diesters in water samples. J Chromatogr Sci 53(2):385–391. https://doi.org/10.1093/chromsci/bmu058
Jiang J, Zhang X, Gao S, Li M, Hou H (2022) Effects of adding methods and modification types of cellulose on the physicochemical properties of starch/PBAT blown films. Int J Biol Macromol 223(PA):1335–1343. https://doi.org/10.1016/j.ijbiomac.2022.11.118
Jiang Z, Ngai T (2022) Recent advances in chemically modified cellulose and its derivatives for food packaging applications: a review. Polymers 14(8). https://doi.org/10.3390/polym14081533
Kallakas H, Kattamanchi T, Kilumets C, Tarasova E, Krasnou I, Savest N, Ahmadian I, Kers J, Krumme A (2023) Tensile and surface wettability properties of the solvent cast cellulose fatty acid ester films. Polymers 15(12). https://doi.org/10.3390/polym15122677
Karkhanis SS, Stark NM, Sabo RC, Matuana LM (2017) Performance of poly(lactic acid)/ cellulose nanocrystal composite blown films processed by two different compounding approaches. Polym Eng Sci 58(11):1964–1974. https://doi.org/10.1002/pen
Katare Y, Prashant S, Mahipal Singh S, Muskan S, Ekta BJ, Kapil P, Bhagyashri TN, Ashutosh T, Leena B (2022) Microplastics in aquatic environments: sources, ecotoxicity, detection & remediation. Biointerface Res Appl Chem 12(3):3407–3428. https://doi.org/10.1007/978-3-031-10729-0_8
Kato LS, Conte-Junior CA (2021) Safety of plastic food packaging: the challenges about non-intentionally added substances (NIAS) discovery, identification and risk assessment. Polymers 13(13):1–43. https://doi.org/10.3390/polym13132077
Kaushal J, Khatri M, Arya SK (2021) Recent insight into enzymatic degradation of plastics prevalent in the environment: a mini—review. Clean Eng Technol 2(March):100083. https://doi.org/10.1016/j.clet.2021.100083
Kaya EC, Yucel U (2023) Advances in cellulose-based packaging films for food products. Cellul Fundam Convers Biofuel Useful Chem 1–21. https://doi.org/10.5772/intechopen.110817
Kehinde O, Ramonu OJ, Babaremu KO, Justin LD (2020) Plastic wastes: environmental hazard and instrument for wealth creation in Nigeria. Heliyon 6(10):e05131. https://doi.org/10.1016/j.heliyon.2020.e05131

Kramar A, Luxbacher T, Moshfeghi Far N, González-Benito J (2023) Active cellulose acetate/chitosan composite films prepared using solution blow spinning: structure and electrokinetic properties. Polymers 15(15). https://doi.org/10.3390/polym15153276

Kumari S, Rao A, Kaur M, Dhania G (2023) Petroleum-based plastics versus bio-based plastics: a review. Nat Environ Pollut Technol 22(3):1111–1124. https://doi.org/10.46488/NEPT.2023.v22i03.003

Lakhiar IA, Yan H, Zhang J, Wang G, Deng S, Bao R, Zhang C, Syed TN, Wang B, Zhou R, Wang X (2024) Plastic pollution in agriculture as a threat to food security, the ecosystem, and the environment: an overview. Agronomy 14(3). https://doi.org/10.3390/agronomy14030548

Landrigan PJ, Raps H, Cropper M, Bald C, Brunner M, Canonizado EM, Charles D, Chiles TC, Donohue MJ, Enck J, Fenichel P, Fleming LE, Ferrier-Pages C, Fordham R, Gozt A, Griffin C, Hahn ME, Haryanto B, Hixson R, Dunlop S (2023) The Minderoo-Monaco commission on plastics and human health. Ann Glob Health 89(1). https://doi.org/10.5334/aogh.4056

Lebreton LCM, Van Der Zwet J, Damsteeg JW, Slat B, Andrady A, Reisser J (2017) River plastic emissions to the world's oceans. Nat Commun 8:1–10. https://doi.org/10.1038/ncomms15611

Li T, Chen C, Brozena AH, Zhu JY, Xu L, Driemeier C, Dai J, Rojas OJ, Isogai A, Wågberg L, Hu L (2021) Developing fibrillated cellulose as a sustainable technological material. Nature 590(7844):47–56. https://doi.org/10.1038/s41586-020-03167-7

Li X, Wan C, Tao T, Chai H, Huang Q, Chai Y, Wu Y (2024) An overview of the development status and applications of cellulose-based functional materials. In: Cellulose, vol 31, no 1. Springer Netherlands. https://doi.org/10.1007/s10570-023-05616-8

Liu X, Qin Z, Ma Y, Liu H, Wang X (2023) Cellulose-based films for food packaging applications: review of preparation, properties, and prospects. J Renew Mater 11(8):3203–3225. https://doi.org/10.32604/jrm.2023.027613

Lopes AC, Barcia MK, Veiga TB, Yamashita F, Grossmann MVE, Olivato JB (2020) Eco-friendly materials produced by blown-film extrusion as potential active food packaging. Polym Adv Technol 32(2):779–788. https://doi.org/10.1002/pat.5130

Lozano YM, Lehnert T, Linck LT, Lehmann A, Rillig MC (2021) Microplastic shape, polymer type, and concentration affect soil properties and plant biomass. Front Plant Sci 12(February):1–14. https://doi.org/10.3389/fpls.2021.616645

Madhushree M, Vairavel P, Mahesha GT, Bhat KS (2024) A comprehensive review of cellulose and cellulose-based materials: extraction, modification, and sustainable applications. J Nat Fibers 21(1). https://doi.org/10.1080/15440478.2024.2418357

Madiraju SVH, Pamula ASP, Darsi BP (2024) Theoretical Review on microplastic pollution: a multifaceted threat to marine ecosystems, human health, and environment. IgMin Res 2(6):460–468. https://doi.org/10.61927/igmin203

Magalhaes S, Fernandes C, Pedrosa JFS, Medronho B, Ferreira PJT, Rasteiro G (2023) Eco-friendly methods for extraction and modification of cellulose: an overview solange. Eco-Friendly Methods Extr Modif Cellul Overview Solange 15:3138

Magesh N, Ajith KV (2024) Microplastics contamination and risk assessment in bivalves of economic importance from Beypore estuary, Southern India. Environ Res 261(November). https://doi.org/10.1016/j.envres.2024.119711

Manoli E, Voutsa D (2019) Food containers and packaging materials as possible source of hazardous chemicals to food. In: Handbook of environmental chemistry, vol 78, pp 19–50. https://doi.org/10.1007/698_2016_121

Masek A, Kosmalska A (2022) Technological limitations in obtaining and using cellulose biocomposites. Front Bioeng Biotechnol 10(August):1–17. https://doi.org/10.3389/fbioe.2022.912052

Massahi T, Omer AK, Kiani A, Soleimani H, Fattahi N, Sharafi K (2025) Assessing the effect of sunlight exposure and reuse of polyethylene terephthalate bottles on phthalate migration. 962(January)

Matavos-aramyan S (2024) Addressing the microplastic crisis: a multifaceted approach to removal and regulation. Environ Adv 17(August):100579. https://doi.org/10.1016/j.envadv.2024.100579

Mohammadi MA, Hosseini SM, Yousefi M (2020) Application of electrospinning technique in development of intelligent food packaging: a short review of recent trends. Food Sci Nutr 8(9):4656–4665. https://doi.org/10.1002/fsn3.1781

Mohanan N, Montazer Z, Sharma PK, Levin DB (2020) Microbial and enzymatic degradation of synthetic plastics. Front Microbiol 11(November). https://doi.org/10.3389/fmicb.2020.580709

Mori R (2023) Replacing all petroleum-based chemical products with natural biomass-based chemical products: a tutorial review. RSC Sustain 1(2):179–212. https://doi.org/10.1039/d2su00014h

Motaung TE (2021) Mat A Crit Rev 55:1–12

Mu Y, Lv S, Liu J, Tong J, Liu L, Wang J, He T, Wei D (2025) Recent advances in research on biomass-based food packaging film materials. Compr Rev Food Sci Food Saf 24(1):2025. https://doi.org/10.1111/1541-4337.70093

Nchimbi AA, Shilla DA, Kosore CM, Shilla DJ, Shashoua Y, Khan FR (2022) Microplastics in marine beach and seabed sediments along the coasts of Dar es Salaam and Zanzibar in Tanzania. Mar Pollut Bull 185(PA):114305. https://doi.org/10.1016/j.marpolbul.2022.114305

Ncube LK, Ude AU, Ogunmuyiwa EN, Zulkifli R, Beas IN (2020) Environmental impact of food packaging materials: a review of contemporary development from conventional plastics to polylactic acid based materials. Materials 13(21):1–24. https://doi.org/10.3390/ma13214994

Ncube LK, Ude AU, Ogunmuyiwa EN, Zulkifli R, Beas IN (2021) An overview of plasticwaste generation and management in food packaging industries. Recycling 6(1):1–25. https://doi.org/10.3390/recycling6010012

Piccinno F, Hischier R, Seeger S, Som C (2015) Life cycle assessment of a new technology to extract, functionalize and orient cellulose nanofibers from food waste. ACS Sustain Chem Eng 3(6):1047–1055. https://doi.org/10.1021/acssuschemeng.5b00209

Pilapitiya PGCNT, Ratnayake AS (2024) The world of plastic waste: a review. Clean Mater 11(January):100220. https://doi.org/10.1016/j.clema.2024.100220

Rashidi M, Seyyedi Mansour S, Mostashari P, Ramezani S, Mohammadi M, Ghorbani M (2021) Electrospun nanofiber based on Ethyl cellulose/Soy protein isolated integrated with bitter orange peel extract for antimicrobial and antioxidant active food packaging. Int J Biol Macromol 193(PB):1313–1323. https://doi.org/10.1016/j.ijbiomac.2021.10.182

Rebec KM, Turk J, Kunaver M (2024) Quantifying the environmental implication of cotton-fiber-based nanocrystalline cellulose: a life-cycle assessment. Sustain Prod Consum 45(December 2023):192–202. https://doi.org/10.1016/j.spc.2023.12.026

Ren Y, Ling Z, Huang C, Lai C, Yong Q (2023) Layer-by-layer assembly induced strong, hydrophobic and antibacterial TEMPO oxidized cellulose nanofibrils films for highly efficient UV-shielding and oil-water separation. Int J Biol Macromol 253(P1):126486. https://doi.org/10.1016/j.ijbiomac.2023.126486

Rezaei A, Nasirpour A, Fathi M (2015) Application of cellulosic nanofibers in food science using electrospinning and its potential risk. Compr Rev Food Sci Food Saf 14(3):269–284. https://doi.org/10.1111/1541-4337.12128

Righi S, Morfino A, Galletti P, Samorì C, Tugnoli A, Stramigioli C (2011) Comparative cradle-to-gate life cycle assessments of cellulose dissolution with 1-butyl-3-methylimidazolium chloride and N-methyl-morpholine-N-oxide. Green Chem 13(2):367–375. https://doi.org/10.1039/c0gc00647e

Romao S, Bettencourt A, Ribeiro IAC (2022) Novel features of cellulose-based films as sustainable alternatives for food packaging. Polymers 14(22). https://doi.org/10.3390/polym14224968

Safahieh A, Oveidizadeh H, Salamat N (2025) Microplastic occurrence and abundance in the subtidal sediments 2 from the northwest Persian Gulf, pp 1–20. https://doi.org/10.2139/ssrn.4998389

Saha G, Saha SC (2024) Tiny particles, big problems: the threat of microplastics to marine life and human health. Processes 12(7). https://doi.org/10.3390/pr12071401

Saini A, Sharma D, Xia Y, Saini A, You X, Su Y, Chen L, Yadav C, Li X (2021) Layer-by-layer assembly of cationic guar gum, cellulose nanocrystals and hydroxypropyl methylcellulose

based multilayered composite films. Cellulose 28(13):8445–8457. https://doi.org/10.1007/s10570-021-04064-6
Sambyal P, Najmi P, Sharma D, Khoshbakhti E, Hosseini H, Milani AS, Arjmand M (2025) Plastic recycling: challenges and opportunities. Can J Chem Eng 103(6):2462–2498. https://doi.org/10.1002/cjce.25531
Shen L, Patel MK (2008) Life cycle assessment of polysaccharide materials: a review. J Polym Environ 16(2):154–167. https://doi.org/10.1007/s10924-008-0092-9
Simoes A, Ramos A, Domingues F, Luís Â (2024) Pullulan-Tween 40 emulsified films containing geraniol: production and characterization as potential food packaging materials. Eur Food Res Technol 250(6):1721–1732. https://doi.org/10.1007/s00217-024-04514-y
Singh N, Walker TR (2024) Plastic recycling: a panacea or environmental pollution problem. Npj Mater Sustain 2(1):1–7. https://doi.org/10.1038/s44296-024-00024-w
Sulaiman B, Woodward JC, Shiels HA (2023) Riverine microplastics and their interaction with freshwater fish. Water Biol Secur 2(4):100192. https://doi.org/10.1016/j.watbs.2023.100192
Tajeddin B (2014) Cellulose-based polymers for packaging applications. Wiley Blackwell 6, 9781118773, pp 477–498. https://doi.org/10.1002/9781118773949.ch21
Thapa K, Vermeulen WJV, De Waal MM, Deutz P, Nguyễn HQ (2024) Towards a just circular economy transition: the case of european plastic waste trade to Vietnam for recycling. Circ Econ Sustain 4(2):851–876. https://doi.org/10.1007/s43615-023-00330-w
Tiwari E, Sistla S (2024) Agricultural plastic pollution reduces soil function even under best management practices. PNAS Nexus 3(10):1–10. https://doi.org/10.1093/pnasnexus/pgae433
Topuz F, Uyar T (2024) Electrospinning of sustainable polymers from biomass for active food packaging. Sustain Food Technol 2(5):1266–1296. https://doi.org/10.1039/d4fb00147h
Ungureanu EL, Alimentare IDB, Popa ME (2019) An overview on human potential exposure to bisphenol a from food-contact materials used in fruits packaging and processing. Sci Bull Ser F. Biotechnol 199–205
Wagberg L, Erlandsson J (2021) The use of layer-by-layer self-assembly and nanocellulose to prepare advanced functional materials. Adv Mater 33(28):1–13. https://doi.org/10.1002/adma.202001474
Wang J, Liu X, Li Y, Powell T, Wang X, Wang G, Zhang P (2019) Microplastics as contaminants in the soil environment: a mini-review. Sci Total Environ 691:848–857. https://doi.org/10.1016/j.scitotenv.2019.07.20
Wang J, Wang L, Gardner DJ, Shaler SM, Cai Z (2021). Towards a cellulose-based society: opportunities and challenges. In: Cellulose, vol 28, no 8. Springer Netherlands. https://doi.org/10.1007/s10570-021-03771-4
Wang X, Nag R, Brunton NP, Siddique MAB, Harrison SM, Monahan FJ, Cummins E (2023) A probabilistic approach to model bisphenol A (BPA) migration from packaging to meat products. Sci Total Environ 854(September 2022):158815. https://doi.org/10.1016/j.scitotenv.2022.158815
Wang YL, Lee YH, Chiu IJ, Lin YF, Chiu HW (2020) Potent impact of plastic nanomaterials and micromaterials on the food chain and human health. Int J Mol Sci 21(5):1727. https://doi.org/10.3390/ijms21051727
WHO (2025) WHO. Addressing plastic pollution: a call to action on world environment day
Winiarska E, Jutel M, Zemelka-Wiacek M (2024) The potential impact of nano- and microplastics on human health: understanding human health risks. Environ Res 251(P2):118535. https://doi.org/10.1016/j.envres.2024.118535
Xie F (2023) Biopolymer-based multilayer films and coatings for food preservation: an update of the recent development. Curr Food Sci Technol Rep 1(1):1–12. https://doi.org/10.1007/s43555-023-00002-8
Yang X, Guo X, Huang S, Xue S, Meng F, Qi Y, Cheng W, Fan T, Lwanga EH, Geissen V (2020) Microplastics in soil ecosystem: insight on its fate and impacts on soil quality. In: Handbook of environmental chemistry, vol 95, pp 245–258. https://doi.org/10.1007/698_2020_458

Yu J, Adingo S, Liu X, Li X, Sun J, Zhang X (2022) Micro plastics in soil ecosystem—a review of sources, fate, and ecological impact. Plant, Soil Environ 68(1):1–17. https://doi.org/10.17221/242/2021-PSE

Yu RS, Singh S (2023) Microplastic pollution: threats and impacts on global marine ecosystems. Sustainability (Switzerland) 15(17). https://doi.org/10.3390/su151713252

Zettler ER, Mincer TJ, Amaral-Zettler LA (2013) Life in the "plastisphere": microbial communities on plastic marine debris. Environ Sci Technol 47(13):7137–7146. https://doi.org/10.1021/es401288x

Zhan Z, Feng Y, Zhao J, Qiao M, Jin Q (2024) Valorization of seafood waste for food packaging development. Foods 13(13). https://doi.org/10.3390/foods13132122

Zhang C, Shao Y, Zhu L, Wang J, Wang J, Guo Y (2017) Acute toxicity, biochemical toxicity and genotoxicity caused by 1-butyl-3-methylimidazolium chloride and 1-butyl-3-methylimidazolium tetrafluoroborate in zebrafish (Danio rerio) livers. Environ Toxicol Pharmacol 51:131–137. https://doi.org/10.1016/j.etap.2017.02.018

Zhang Y, Min T, Zhao Y, Cheng C, Yin H, Yue J (2024) The developments and trends of electrospinning active food packaging: a review and bibliometrics analysis. Food Control 160. https://doi.org/10.1016/j.foodcont.2024.110291

Zhang Y, Zhang C, Wang Y (2021) Recent progress in cellulose-based electrospun nanofibers as multifunctional materials. Nanoscale Adv 3(21):6040–6047. https://doi.org/10.1039/d1na00508a

Zhao X, Cornish K, Vodovotz Y (2020) Narrowing the gap for bioplastic use in food packaging: an update. Environ Sci Technol 54(8):4712–4732. https://doi.org/10.1021/acs.est.9b03755

Zhao ZY, Wang PY, Xiong XB, Irum M, Mo F, Cao J, Wang N, Tao HY, Hao M, Wang S, Xiao KW, Yang FK, Wang Y, Khan A, Wang WY, Grace UA, Xiong YC (2024) Size and concentration effects of polythene plastic residues on soil physiochemical traits and maize productivity: A three-year appraisal. J Clean Prod 435(January). https://doi.org/10.1016/j.jclepro.2024.140572

Zhou H, Tong H, Lu J, Cheng Y, Qian F, Tao Y, Wang H (2021) Preparation of bio-based cellulose acetate/chitosan composite film with oxygen and water resistant properties. Carbohyd Polym 270(March):118381. https://doi.org/10.1016/j.carbpol.2021.118381

Invisible Intruders: Microplastic Contamination in Soil and Its Influence on Microbial Communities and Degradation Mechanisms

Jaya Jaysudha, Pardeep Kaur, Gaurav Mudgal, and Sonia Kumari

Abstract The widespread production of plastics, coupled with inadequate disposal and degradation practices, has led to pervasive environmental contamination. Plastics, composed of diverse synthetic polymers, gradually fragment into microplastic particles smaller than 5 mm through physical, chemical, and biological processes. These pollutants originate from two primary sources: primary microplastics, intentionally manufactured for commercial use, and secondary microplastics, formed through the breakdown of larger plastic debris. While most research has focused on marine microplastic pollution, terrestrial ecosystems—particularly soil—remain underexplored despite being significant sinks for microplastic accumulation. Sources such as sewage sludge, landfills, agricultural mulching films, fertilizers, and municipal wastewater contribute to the infiltration of microplastics into soil matrices. Soil ecosystems host complex and diverse microbial communities, which are sensitive to microplastic presence. The interaction between microplastics and soil microbiota can alter microbial abundance, diversity, and functional dynamics, depending on the type and concentration of microplastics involved. Notably, certain microorganisms exhibit the capacity to degrade microplastics into environmentally benign byproducts, offering potential pathways for bioremediation. This chapter provides a comprehensive overview of microplastic contamination in soil, emphasizing its ecological impact on microbial communities and the mechanisms of microbial degradation. It also highlights critical knowledge gaps and proposes directions for future research to better understand and mitigate microplastic pollution in terrestrial environments.

Keywords Microplastic · Pollution · Primary · Secondary · Microbiota · Remediation · Degradation · Adversity

J. Jaysudha · P. Kaur · G. Mudgal · S. Kumari (✉)
University Institute of Biotechnology, Chandigarh University, Mohali, Punjab, India
e-mail: soniyabt17@gmail.com

P. Kaur
e-mail: Pardeepkaur2108@gmail.com

K. K. Kesari et al. (eds.), *Smart Sustainable Materials Advancing Re-manufacturing and Technology*, World Sustainability Series, https://doi.org/10.1007/978-3-032-17907-4_8

1 Introduction

Plastics, or synthetic organic polymers, have become an integral part of our modern world, making it challenging to envision a day without them. The mass production of plastics began around 1950 on a significant scale. Synthetic plastics, such as Bakelite, were first utilized in the early twentieth century. After the occurrence of World War II, there was a vast use of plastics outside of the military. The subsequent rapid rise in the manufacturing of plastics is exceptional, outpacing most other synthetic substances. Steel and cement are notable exemptions as they are widely used in the construction industry (World Steel Association (WSA) 1978; Cement Statistics and Information n.d.). Most of the monomers used in plastic manufacturing are derived from propylene or ethylene, the source of both can be attributed to fossil hydrocarbons. The plastics commonly used in our daily lives are predominantly non-biodegradable, leading to significant environmental concerns due to their contribution to ecosystem pollution (Chen and Patel 2012). Microplastics are a common term that is used in association with synthetic solid plastic particle or polymeric matrices, which ranges below the size of 5 mm (Frias and Nash 2019). Concerns have been raised regarding the distribution of microplastics in the global biosphere, given the numerous risks associated with their environmental presence (Ajith et al. 2020). Microplastics can be categorized into two types based on their sources: primary and secondary. Primary sources involve the direct production of small plastic particles to meet industrial requirements, such as in cosmetics, polishing agents, toothpaste, and detergents. All of these contain a certain amount of plastic in them, which is not biodegradable (Duis and Coors 2016; Cincinelli et al. 2019). Secondary sources include microplastic contamination from plastic fragments that break down into tiny particles due to the effects of water and wind. The source also includes the plastic, which is exposed to chemical and biological degradation after the plastic enters the environment. It might be through light, heat, or microorganisms (Khalid et al. 2021). Several studies have reported that microplastics have been found in human organs, food, and drinking water. This is due to the deposition of microplastics in soil and aquatic systems, which eventually enter the food cycle of humans and other organisms (Andrady 2017a; Vethaak and Legler 2021a; Ragusa et al. 2021a).

Soil is the uppermost layer of the Earth's crust formed due to the weathering of mineral rocks and serves as a ground for the growth of plants, and is a habitat of numerous microorganisms that help balance the biogeochemical cycle. But lately, soil has become a repository for microplastics in the terrestrial ecosystem (Rillig n.d.; Yang et al. 2021a). Many reports have suggested that huge amounts of microplastic debris are present in compost and sewage sludge (Li et al. 2018a; Mahon et al. 2017a; Zhang et al. 2020a). Microplastics enter the soil through various activities like the dumping of plastics in landfills, sewage sludge used in agricultural productions, and atmospheric depositions (Shen et al. 2022a; Zhang et al. 2019a; Rochman 2018a). Microplastics, once introduced into the soil environment, tend to accumulate over extended periods, ultimately affecting soil organisms (Zhou et al. 2020a). Microplastics can be harmful to the soil ecosystem as they act as carriers of toxic substances;

the contaminants can degrade the quality as well as function of the soil (Guo et al. 2020a).

Research in the past has mainly concentrated on investigating the impacts of microplastics in aquatic environments, rather than on terrestrial ecosystems (Monkul and Ozhan 2021; Hur and Jho 2021). As the soil is crucially important for a variety of purposes like plant growth, agricultural activities, nutrient cycling, and maintaining the biodiversity for microorganisms, it is important to evaluate the risk factors associated with microplastic pollution in the soil environment (Hur and Jho 2021; Zhu et al. 2019a; Hasan and Jho 2022). The expected interactions between both soil particles and plastic polymers are predicted to happen at the interface known as the Plastisphere. This interface emerges as a consequence of the influence exerted by microplastic particles. The physicochemical properties of microplastics could induce changes in the diversity and activities of soil microbial communities at the soil-plastic interface (Kuzyakov and Blagodatskaya 2015). Plastics have unique physical properties that make them susceptible to degradation and hence lengthen their shelf life. The rate of degradation of plastics is attributed to several factors, like degree of crystallinity, chemical structure, and molecular weight (Rajendran et al. 2015). Presently, there are few methods for plastic degradation, but they too contribute to environmental deterioration in some or the other way. These processes include incineration, landfilling, recycling, etc. There are consequences related to all these processes, like the production of carcinogenic gases from the process of incineration, there is cost ineffectiveness in the process of recycling, and landfilling processes occupy huge amounts of land (Verma et al. 2016). Due to these problems, researchers are looking for better alternative ways that are non-hazardous for the environment as well as efficient for the degradation of microplastics. A better alternative for the sustainable degradation of microplastics can be microorganisms. Various microbial species have been identified for their ability to degrade microplastics, indicating their potential in the degradation process. Recent reports have also emerged suggesting that certain insects can break down plastics during their larval stage. These include superworms, waxworms, and mealworms. These worms are believed to eat, degrade, and mineralize plastics because of the presence of microorganisms in their gut (Zhang et al. 2020b). Investigating the interactions between microplastics and the soil microbiome, as well as identifying potential soil microorganisms capable of degrading microplastics, is crucial. The chapter aims to evaluate the impacts of microplastics on the soil microbiome and the microbial degradation of microplastics. The chapter comprehensively describes the factors influencing the structure, composition, diversity, and functionality of soil microbial communities. The identification of challenges that need to be addressed to fill the knowledge gaps is presented, along with potential opportunities for future research.

2 Soil- Its Flora and Fauna

Soil stands as an extremely valuable resource to mankind as well as to a group of organisms whose viability depends on it. Human fate is incredibly connected to the existence of soil. From the source of our food to the source of our clothes, it serves for our living. Soil is a complex of air, water, weathered mineral particles, moisture, and organic matter. These compounds are present at varying levels according to the soil type. There are numerous soil organisms present in the soil that help in its transformation and replenishment. The soil medium gains energy through the decomposition of organic matter, with microbial biomass regulating the rate of decomposition. Soil serves as a habitat for many living components due to its biologically diverse nature. Flora and fauna represent living components within an ecosystem. There can be 20 different lineages of arthropods that represent the faunal group, and it's the most diversified phylum among the living organisms. They are believed to be present in various ecosystems depending on the soil type and texture (Anderson 1988). Apart from the fauna of arthropods, there were various other species found in the soil, including the recognizable macro fauna like termites, centipedes, earthworms, and beetles, as well as the numerous invisible germs, bacteria, and fungi. Plant roots, with their symbiotic relationships and interactions with other soil components, can be considered as soil organisms. The various organisms within the ecosystem interact with diverse plants and animals, contributing to the formation of a complex network of biological activity. Environmental factors such as temperature, moisture, and acidity, as well as anthropogenic actions like management practices, tend to influence the soil biological community and its functions to varying degrees (Frampton et al. 2000; Siira-Pietikäinen et al. 2001).

The interaction between soil organisms and human interventions in land management for agriculture and forestry profoundly influences soil health and quality. The significance of soil fauna becomes more pronounced in tropical nations like India, characterized by rapid population growth and low fertility owing to highly weathered soil. The diversity of soil organisms, particularly beneficial ones, plays a pivotal role in preserving and enhancing soil fertility (Warren and Zou 2002). The health and quality of soil are intimately tied to its soil biota, which includes a diverse array of microbial communities, their food web structure, and the various functions they perform. While soil biodiversity and abundance may not be deemed essential for production, they may be essential for the soil's continued ability to support production. The presence of microflora in the soil is essential for the mineralization of organic matter within the soil. Furthermore, within agrosystems, the soil microflora assumes a significant function in the nutrition and development of plants (Bérard et al. 2015). Fungi and bacteria are the starters of the food web in the soil that encourage the viability of other organisms. Bacteria represent the most abundant group of microorganisms found in soil, whereas the fungal population in soils encompasses a diverse array of organisms. Of all the different kinds of fungal soil microorganisms, *Penicillium, Fusarium, Verticillium, Aspergillus,* and *Trichoderma* species are the most common. *Aspergillus niger* boasts the largest fungal colonies found in soil (Rajik et al. 2011). The microbial communities found in the soil of various cultivated soils are relatively similar. For example, the microbes found in mowed grasslands and cultivated deciduous forests are typically the same. Cultivated lands exhibit varying soil properties and vegetation conditions; however, they all harbor a shared microbial population. Nevertheless, some reports indicate that the transformation of deciduous forests into tilled croplands has led to alterations in soil characteristics, consequently affecting soil microflora. Consequently, in addition to vegetation and soil characteristics, past land use plays a significant role in determining soil microflora (Jangid et al. 2011).

3 Microplastics and Its Sources

- **Plastic**

Synthetic organic polymers, derived from polymerized monomers extracted from petroleum sources, constitute the building blocks of plastics. These advantageous material properties—inertness, durability, lightweight character, and corrosion resistance—have driven the extensive manufacture of plastics, making them a material of choice for numerous applications. The characteristics of plastics have resulted in their widespread utilization across a variety of applications that seem limitless (Cole et al. 2011). The plastics' chemical inertness increased their resistance to full degradation through the usual process, enabling them to endure for centuries without breaking down entirely (Thompson et al. 2004). Each year, a vast amount of plastic is manufactured worldwide for various uses. It is alarming to discover that a substantial percentage of these plastics are disposed of once they fulfill their intended function (Geyer et al. 2017). Based on the findings of Masiá et al. (2019), only 9% of discarded plastics undergo recycling. Approximately 12% of these materials are subjected to incineration, leaving the remaining 79% to be deposited into the environment. Plastic waste can build up as a result of intentional disposal on land or in the ocean, as well as from being carried away from landfill sites by wind, or due to spillages during transportation. This accumulation of plastics ultimately leads to pollution in marine environments worldwide. The primary issues surrounding the build-up of these plastics are their non-biodegradability and ability to persist in the environment for centuries (Geyer et al. 2017; Andrady 2011). In industrial applications, a variety of plastics are utilized, including Polyethylene, Polypropylene, Polystyrene, Poly(ethylene terephthalate), and Poly(vinyl chloride) (Barnes et al. 2009). The production scale has not only grown, but the durability of the plastics being manufactured has also increased. Moreover, the combination of low recycling rates and inadequate plastic management practices has resulted in the accumulation of plastic debris on a global scale (Cole et al. 2011; Xiang et al. 2022; Ragusa et al. 2021b). Recent studies indicate that despite being durable and inert, these plastics could potentially break down under specific conditions. The presence of atmospheric agents, including ultraviolet radiation, wave action, corrosion, and photo-oxidation, along with bacterial activity, contributes to the degradation of these plastics, leading to their fragmentation into micro- and nano-scale particles (Derraik 2002). Due to their diminutive size and minimal density, these microplastics can endure in the environment for extended durations, presenting an escalating environmental threat (Weithmann et al. 2018a). Microplastics have become a worldwide issue, focusing primarily on the necessity to eliminate these tiny particles from the environment (EFSA Contam Panel 2016).

Microplastics consist of a wide variety of substances, such as broken pieces, fibers, spheres, particles, pellets, thin layers, or small balls, which display a variety of forms. These minute particles have sizes that span from 0.1 to 5,000 μm (Abu-Hilal and Al-Najjar 2009). Microplastic pollution poses a significant risk to ecological systems and human health. Nonetheless, the level of uncertainty surrounding this issue is substantial. To determine the probable hazards posed by microplastics to both the environment and human health, data must be collected on both exposure levels and their effects. Microplastics are typically divided into two primary categories depending on their origins: primary microplastics and secondary microplastics (Boucher and Friot 2017). Primary microplastics refer to tiny plastic particles deliberately created for industrial use, such as in cosmetics, toothpaste, detergents, and specific abrasive polishing products. These products consciously include a certain amount of plastic particles. Secondary microplastics, meanwhile, are plastic waste fragments that are formed as a result of biological and chemical breakdown caused by exposure to light, heat, and microorganisms after it has been introduced into the environment, along with mechanical erosion from wind and water (Yurtsever 2019; Laskar and Kumar 2019) (Fig. 1).

A multitude of research investigations have extensively documented the presence of various types of microplastics in the environment, as well as in consumable items

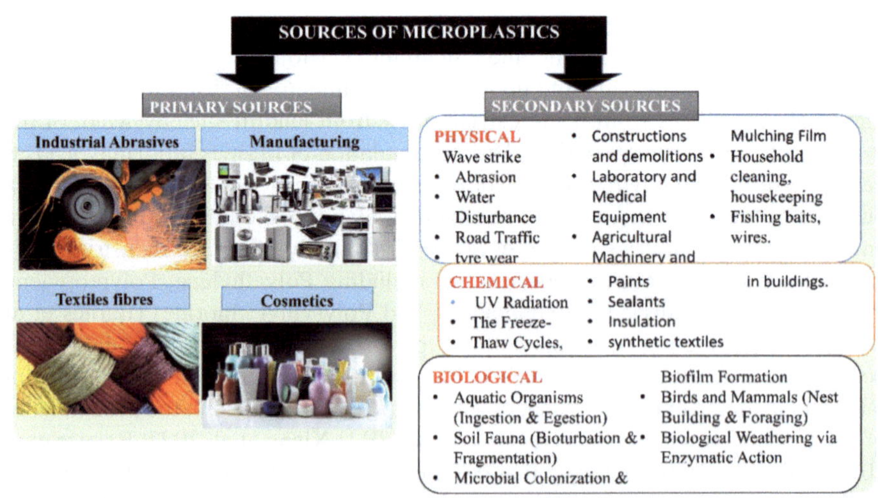

Fig. 1 The diverse sources of microplastics

and even within the human body (Andrady 2017b; Vethaak and Legler 2021b; Ragusa et al. 2021c) (Fig. 2). At present, the primary focus of microplastics research revolves around the aquatic ecosystem. Moreover, substantial evidence indicates the presence of microplastics within terrestrial ecosystems. It is noteworthy that approximately 80% of global plastic waste is deposited in landfills, highlighting the pivotal role of soil as a primary repository for microplastics. Although more and more researchers are turning their attention to the land environment, research into microplastics in the soil is still deficient and requires greater emphasis. Soil, the unconsolidated layer enveloping the surface of the Earth, fulfills an essential function in facilitating the development of plants and assumes a pivotal responsibility for the protection of the environment, ensuring energy stability and conservation of biodiversity. Nevertheless, it is crucial to recognize that soil serves as a key storage site for microplastics within the terrestrial ecosystem (Rillig n.d.; Yang et al. 2021b). According to a report, approximately 90% of plastic waste on land finds its way into the soil environment, either through direct deposition or indirect pathways (PlasticsEurope and Plastics—The Facts 2019). Several previous investigations have underscored the presence of significant quantities of microplastic fibers or waste in sewage sludge and compost (Li et al. 2018b; Mahon et al. 2017b; Zhang et al. 2020c). Given that sewage sludge is utilized in agricultural production, a considerable influx of microplastics is observed to enter the soil environment (Nizzetto et al. 2016). Microplastic contamination of soil environments arises from multiple sources. These include the breakdown of agricultural mulch film, the fragmentation of plastic waste in landfills, and the atmospheric deposition of microplastics (Shen et al. 2022b; Zhang et al. 2019b; Rochman 2018b). Once introduced into the soil, these microplastics persist and accumulate over an extended period, ultimately impacting soil organisms (Chae and An 2018). In addition, microplastics can play a role as vehicles for different toxic contaminants, potentially causing harm to the soil ecosystem if these contaminants are transferred (Zhou et al. 2020b; Guo et al. 2020b). This kind of damage can induce severe consequences for soil health and function. Most of the existing research on environmental microplastics predominantly focuses on aquatic ecosystems, neglecting the critical issue of microplastics in soil. Therefore, it is crucial to comprehend recent advancements in this field to facilitate future research endeavors on this matter. Currently, a growing body of evidence indicates that microplastics are widely distributed in soil, with several studies specifically examining their presence in this environment (Zhu et al. 2019a, 2019b; Guo et al. 2020b; Qi et al. 2020; Xu et al. 2020; He et al. 2018; Chia et al. 2021, 2022; Wang et al. 2019, 2020a; Helmberger et al. 2020; Möller et al. 2020; Li et al. 2020; Kim et al. 2021; Zhang et al. 2021, 2020d; Thomas et al. 2020).

4 The Dispersion of Microplastics Within the Soil Matrix

Numerous researchers are investigating the dispersion of microplastics in the land environment. Nevertheless, due to the absence of an established analytical test standard, numerous researchers have primarily concentrated on extracting, identifying,

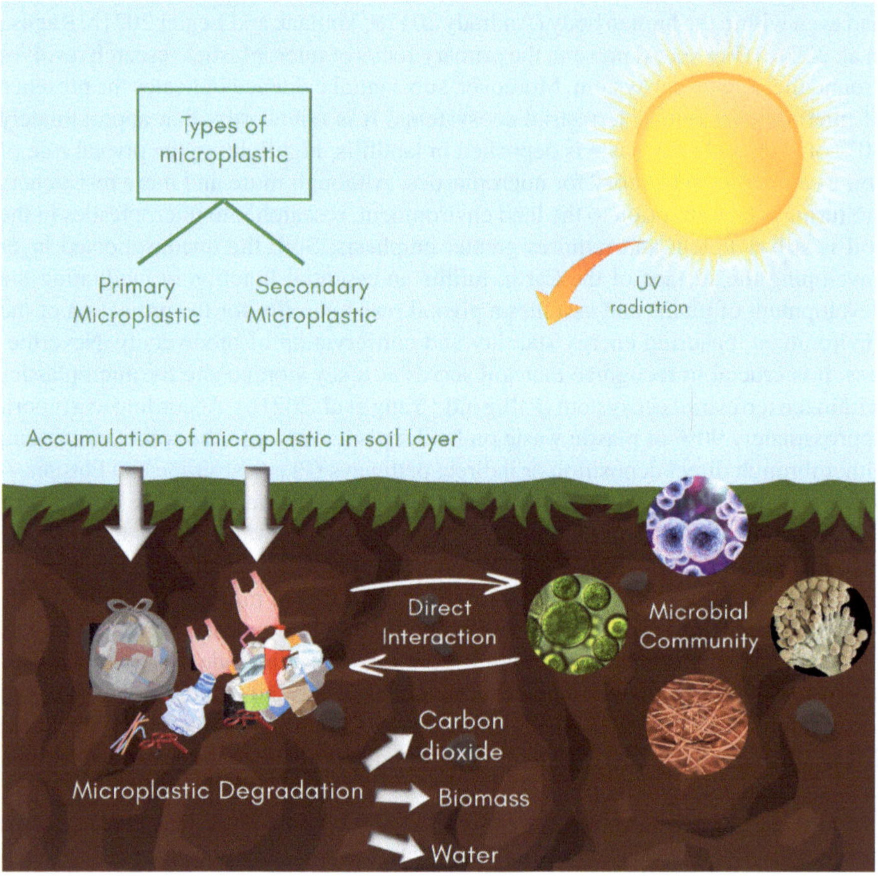

Fig. 2 Types of Microplastics and their effect on soil

and mapping microplastics (Fuller and Gautam 2016; Piehl et al. 2018; Scheurer and Bigalke 2018; Weber and Opp 2020; Corradini et al. 2019; Huerta Lwanga et al. 2016; Berg et al. 2020), predominantly found within agricultural soils. In the farmland soil of southeastern Germany, the mean concentration of microplastics was measured at a low 0.34 ind·kg^{-1} (Piehl et al. 2018). However, the soil in a conventional residential garden found in southeastern Mexico demonstrated notably elevated levels, with a few measurements reaching as high as 870 ind·kg^{-1} (Huerta Lwanga et al. 2017). The difference in mean abundance between these two sites equated to around 2500-fold. The levels of soil microplastics vary significantly across different land types, even within the same city (Chen et al. 2020; Zhou et al. 2019). Furthermore, the estuarine intertidal zone is situated in the transitional region connecting the land and the ocean, making it vulnerable to the contamination of microplastics. (Qian et al. 2016; Zhou et al. 2018). Therefore, the tidal flat soil in the coastal region has become a key area of interest. Several research studies have provided evidence supporting the widespread presence and significant quantities of microplastics. Additionally, these studies have identified a correlation between decreasing particle size and increasing levels of microplastics. It is important to highlight that there has been a notable rise in the proportion of microplastics exhibiting a diminutive particle size, measuring less than 1 mm. The city of Wuhan in China has been identified as having an unusually high concentration of microplastics, with levels ranging from 2.2 × 104 to 6.9 × 105 ind·kg^{-1} (Zhou et al. 2019). This abundance surpassed levels observed in other regions, suggesting a strong correlation with human activities in the study area. Film plastic is extensively utilized in agricultural fields due to its ability to enhance soil temperature, retain moisture, and facilitate seed germination and the rapid growth of young plants. Consequently, the investigation of soil microplastics has placed significant emphasis on farmland covered with plastic films (Huang et al. 2020; Zhou et al. 2020c). Several research studies have demonstrated a direct correlation between the duration of farmland coverage with plastic film and the increased presence of microplastics in the soil. This can be attributed to the fact that plastic film does not readily degrade and instead accumulates over time. Furthermore, it has been observed that certain regions, both agricultural and non-agricultural, exhibit a higher abundance of microplastics in their soils, as evidenced by studies conducted in various locations (Fuller and Gautam 2016; Piehl et al. 2018; Scheurer and Bigalke 2018; Weber and Opp 2020; Corradini et al. 2019; Huerta Lwanga et al. 2016, 2017; Berg et al. 2020; Chen et al. 2020; Zhou et al. 2019, 2018, 2020c; Qian et al. 2016; Huang et al. 2020; Zhang and Liu 2018; Ding et al. 2020). The reported levels of microplastics in soils vary significantly across different countries. Not only is this connected to varying land use patterns, but it also pertains to the inconsistent techniques employed for extracting, isolating, and analyzing microplastics in soil. It is important to highlight that the majority of scholars in current soil microplastics research have solely concentrated on surface soil, and there is still a lack of understanding regarding the overall accumulation of microplastics in soil. Moving forward, it is crucial to establish unified and comparable research methods to standardize the measurement of microplastic abundance.

5 Impact of Microplastics on the Soil Environment

Soil is a sophisticated amalgamation of minerals, organic matter, water, and air (Monreal et al. 2010). The occurrence of microplastics in soil is impacted by a range of factors, such as human activities, inherent soil attributes, and the specific physical and chemical properties of the plastics.

5.1 Human Activities

Microplastics in soil can arise from multiple sources, such as the residues left behind by agricultural film, the recycling of sludge, the practice of sewage irrigation, and the application of fertilizers, among other contributing factors (Zhang et al. 2020d). Human activities have a substantial impact on the distribution of microplastics in the soil, as they introduce these particles into the environment. Certain environmental conditions that are external to the soil have a major impact on the prevalence of microplastics there. Consequently, these specific characteristics are what primarily influence the presence of microplastics in soil. In fields covered with film, the quantity of microplastics detected in the soil was highly impacted by the duration of the film mulching application. Zhou et al.'s study (2020c) indicated that there were more microplastics in the mulched farmland near Hangzhou Bay's soil than in the unmulched agricultural area. In Shihezi, Xinjiang, the cotton fields covered with film have shown varying levels of microplastic abundance in the soil over different periods. After 5 years of film coverage, the mean abundance of microplastics was recorded at 80.3 ind·kg^{-1}. This increased to 308 ind·kg^{-1} after 15 years of film coverage. However, the microplastic abundance reached a significantly higher level of 1075.6 ind·kg^{-1} after 24 years of continuous film coverage in the cotton fields (Yurtsever 2019). As the farmland coated with film increases, so does the amount of microplastics present in the soil. This discovery highlights even more how important agricultural film residue is as a major factor in the ongoing deposition of microplastics in soil. The introduction of sludge into the soil ecosystem brings forth a significant quantity of microplastics. Between 0.63 and 4.30 million tons of microplastics are thought to enter the soil in Europe as a result of sludge application, and between 0.44 and 3.00 million tons are thought to enter the soil in North America (Ng et al. 2018). In China, an estimated 1.56 × 1014 microplastics are introduced into the environment annually through sludge (Li et al. 2018b). As a result, how much sludge is used has a big influence on how much microplastics accumulate in the soil. Moreover, a major external source of microplastics in the soil is the extended use of organic fertilizer in agricultural areas. According to a study, there were anywhere between 14 and 895 ind·kg–1 plastic pieces bigger than 1 mm in organic fertilizers (Weithmann et al. 2018b). The removal of microplastic particles from irrigation water is mostly dependent on filtration in regions where sewage irrigation is used. The quantity of microplastics released into agricultural soil is directly impacted by the efficacy of this filtering procedure. For instance, sewage still releases over 30 billion microplastics into the environment each year, even after being treated at a treatment plant in Vancouver, Canada. In coastal tidal flat soils, the distribution and composition of microplastics are altered by intensive human activities such as development, port operations, tourism, and mariculture (Zhou et al. 2018).

5.2 *Impact of Physicochemical Properties of Soil*

After entering the soil, a variety of soil variables have a significant impact on how microplastics disperse within the soil ecosystem. These elements include organic matter content, soil texture, pH, and the existence of soil organisms (Wang et al. 2020a, 2021, 2022; Yu et al. 2021; Cao et al. 2021). By competing with H + ions, the pH of the soil, in particular, affects the soil's capacity to adsorb microplastics. The soil's surface potential and particular adsorption sites are altered by this competition. A recent study found that compared to neutral soil, acidic soil had much less microplastics (Cao et al. 2021). The soil texture impacts the distribution of microplastics by influencing their mobility within the soil layers (Thomas et al. 2020; Guo et al. 2022). Microplastics are more likely to travel into deeper layers in sandy soils, whereas in clay soils, the cohesiveness of the clay particles can prevent microplastics from moving into the deeper layers (Cao et al. 2021).

According to a recent study, sand loam soils have a much greater concentration of microplastics than other soil textures (Yu et al. 2021). To ascertain how microplastics affect the hydraulic properties of soil, Guo et al. (2022) stressed the significance of soil texture. They found that the size of the particles affected how negatively microplastics affected the infiltration characteristics of loam, clay, and sandy soils. Larger particles notably had the least effect. Furthermore, through consumption or excretion, soil fauna contribute to the spread of microplastics within soil strata. Moreover, their displacement and mobility may cause microplastics to migrate between various soil levels (Zhang et al. 2022; Kumar et al. 2020). In research by Yu et al. (2019), it was shown that earthworms may eat microplastics found in topsoil and carry them down to the soil's deeper layers. In a similar vein, Zhu et al. (2018) have noted that surface attachment, gripping, and pushing by minute soil fauna like mites and collembolans can facilitate the mobility of microplastics inside the soil. Microplastic movement in the soil was found to be facilitated by the presence of collembola, but migratory capacity varied throughout collembola species. Microplastics attach themselves to the cuticles of these microarthropods and are then carried by their extra motions, which is the mechanism underlying this phenomenon (Maaß et al. 2017). Furthermore, several elements, including their chemical and physical characteristics, affect how microplastics are distributed in the soil environment. Consider polyethylene, a long-chain polymer made up of several ethylene (–CH_2–) monomers. Because of its hydrophobic properties and large molecular weight, this polyethylene terephthalate is chemically inert and has limited degradability, which causes it to accumulate extensively in the soil environment (Yoshida et al. 2016). The widespread production and utilization of this material have contributed to its substantial presence. Furthermore, several variables, including temperature, UV radiation, and microfauna populations, might affect the dispersion of soil microplastics (Xi et al. 2022; Lin et al. 2022). Nevertheless, the impact of each polymer's refractory properties may not be immediately significant. A particular polymer poses significant challenges in terms of degradation (Chiellini et al. 2003; Nanda and Sahu 2010). Polyethylene terephthalate's chemical inertness makes it difficult to degrade, and its widespread use has resulted in significant accumulation in soil (Yoshida et al. 2016). Other factors, including temperature, ultraviolet rays, and microfauna communities, can impact soil microplastic distribution (Xi et al. 2022; Lin et al. 2022).

6 Interaction Between Soil Microbial Community and Microplastics

Microorganisms engage with microplastics in diverse settings, such as soil (Mammo et al. 2020). The introduction of more MPs in soil raises concerns about their ecological impact, particularly on soil microbiota. Plastic mulch films were used in agriculture at a rate of 7.4 million tonnes in 2019 compared to 4.4 million tonnes in 2012 (Batista et al. 2022). Consequently, the amount of microplastics that have contaminated the terrestrial environment has increased significantly (Huang et al. 2020). A variety of microorganisms have been found to adhere to MP surfaces (Masó et al. 2003; Tender et al. 2017). Previous research has shown that various MP types can specifically promote the establishment of particular soil microbial taxa within a community. Exposing soil to various plastic types enriched specific bacterial and fungal taxa. It is intriguing to record that the diversity and richness of microbial communities on the surfaces of microplastics are found to be unique and not as diverse as compared to those that are found colonizing materials like wood and rhizospheric soil (Oberbeckmann et al. 2017; Yu et al. 2021).

Microplastispheres or plastispheres are the terms used to describe the microbial communities found on microplastics, primarily due to their exceptional characteristics (Zettler et al. 2013). Several investigations have shown inconsistent results about the effects of MPs on soil microbial populations. While other research has not shown any conclusive evidence of a major impact on the composition of soil microbial populations (Huang et al. 2019; Yan et al. 2021; Judy et al. 2019), others have revealed noteworthy changes in the abundance and diversity of these communities, ultimately affecting the overall functioning of the soil (Gao et al. 2021). Yi et al.'s (2019) investigation found that soil microbial populations were significantly impacted by membrane-like PE and fibrous PP. As compared to Deinococcus-Thermus and Chloroflexi, the quantity of Acidobacteria and Bacteroidetes increased (Yi et al. 2021). Another analysis highlighted the effect of MPs on microbial activities by showing that they reduced substrate-induced respiration and increased the rates at which arbuscular mycorrhizal fungus colonized their roots (Judy et al. 2019). In soil treated with 0.007% low-density polyethylene (LDPE), the rate of turnover of the bacterial population was three times higher than in soil that was not treated. As exposure duration rose, so did the divergence of soil microbial populations (Gao et al. 2021). According to Ng et al. (Yi et al. 2021), the growth medium's microbial activities and the dispersion and variety of soil microorganisms were enhanced by the addition of LDPE (3% w/w) and PET (0.2% and 0.4% w/w). Furthermore, it was shown that 14 different bacterial genera were both more common and enriched. While de

Souza Machado et al. (Ng et al. 2021) and Awet et al. (Souza Machado et al. 2018) observed that polyester (0.05–0.4%), polyacrylic (0.05–0.4%), and PS (1 mg kg-1) had a substantial detrimental influence on soil microbial activity, Liu et al. (2017) Wang et al. (2020b) discovered that PP (7% and 28%) had a favorable impact on soil microbiota. Because MP kinds, sizes, forms, concentrations, and environmental circumstances vary throughout research, it is challenging to generalize about the influence of MPs on soil microbial populations.

The different physical, chemical, and biological characteristics of the soil types employed in these studies may have affected the results.

7 Factors Influencing the Interaction of Microplastics with the Microbial Community

The composition, richness, and diversity of soil microbial communities are strongly influenced by MP-related characteristics. The effects of MP size on soil microbial communities—including attachment, colonization, dispersion, and diversity—have not received much attention in the literature. Different impacts on bacterial diversity at the genus level were seen with LDPE powder (150–250 μm) at 2% and 7% (w/w) (Rong et al. 2021), although LDPE particles (30 μm) at 0.2% (w/w) had no discernible influence on the microbial community (Li et al. 2011). The richness and composition of the microbial communities adhering to MPs were shown to be independent of MP size or type, according to earlier studies by Frère et al. (2018). In contrast to these results, Guo et al. (Guo et al. 2021) observed a considerable increase in soil microbial community richness in the presence of PE microfibers that are less than 2 mm in length. Compared to macroplastics, actinobacteria were shown to be more prevalent in PE, PS, and PP microplastics (Zhang et al. 2017). These results imply that microbial diversity may be influenced by MP size. To learn more about how MP size impacts soil microbial populations, further investigation is required. Different types of MPs have different effects on soil microbial communities than merely size in terms of composition, richness, and diversity. In comparison to soils treated with PS, soil samples treated with PE, PS, and PP of identical size at 150 μm and 1% w/w concentration showed distinct microbial communities. However, the variety of microorganisms among the MP kinds did not significantly differ (Sun et al. 2022). The application of PS (0.33–0.64 μm) and LDPE (150–250 μm) particles to the soil did not substantially alter the variety of bacteria (Rong et al. 2021; Xu et al. 2021). The application of LDPE (678 μm) and PVC (15 μm) particles at the same concentration to the soil, however, led to a reduction in the microbial communities' Shannon diversity indices (Fei et al. 2020). In comparison to soils treated with PS, those treated with PP and PE exhibited greater relative abundances of Bacteroidetes and Gemmatimonadetes (Yi et al. 2021). The capacity of soil microorganisms to react to perturbations varies, which causes differences in results (Lee et al. 2017). Furthermore, soil microorganisms can have complicated reactions to a single artificial pressure (Rillig et al. 2019). Different microbial communities can result from microbial adhesion and colonization on MP surfaces depending on their roughness and hydrophobicity. EPS, which function as bioadhesives and improve microorganism adhesion to the surface, are secreted by colonizing microorganisms on MP surfaces (Ogonowski et al. 2018). The connection between soil microbial community interactions and MP surface hydrophobicity has not received much attention. On the hydrophilic carrier contact, marine bacteria create biofilms in the early colonization stage (Pinto et al. 2019). MP's hydrophobic characteristics significantly decreased in saltwater as a result of an increase in hydrophilic groups (C-O and C=O) on the PE surface (Tu et al. 2020). During the creation and maturation of biofilms, microbial adherence to MP surfaces is enhanced by bacterial hydrophobicity (Pompilio et al. 2008). Aging can cause changes to MP's surface topography, polarity, roughness, and surface area (Mao et al. 2020). Microbial communities can be impacted by structural and physical changes, which can modify their composition and diversity. It was discovered that gammaproteobacteria predominated on rough PE surfaces and betaproteobacteria on smooth PS surfaces (Piehl et al. 2018). Positive correlations were observed between the growth rate and density of adherent microorganisms and a rougher MP surface (Pompilio et al. 2008). In the first stages of biofilm colonization, *Vibrio crassostreae* colonized the smooth PS surface more quickly than the rough PS surface, taking less than 10 h (6 d). But the even PS surface likewise decolonized fast, zero in less than 24 h (Foulon et al. 2016). The quantity and variety of microbial communities are impacted by MPs' surface charge. The ideal range for microbial colonization is 31–43 mN m-1 for plastic surface energy (Nauendorf et al. 2016). Plastic additives such as citrate esters, phthalate esters, fatty acid esters, glycerides, and polyhydric alcohols can leak out during the weathering process (Hahladakis et al. 2018). Over time, their presence in the soil may increase, and they may cause numerous changes to the microbial communities in the soil. The effects of plastic additions on microcosm activity and soil microbial communities have not been well studied (Kong et al. 2012, 2018; Cartwright et al. 2000). The structure and functional diversity of soil microbial communities were not significantly affected by soil treated with DEP and DEHP at 0.1 mg g-1 (Cartwright et al. 2000). Bacterial diversity declined as soil dibutyl phthalate (DBP) concentration rose, and as DBP breakdown progressed, so did the composition, abundance, and structure of soil bacterial communities (Kong et al. 2018). To get further insight into how MPs colonize natural habitats, more studies on the effects of plastic additives on soil microbiomes at relevant quantities are required (Cartwright et al. 2000). Aside from MP characteristics, other variables may also affect the structure, diversity, and activity of soil microbial communities. The variety and makeup of soil microbial populations can be impacted by the exposure of these organisms to MPs (Xu et al. 2019). Microorganism colonization is enhanced by the adsorption of substances such as carbohydrates, proteins, lipids, nucleic acids, and fatty acids onto MPs (Keswani et al. 2016). The composition, diversity, and activity of microbial communities associated with MPs are influenced by several environmental parameters, including salinity, biotic factors, temperature, pH, light intensity, nutrition availability, ionic strength, and pollution (Oberbeckmann et al. 2017; Kesy et al. 2019). The diversity of microbial communities linked to MPs can be strongly impacted by seasonal fluctuations and geographic location (Kesy et al. 2019; Curren and Leong 2019). It is yet unknown how microbiological colonization on MP surfaces operates. Due to natural weathering, plastic can alter the physicochemical composition of soil (Pickett and Kutz 2018), which may have an impact on microbial populations. Virgin plastics usually have a low surface charge and a smooth, homogeneous structure (Fotopoulou and Karapanagioti 2015). In plastic polymers exposed to sunlight, photooxidation is caused by UV and blue photons striking chromophores, which starts the process of photooxidative breakdown (Pickett and Kutz 2018). Plastic polymers experience chain scissions as a result of radicals produced by the photooxidative weathering process. The surface of the polymer gets oxidized when these radicals react with oxygen. As a result, plastic surfaces have more hydrophilic and carbonyl groups, which promote greater microbial community development and biodegradation (Oberbeckmann et al. 2017; Judy et al. 2019; Nauendorf et al. 2016; Hahladakis et al. 2018; Kong et al. 2012, 2018; Cartwright et al. 2000; Kirstein et al. 2019; Xu et al. 2019; Keswani et al. 2016; Kesy et al. 2019; Curren and Leong 2019; Pickett and Kutz 2018; Fotopoulou and Karapanagioti 2015; Wei and Zimmermann 2017; Dong et al. 2020). Plastic additives like plasticizers, UV stabilizers, heat stabilizers, flame retardants, and pigments are mostly broken down during the photooxidation process (Pickett and Kutz 2018). It is essential for determining the makeup, variety, and quantity of microbial communities connected to the MPs found in soil (Yan et al. 2021; Wei and Zimmermann 2017). At high temperatures, chain scissions during thermal weathering generate radicals, just like in photooxidation. Consequently, MP surfaces are more favourable for microbial colonization, biofilm development, and biodegradation (Volke et al. 2001). Increased microbial colonization and the development of plastic-associated microbial communities might result from mechanical stress and biogeochemical assaults that lessen MPs' hydrophobicity, brittleness, and stiffness [173, 366].

8 Microbial Degradation of Microplastics

There is a possibility that long stretches of synthetic plastic will remain in the environment without degrading. The breakdown of MPs in nature is a multi-step process involving physical, chemical, and biological factors (Ammalaa et al. 2010). Microbial strains found in diverse environments, including plastic-dumping sites, mulch film waste, marine water, soils contaminated with crude oil, landfills, sewage sludge, and insects that consume plastic, can biodegrade MPs even though they are not as easily degraded as other biodegradable materials (Krueger et al. 2015). The tolerance of microorganisms for harsh environmental conditions and their ability to grow on a variety of organic and inorganic substrates allow them to degrade a broad spectrum of environmental pollutants. As a result, they offer an environmentally viable and long-lasting solution to the problem of plastic pollution. By facilitating colonization and supplying resources for their nutritional requirements, MPs enable microorganisms to occupy diverse and new ecological niches. MPs experience structural deformation and property loss because of microbial enzymatic activity (Miao et al. 1016). In the following sections role of different types of microorganisms in the degradation of MP is discussed as also tablulated ahead (Table 1).

Table 1 Diverse microorganisms and their role in the degradation of MPs

S. no	Type of polymer	Microbial species	Rate of degradation (%)	References
1	PE	*Rhodococcus ruber* C208	7.5	Sivan et al. (2006)
	HDPE	*Streptomyces* sp	18.26	Farzi et al. (2017)
	LDPE	*Streptomyces coelicoflavus* NBRC 15399T	30.0	Duddu et al. (2015)
	PP	*Rhodococcus* sp. strain 36	6.4	Auta et al. (2018)
	PLA	*Rhodococcus* sp. strain 36	97.0	Jarerat et al. (2003)
6	PS-PLA Nanocomposites	Pseudomonas aeruginosa	9.9	Shimpi et al. (2012)
7	PS (Styrofoam)	Pestalotiopsis sp.	74.43	Yanto et al. (2019)
	PU	Cladosporium pseudocladosporioides	87.0	Álvarez-Barragán et al. (2016)
9	polycyclic aromatic hydrocarbons (PAH)	Mortierellomycota	NR	Wang et al. (2020)
10	High-density polyethylene	*Aspergillus flavus* PEDX3	3.9	Zhang et al. (2020)
11	"Nylon 66 and nylon 6"	" *Bacillus cereus* and *Bacillus sphaericus*"	2–7	Sudhakar et al. (2007)
12	Polyurethane (PUR)	*Comamonas acidovorans*	100	Nakajima-Kambe et al. (1997)
13	Plastic fragments	Mortierellomycota and Ascomycota	NR	Temporiti et al. (2022)
14	LLDPE	*Alcaligenes faecalis* (MK517568)	3.5	Tareen et al. (2022)
15	Polyethylene terephthalate (PET)	*Vibrio*	35	Sarkhel et al. (2020)

NR-Not Reported

8.1 *Bacterial-Based Degradation of Microplastics*

Bacteria are the most prevalent type of microbes present in diverse environmental habitats, including soil, water, and atmosphere. Various species of bacteria have been reported for their potential role in breaking down complex polymers and addressing environmental issues by remediating pollutants like crude oil, antibiotics, metal compounds, plastic, and other harmful substances (Bakir et al. 2014). Bacterial strains possess a range of metabolic activities and diversity that enable them to adsorb, desorb, and degrade MPs (Auta et al. 2018; Auta et al. 2017). In minimal media, it has been discovered that different strains of bacteria rely primarily on MPs as their source of carbon, rendering pronounced changes in their morphology and chemical constituents, besides an appreciable weight loss (Janssen et al. 2002). Within the natural environment, different bacterial strains exist in the form of creating a consortium which helps them degrade the crystalline structure of MP even better (Shah et al. 2008). It is essential to conduct further research to uncover the potential of various bacterial consortia in the biodegradation of microplastics (MP) to further augment their degradation capability.

Actinomycetes are filamentous bacteria with Gram-positive characteristics and are found in diverse habitats, including freshwater, soil, plant tissues, and marine environments (Omidoyin and Femi-Ola 2020). Actinomycetes play important roles in organic carbon recycling, chitin, and cellulose breakdown, and the degradation of recalcitrant structures (Gohain et al. 2020). They can grow on and break down various complex polymers, such as MPs, due to the action of their hydrolytic enzymes and bioactive metabolites (Sharma et al. 2014; Jabloune et al. 2020). Actinomycetes are known to exude biopolymers, including glycogen, levan, and N-acetylglucosamine-rich dextran slime. These biopolymers are thought to increase medical implant surfaces' adherence, which promotes further microbial processes (Petar et al. 2015). Various important Actinomycetes involved in plastic degradation are *Streptomyces, Actinomadura, Rhodococcus,* and thermophilic species of *Thermoactinomycetes* (Jabloune et al. 2020; Auta et al. 2018). Only a few actinomycete genera have been recognized for their potential to degrade MPs, such as *Streptomyces, Rhodococcus, and Micrococcus.* Studies indicate that certain actinomycetes are capable of forming biofilms, similar to various bacterial strains, to enhance their colonization and survival on MPs (Gilan and Sivan 2013).

8.2 *Algae-Based Degradation of Microplastics*

Extensive research has been conducted and confirmed the capacity of various photosynthetic and heterotrophic algae to break down environmental contaminants, encompassing both organic and inorganic substances (Kathirvel et al. 2020; Correa et al. 2020; Mustafa et al. 2021). The research work in the field is focused on the capacities of bacteria and fungi to break down environmental contaminants by adsorbing, accumulating, or metabolizing them into less toxic forms (Correa et al. 2020; Mustafa et al. 2021). However, there has been a limited amount of research on the exploitation of algae for degrading MPs, with current studies shifting towards exploring algae's role in green plastic production (Shanmugam et al. 2021). Numerous algae species, including *Oscillatorian, Spirogyra, Spirulina, Anabaena,* and *Chlorella,* have been found to colonize terrestrial settings, including MP surfaces. However, there is currently insufficient information to draw concrete conclusions regarding these algae role to metabolizing MPs (Sarmah and Rout 2018; Wang et al. 2020c). Khoironi et al. (2019) reported Polypropylene (PP) and polyethylene terephthalate (PET) degradation by *Spirulina* sp. in 112 days of incubation, exhibiting a decrease in tensile strength of 0.1977 MPa and 0.9939 MPa, respectively. In comparison to fungi and bacterial-mediated degradation, the rate of polyethylene (PE) degradation by algae is relatively lower. For instance, during a 30-day incubation period, thermophilic bacteria called *Brevibacillus borstelensis* demonstrated a 30% deterioration in PE films (Hadad et al. 2005). After 28 days of incubation, *Zalerion maritumum,* a marine fungus, managed to break down 43% of PE pellets (Paço et al. 2017). Algae are primarily photoautotrophic organisms that use ambient CO_2 as their major carbon source and sunlight as their primary energy source, which accounts for their slower rate of MP breakdown as compared to bacteria and fungi (Dineshbabu et al. 2020). In heterotrophic environments, when light is scarce, certain kinds of algae may survive by obtaining their energy and carbon from outside sources (Jareonsin and Pumas 2021). While these algae can successfully colonize and absorb MPs, they lack the metabolic ability to break them down (Gopal 2017). Consequently, there is a potential for plastic accumulation within algae, which could then enter the food chain (Hoffmann et al. 2020).

8.3 Fungi-Based Degradation of Microplastics

Fungi are renowned for their extensive metabolic capabilities, which encompass the synthesis of extracellular multienzyme complexes (Matavuly and Peter 2009). Consequently, they have become important player in microbial-mediated biodegradation. Various fungal strains inhabit diverse natural habitats and contribute significantly to the maintenance of biogeochemical cycles, as well as the conversion of various substances (Chen et al. 2016). Multiple fungal species and consortia have been studied for their role in breaking down MPs. These species primarily belong to the genera *Penicillium, Aspergillus, Mucor, Trichoderma,* and *Cladosporium.* Furthermore, specific types of edible fungi such *as Pleurotus abalones, Pleurotus ostreatus,* and *Agaricus bisporus* have been reported to use PE and polystyrene (PS) for their growth and development activities (Ong and Ling 2019; Mitik-Dineva et al. 2009). Through their metabolic activities, fungi potentially mediate changes in the hydrophobic properties of microplastics by acting on various groups of MPs including carboxyl, carbonyl, and ester (Yuan et al. 2020).

8.4 Biofilm-Mediated Degradation of MP

A variety of microorganisms, including viruses, bacteria, actinomycetes, fungi, protists, and algae can grow biofilms on the surface of the MP (Yuan et al. 2020). The interaction between biofilms and microplastics can modify the chemical and physical properties of the polymer surface, resulting in the biological breakdown of microplastics. The primary factor influencing this phenomenon is the time required for microorganisms to alter the intrinsic characteristics of microplastics. Upon the introduction of MPs into the biofilm, several extracellular oxidases and hydrolases decompose the substantial macromolecular polymers into smaller oligomers and monomers. The proliferation of biofilms on microplastic surfaces is affected by variables such as pH, salinity, temperature, and ultraviolet exposure. The biofilm-mediated degradation can be increased by supplementation of glucose as an external carbon source Significant deterioration of MPs occurred by biofilm treatment, which is facilitated by high concentration (120,000 ppm) of methane gas in an environment.

Biofilm formation and degradation of microplastics (MP) involve four stages: adhesion, enzymatic action, fragmentation, and penetration, with microbial filaments and water serving as nutrient sources (Yuan et al. 2020). Adhesion involves bacterial and fungal colonization on MP surfaces, forming biofilms that enhance nutrient availability and provide a stable environment for microbial growth (Sahu et al. 2024). The physicochemical properties of MPs are altered by biofilm formation, facilitating microbial attachment and subsequent degradation (Ventura et al. 20241). Further enzymatic action and fragmentation involves bacteria like Bacillus and Pseudomonas producing hydrolases enzymes that break down the complex structures of MPs into simpler compounds (Silva et al. 2024). This enzymatic activity leads to fragmentation, making MPs more susceptible to further microbial degradation (Emisha et al. 2023). The biofilm matrix allows for the penetration of microbial filaments by utilization water and other nutrients to sustain metabolic processes, thereby enhancing degradation efficiency (Chettri et al. 2023). According to recent review few studies on biofilm forming bacterial species *Rhodococcus* sp., *Bacillus* sp. *Stenotrophomonas panacihumi* assists the detoration of PP, *Pseudomonas aeruginosa* NB26 and Bacillus sp. NB6 for PS and Microbacterium sp. NA23 for PVC (Ali et al. 2024). Rhodococcus ruber is reported to colonize the polyethylene surfaces and making the biofilm, consequently resulting in a decrease in their average molecular weight reductions of 14% (Hadad et al. 2005). Biofilm formation by *erythrobacter* sp. and *Alcanivorax borkum* has been implicated in the biodegradation of LDPE and its subsequent removal from aquatic ecosystems (Yang et al. 2020). While microbial degradation of MPs shows promise, challenges remain in fully understanding the mechanisms and optimizing conditions for effective bioremediation. Further research is essential to address these gaps and improve strategies for mitigating MP pollution.

9 Conclusion

Microplastics have been a major risk for a prolonged time, demanding an in-depth investigation of their sources, structure, and interactions with various biomes. Microplastics are detrimental to all living beings. Microplastics are hazardous to all organisms. Microplastics serve as transporters for heavy metals and pathogens, demonstrating various toxicities and triggering numerous pathophysiological processes. These pathophysiological mediators later lead to various physiological diseases and chronic consequences. A diverse array of bacteria directly interacts with microplastics in soil, facilitating their rapid breakdown into various by-products. Scientific research has primarily concentrated on investigating the ecological effects of microplastic contamination in aquatic ecosystems, especially in marine settings. To date, there has been a paucity of study on soil pollution caused by microplastics. The soil possesses unique physiochemical characteristics, making it difficult for researchers to study the interaction between the soil microbiota and microplastics. A knowledge gap remains on the mechanism of microbial interaction with microplastics. Future research should focus on identifying the exact sources of microplastics in distinct soil types and under varied situations. Research should uncover more microbial species with enhanced degradation capabilities and utilise them efficiently to mitigate microplastic accumulation. A significant objective for researchers is to examine the correlation between microplastic concentration in soil and its impact on the local microbiota.

References

Abu-Hilal AH, Al-Najjar TH (2009) Plastic pellets on the beaches of the Northern Gulf of Aqaba, Red Sea. Aquat Ecosyst Health Manage 12:461–470. https://doi.org/10.1080/14634980903361200

Ajith N et al (2020) Global distribution of microplastics and its impact on marine environment—a review. Environ Sci Pollution Res 27:25970–25986

Ali SS, Elsamahy T, Al-Tohamy R, Sun J (2024) A critical review of microplastics in aquatic ecosystems: degradation mechanisms and removing strategies. Environ Sci Ecotechnol 25:100427

Álvarez-Barragán J et al (2016) Biodegradative activities of selected environmental fungi on a polyester polyurethane varnish and polyether polyurethane foams. Appl Environ Microbiol 82(17):5225–5235

Ammalaa A, Batemana S, Deana K et al (2010) An overview of degradable and biodegradable polyolefins. Prog Polym Sci 36:1015–1049. https://doi.org/10.1016/j.progpolymsci.2010.12.002

Anderson JM (1988) Invertebrate-mediated transport processes in soils. Agr Ecosyst Environ 24(1–3):5–19

Andrady AL (2011) Microplastics in the marine environment. Mar Pollut Bull 62:1596–1605. https://doi.org/10.1016/j.marpolbul.2011.05.030

Andrady AL (2017a) The plastic in microplastics: a review. Mar Pollut Bull 119:12–22

Andrady AL (2017) The plastic in microplastics: a review. Mar. Pollut Bull 119:12–22. [CrossRef]

Auta HS, Emenike CU, Fauziah SH (2017) Screening of Bacillus strains isolated from mangrove ecosystems in Peninsular Malaysia for microplastic degradation. Environ Pollut 231:1552–1559. https://doi.org/10.1016/j.envpol.2017.09.043

Auta HS, Emenike CU, Jayanthi B, Fauziah SH (2018) Growth kinetics and biodeterioration of polypropylene microplastics by Bacillus sp. and Rhodococcus sp. isolated from mangrove sediment. Mar Pollut Bull 127:15–21. https://doi.org/10.1016/j.marpolbul.2017.11.036

Auta HS et al (2018) Growth kinetics and biodeterioration of polypropylene microplastics by Bacillus sp. and Rhodococcus sp. isolated from mangrove sediment. Marine Pollut Bull 127:15–21

Bakir A, Rowland SJ, Thompson RC (2014) Enhanced desorption of persistent organic pollutants from microplastics under simulated physiological conditions. Environ Pollut 185:16–23. https://doi.org/10.1016/j.envpol.2013.10.007

Barnes DKA, Galgani F, Thompson RC, Barlaz M (2009) Accumulation and fragmentation of plastic debris in global environments. Philos Trans Royal Soc B: Biol Sci 364:1985–1998. https://doi.org/10.1098/rstb.2008.0205

Batista T, Cansado IP, Tita B et al (2022) Dealing with plastic waste from agriculture activity. Agronomy 12:1–12. https://doi.org/10.3390/agronomy12010134

Bérard A, Ruy S, Coronel A, Toussaint B, Czarnes S, Legendre L, Doussan C (2015) Rhizosphere: a leverage for tolerance to water deficits of soil microflora?. InEGU general assembly conference abstracts, p 9481

Berg P Van den, Huerta-Lwanga E, Corradini F, Geissen V (2020) Sewage sludge application as a vehicle for microplastics in eastern Spanish agricultural soils. Environ Pollut 261:114198. [CrossRef]

Boucher J, Friot D (2017) Primary Microplastics in the oceans: a global evaluation of sources. Iucn: Gland, Switzerland, Vol 10

Cao L, Wu D, Liu P, Hu W, Xu L, Sun Y, Wu Q, Tian K, Huang B, Yoon SJ (2021) Occurrence, distribution and affecting factors of microplastics in agricultural soils along the lower reaches of Yangtze River, China. Sci Total Environ 794:148694. [CrossRef]

Cartwright C, Thompson I, Burns R (2000) Degradation and impact of phthalate plasticizers on soil microbial communities. Environ Toxicol Chem 19:1253–1261. https://doi.org/10.1002/etc.5620190506

Cement Statistics and Information | U.S. Geological Survey [Internet]. Available from: https://www.usgs.gov/centers/national-minerals-information-center/cement-statistics-and-information

Chae Y, An Y-J (2018) Current research trends on plastic pollution and ecological impacts on the soil ecosystem: a review. Environ Pollut 240:387–395. [CrossRef]

Chen G-Q, Patel MK (2012) Plastics derived from biological sources: present and future: a technical and environmental review. Chem Rev 112(4):2082–2099

Chen Y, Stemple B, Kumar M, Wei N (2016) Cell surface display fungal laccase as a renewable biocatalyst for degradation of persistent micropollutants bisphenol A and sulfamethoxazole. Environ Sci Technol 50:8799–8808. https://doi.org/10.1021/acs.est.6b01641

Chen Y, Leng Y, Liu X, Wang J (2020) Microplastic pollution in vegetable farmlands of suburb Wuhan, central China. Environ Pollut 257:113449. [CrossRef]

Chettri D, Pati T, Verma AK (2023) Microbe-mediated biodegradation of microplastics from wastes. Water Environ J 37(4):671–685

Chia RW, Lee J-Y, Kim H, Jang J (2021) Microplastic pollution in soil and groundwater: a review. Environ Chem Lett 19:4211–4224. [CrossRef]

Chia RW, Lee J-Y, Jang J, Kim H, Kwon KD (2022) Soil health and microplastics: A review of the impacts of microplastic contamination on soil properties. J Soils Sediments 22:1–16. [CrossRef]

Chiellini E, Corti A, Swift G (2003) Biodegradation of thermally-oxidized, fragmented low-density polyethylenes. Polym Degrad Stab 81:341–351. [CrossRef]

Cincinelli A, Martellini T, Guerranti C, Scopetani C, Chelazzi D, Giarrizzo T (2019) A potpourri of microplastics in the sea surface and water column of the Mediterranean Sea. TrAC Trends Anal Chem 110:321–326. https://doi.org/10.1016/j.trac.2018.10.026

Cole M, Lindeque P, Halsband C, Galloway TS (2011) Microplastics as contaminants in the marine environment: a review. Mar Pollut Bull 62:2588–2597. https://doi.org/10.1016/j.marpolbul.2011.09.025

Corradini F, Meza P, Eguiluz R, Casado F, Huerta-Lwanga E, Geissen V (2019) Evidence of microplastic accumulation in agricultural soils from sewage sludge disposal. Sci Total Environ 671:411–420. [CrossRef] [PubMed]

Correa D, Beyer H, Possingham H, Fargione J, Hill J, Schenk P (2020) Microalgal biofuel production at national scales: Reducing conflicts with agricultural lands and biodiversity within countries. Energy 215:1–12. https://doi.org/10.1016/j.energy.2020.119033

Curren E, Leong SCY (2019) Profiles of bacterial assemblages from microplastics of tropical coastal environments. Sci Total Environ 655:313–320. https://doi.org/10.1016/j.scitotenv.2018.11.250

da Silva MR, Souza KS, Motteran F, de Araújo LC, Singh R, Bhadouria R, de Oliveira MB (2024) Exploring biodegradative efficiency: a systematic review on the main microplastic-degrading bacteria. Front Microbiol 18(15):1360844

de Souza Machado AA, Lau CW, Till J et al (2018) Impacts of microplastics on the soil biophysical environment. Environ Sci Technol 52:9656–9665. https://doi.org/10.1021/acs.est.8b02212

De Tender C, Devriese LI, Haegeman A et al (2017) Temporal dynamics of bacterial and fungal colonization on plastic debris in the North sea. Environ Sci Technol 51:7350–7360. https://doi.org/10.1021/acs.est.7b00697

Derraik JGB (2002) The pollution of the marine environment by plastic debris: a review. Mar Pollut Bull 44:842–852. https://doi.org/10.1016/S0025-326X(02)00220-5

Dineshbabu G, Uma V, Mathimani T, Prabaharan D (2020) Elevated CO_2 impact on growth and lipid of marine cyanobacterium Phormidium valderianum BDU 20041—towards microalgal carbon sequestration. Biocatal Agric Biotechnol 25:1–7. https://doi.org/10.1016/j.bcab.2020.101606

Ding L, Zhang S, Wang X, Yang X, Zhang C, Qi Y, Guo X (2020) The occurrence and distribution characteristics of microplastics in the agricultural soils of Shaanxi Province, in north-western China. Sci Total Environ 720:137525. [CrossRef] [PubMed]

Dong M, Zhang Q, Xing X, Chen W, She Z, Luo Z (2020) Raman spectra and surface changes of microplastics weathered under natural environments. Sci Total Environ 739:1–9. https://doi.org/10.1016/j.scitotenv.2020.139990

Duddu MK et al (2015) Biodegradation of low density polyethylene (LDPE) by a new biosurfactant-producing thermophilic Streptomyces coelicoflavus NBRC 15399T. African J Biotechnol 14(4):327–340

Duis K, Coors A (2016) Microplastics in the aquatic and terrestrial environment: sources (with a specific focus on personal care products), fate and effects. Environ Sci Eur 28:1–25. https://doi.org/10.1186/s12302-015-0069-y

EFSA CONTAM Panel (EFSA Panel on Contaminants in the Food Chain) (2016) Statement on the presence of microplastics and nanoplastics in food, with particular focus on seafood. EFSA J 14:e04501. https://doi.org/10.2903/j.efsa.2016.4501

Emisha L, Wilfred N, Kavitha S, Halder G, Haldar D, Patel AK, Singhania RR, Pandey A (2023) Biodegradation of microplastics: advancement in the strategic approaches towards prevention of its accumulation and harmful effects. Chemosphere 9:140661

Farzi A et al (2017) Biodegradation of high density polyethylene using Streptomyces species. J Coast Life Med 5(11):474–479

Fei Y, Huang S, Zhang H et al (2020) Response of soil enzyme activities and bacterial communities to the accumulation of microplastics in an acid cropped soil. Sci Total Environ 707:1–9. https://doi.org/10.1016/j.scitotenv.2019.135634

Fotopoulou KN, Karapanagioti HK (2015) Surface properties of beached plastics. Environ Sci Pollut Res 22:11022–11032. https://doi.org/10.1007/s11356-015-4332-y

Foulon V, Le Roux F, Lambert C, Huvet A, Soudant P, Paul-Pont I (2016) Colonization of polystyrene microparticles by Vibrio crassostreae: light and electron microscopic investigation. Environ Sci Technol 50:10988–10996. https://doi.org/10.1021/acs.est.6b02720

Frampton GK, Brink PJ vanden, den Brink PJ van, van den, Brink PJ, Rusek J (2002) Influence of cropping on the species composition of epigeic Collembola in arable fields. In: Apterygota at the beginning of the third Millenium. X International Colloquium on Apterygota, Ceske Budejovice, 21–24 August 2000, Pedobiologia 46 (3–4), 328–337

Frère L, Maignien L, Chalopin M et al (2018) Microplastic bacterial communities in the Bay of Brest: Influence of polymer type and size. Environ Pollut 242:614–625. https://doi.org/10.1016/j.envpol.2018.07.023

Frias, João PGL, Nash R (2019) Microplastics: finding a consensus on the definition. Marine Pollution Bulletin 138:145–147

Fuller S, Gautam A (2016) A procedure for measuring microplastics using pressurized fluid extraction. Environ Sci Technol 50:5774–5780. [CrossRef]

Gao B, Yao H, Li Y, Zhu Y (2021) Microplastic addition alters the microbial community structure and stimulates soil carbon dioxide emissions in vegetable-growing soil. Environ Toxicol Chem 40:352–365. https://doi.org/10.1002/etc.4916

Geyer R, Jambeck JR, Law KL (2017) Production, use, and fate of all plastics ever made. Sci Adv 3:19–24. https://doi.org/10.1126/sciadv.1700782

Gilan I, Sivan A (2013) Effect of proteases on biofilm formation of the plastic-degrading actinomycete Rhodococcus ruber C208. FEMS Microbiol Lett 342:18–23. https://doi.org/10.1111/1574-6968.12114

Gohain A, Manpoong C, Saikia R, De Mandal S (2020) Actinobacteria: diversity and biotechnological applications. In: De Mandal S, Bhatt P (eds) Recent advancements in microbial diversity. Academic Press, Amsterdam, pp 217–231

Gopal R (2017) Biodegradation of polyethylene by green photosynthetic microalgae. J Bioremediat Biodegradation 8:1–8. https://doi.org/10.4172/2155-6199.1000381

Guo J-J, Huang X-P, Xiang L, Wang Y-Z, Li Y-W, Li H, Cai Q-Y, Mo C-H, Wong M-H (2020a) Source, migration and toxicology of microplastics in soil. Environ Int 137:105263

Guo QQ, Xiao MR, Ma Y, Niu H, Zhang GS (2021) Polyester microfiber and natural organic matter impact microbial communities, carbon-degraded enzymes, and carbon accumulation in a clayey soil. J Hazard Mater 405:1–10. https://doi.org/10.1016/j.jhazmat.2020.124701

Guo J-J, Huang X-P, Xiang L, Wang Y-Z, Li Y-W, Li H, Cai Q-Y, Mo C-H, Wong M-H (2020) Source, migration and toxicology of microplastics in soil. Environ Int 137:105263. [CrossRef] [PubMed]

Guo Z, Li P, Yang X, Wang Z, Lu B, Chen W, Wu Y, Li G, Zhao Z, Liu G (2022) Soil texture is an important factor determining how microplastics affect soil hydraulic characteristics. Environ Int 165:107293. [CrossRef]

Hadad D, Geresh S, Sivan A (2005) Biodegradation of polyethylene by the thermophilic bacterium Brevibacillus borstelensis. J Appl Microbiol 98(5):1093–1100

Hadad D, Geresh S, Sivan A (2005) Biodegradation of polyethylene by the thermophilic bacterium Brevibacillus borstelensis. J Appl Microbiol 98:1093–1100. https://doi.org/10.1111/j.1365-2672.2005.02553.x

Hahladakis JN, Velis CA, Weber R, Iacovidou E, Purnell P (2018) An overview of chemical additives present in plastics: Migration, release, fate and environmental impact during their use, disposal and recycling. J Hazard Mater 344:179–199. https://doi.org/10.1016/j.jhazmat.2017.10.014

Hasan MM, Jho EH (2022) Effect of microplastics on the germination and growth of terrestrial plants. J Korean Soc Environ Eng 44:375–382. https://doi.org/10.4491/KSEE.2022.44.10.375

He D, Luo Y, Lu S, Liu M, Song Y, Lei L (2018) Microplastics in soils: analytical methods, pollution characteristics and ecological risks. TrAC Trends Anal Chem 109:163–172. [CrossRef]

Helmberger MS, Tiemann LK, Grieshop MJ (2020) Towards an ecology of soil microplastics. Funct Ecol 34:550–560. [CrossRef]

Hoffmann L, Eggers SL, Allhusen E, Katlein C, Peeken I (2020) Interactions between the ice algae Fragillariopsis cylindrus and microplastics in sea ice. Environ Int 139:1–9

Huang Y, Zhao Y, Wang J, Zhang M, Jia W, Qin X (2019) LDPE microplastic films alter microbial community composition and enzymatic activities in soil. Environ Pollut 254:1–10. https://doi.org/10.1016/j.envpol.2019.112983

Huang Y, Liu Q, Jia W, Yan C, Wang J (2020) Agricultural plastic mulching as a source of microplastics in the terrestrial environment. Environ Pollut 260:114096. [CrossRef]

Huang Y, Liu Q, Jia W, Yan C, Wang J (2020). Agricultural plastic mulching as a source of microplastics in the terrestrial environment. Environ Pollut 260:1–6. https://doi.org/10.1016/j.envpol.2020.114096

Huerta Lwanga E, Gertsen H, Gooren H, Peters P, Salánki T, Van Der Ploeg M, Besseling E, Koelmans AA, Geissen V (2016) Microplastics in the terrestrial ecosystem: Implications for Lumbricus terrestris (Oligochaeta, Lumbricidae). Environ Sci Technol 50:2685–2691. [CrossRef] [PubMed]

Huerta Lwanga E, Mendoza Vega J, Ku Quej V, Chi JDLA, Sanchez del Cid L, Chi C, Escalona Segura G, Gertsen H, Salánki T, van der Ploeg M (2017) Field evidence for transfer of plastic debris along a terrestrial food chain. Sci Rep 7:1–7. [CrossRef]

Hur J, Jho EH (2021) Current research trends on the effects of microplastics in soil environment using earthworms: mini-review. J Korean Soc Environ Eng 43:299–306. https://doi.org/10.4491/KSEE.2021.43.4.299

Jabloune R, Khalil M, Ben Moussa IE et al (2020) Enzymatic degradation of p-Nitrophenyl esters, polyethylene terephthalate, cutin, and suberin by Sub1, a suberinase encoded by the plant pathogen Streptomyces scabies. Microbes Environ 35:1–7. https://doi.org/10.1264/jsme2.ME19086

Jangid K, Williams MA, Franzluebbers AJ, Schmidt TM, Coleman DC, Whitman WB (2011) Land-use history has a stronger impact on soil microbial community composition than aboveground vegetation and soil properties. Soil Biol Biochem 43(10):2184–2193

Janssen PH, Yates PS, Grinton BE, Taylor PM, Sait M (2002) Improved culturability of soil bacteria and isolation in pure culture of novel members of the divisions Acidobacteria, Actinobacteria, Proteobacteria, and Verrucomicrobia. Appl Environ Microbiol 68:2391–2396. https://doi.org/10.1128/aem.68.5.2391-2396.2002

Jareonsin S, Pumas C (2021) Advantages of heterotrophic microalgae as a host for phytochemicals production. Front Bioeng Biotechnol 9:1–17. https://doi.org/10.3389/fbioe.2021.628597

Jarerat A, Tokiwa Y, Tanaka H (2003) Poly (L-lactide) degradation by Kibdelosporangium aridum. Biotech Lett 25:2035–2038

Judy JD, Williams M, Gregg A et al (2019) Microplastics in municipal mixed-waste organic outputs induce minimal short to long-term toxicity in key terrestrial biota. Environ Pollut 252:522–531. https://doi.org/10.1016/j.envpol.2019.05.027

Kathirvel B, Mathimani T, Rene E, Shanmugam S, Chi N, Pugazhendhi A (2020) Impact of cultivation conditions on the biomass and lipid in microalgae with an emphasis on biodiesel. Fuel 284:1–8. https://doi.org/10.1016/j.fuel.2020.119058

Keswani A, Oliver DM, Gutierrez T, Quilliam RS (2016) Microbial hitchhikers on marine plastic debris: human exposure risks at bathing waters and beach environments. Mar Environ Res 118:10–19. https://doi.org/10.1016/j.marenvres.2016.04.006

Kesy K, Oberbeckmann S, Kreikemeyer B, Labrenz M (2019) Spatial environmental heterogeneity determines young biofilm assemblages on microplastics in Baltic Sea mesocosms. Front Microbiol 10:1–18. https://doi.org/10.3389/fmicb.2019.01665

Khalid N, Aqeel M, Noman A et al (2021) Linking effects of microplastics to ecological impacts in marine environments. Chemosphere 264:1–13. https://doi.org/10.1016/j.chemos

Khoironi A, Anggoro S, Sudarno S (2019) Evaluation of the interaction among microalgae Spirulina sp, plastics polyethylene terephthalate and polypropylene in freshwater environment. J Ecol Eng 20:161–173. https://doi.org/10.12911/22998993/108637

Kim Y-N, Yoon J-H, Kim K-HJ (2021) Microplastic contamination in soil environment—a review. Soil Sci Annu 71:300–308. [CrossRef]

Kirstein IV, Wichels A, Gullans E, Krohne G, Gerdts G (2019) The Plastisphere—Uncovering tightly attached plastic "specific" microorganisms. PLoS One 14:1–17. https://doi.org/10.1371/journal.pone.0215859

Kong S, Ji Y, Liu L et al (2012) Diversities of phthalate esters in suburban agricultural soils and wasteland soil appeared with urbanization in China. Environ Pollut 170:161–168. https://doi.org/10.1016/j.envpol.2012.06.017

Kong X, Jin D, Jin S et al (2018) Responses of bacterial community to dibutyl phthalate pollution in a soil-vegetable ecosystem. J Hazard Mater 353:142–150. https://doi.org/10.1016/j.jhazmat.2018.04.015

Krueger M, Harms H, Schlosser D (2015) Prospects for microbiological solutions to environmental pollution with plastics. Appl Microbiol Biotechnol 99:8857–8874. https://doi.org/10.1007/s00253-015-6879-4

Kumar M, Xiong X, He M, Tsang DC, Gupta J, Khan E, Harrad S, Hou D, Ok YS, Bolan NS (2020) Microplastics as pollutants in agricultural soils. Environ Pollut 265:114980. [CrossRef] [PubMed]

Kuzyakov Y, Blagodatskaya E (2015) Microbial hotspots and hot moments in soil: concept and review. Soil Biol Biochem 83:184–199. https://doi.org/10.1016/j.soilbio.2015.01.025

Laskar N, Kumar U (2019) Plastics and microplastics: a threat to environment. Environ. Technol Innov 14:100352. [CrossRef]

Lee S-H, Sorensen JW, Grady KL, Tobin TC, Shade A (2017) Divergent extremes but convergent recovery of bacterial and archaeal soil communities to an ongoing subterranean coal mine fire. ISME J 11:1447–1459. https://doi.org/10.1038/ismej.2017.1

Li X, Chen L, Mei Q, Dong B, Dai X, Ding G, Zeng EY (2018a) Microplastics in sewage sludge from the wastewater treatment plants in China. Water Res 142:75–85

Li X, Chen L, Mei Q, Dong B, Dai X, Ding G, Zeng EY (2018) Microplastics in sewage sludge from the wastewater treatment plants in China. Water Res 142:75–85. [CrossRef] [PubMed]

Li J, Song Y, Cai Y (2020) Focus topics on microplastics in soil: Analytical methods, occurrence, transport, and ecological risks. Environ Pollut 257:113570. [CrossRef]

Li HZ, Zhu D, Lindhardt JH, Lin SM, Ke X, Cui L (2021) Long-term fertilization history alters effects of microplastics on soil properties, microbial communities, and functions in diverse farmland ecosystem. Environ Sci Technol 55:4658–4668. https://doi.org/10.1021/acs.est.0c04849

Lin Z, Jin T, Zou T, Xu L, Xi B, Xu D, He J, Xiong L, Tang C, Peng J (2022) Current progress on plastic/microplastic degradation: fact influences and mechanism. Environ Pollut 304:119159. [CrossRef]

Liu H, Yang X, Liu G et al (2017) Response of soil dissolved organic matter to microplastic addition in Chinese loess soil. Chemosphere 185:907–917. https://doi.org/10.1016/j.chemosphere.2017.07.064

Maaß S, Daphi D, Lehmann A, Rillig MC (2017) Transport of microplastics by two collembolan species. Environ Pollut 225:456–459. [CrossRef]

Mahon AM, O'Connell B, Healy MG, O'Connor I, Officer R, Nash R, Morrison L (2017a) Microplastics in sewage sludge: effects of treatment. Environ Sci Technol 51:810–818

Mahon AM, O'Connell B, Healy MG, O'Connor I, Officer R, Nash R, Morrison L (2017) Microplastics in sewage sludge: Effects of treatment. Environ Sci Technol 51:810–818. [CrossRef]

Mammo FK, Amoah ID, Gani KM et al (2020) Microplastics in the environment: interactions with microbes and chemical contaminants. Sci Total Environ 743:1–21. https://doi.org/10.1016/j.scitotenv.2020.14051

Mao R, Lang M, Yu X, Wu R, Yang X, Guo X (2020) Aging mechanism of microplastics with UV irradiation and its effects on the adsorption of heavy metals. J Hazard Mater 393:1–9. https://doi.org/10.1016/j.jhazmat.2020.122515

Masiá P, Ardura A, Garcia-Vazquez E (2019) Microplastics in special protected areas for migratory birds in the Bay of Biscay. Mar Pollut Bull 146:993–1001. https://doi.org/10.1016/j.marpolbul.2019.07.065

Masó M, Garcés E, Pagès F, Camp J (2003) Drifting plastic debris as a potential vector for dispersing Harmful Algal Bloom (HAB) species. Sci Mar 67:107–111. https://doi.org/10.3989/scimar.2003.67n110

Matavuly M, Peter M (2009) Marine fungi: degraders of poly-3-hydroxyalkanoate based plastic materials. Zb Matice Srp Prir Nauke 116:253–265. https://doi.org/10.2298/ZMSPN0916253M

Miao L, Wang P, Hou J et al (2019). Distinct community structure and microbial functions of biofilms colonizing microplastics. Sci Total Environ 650:2395–2402. https://doi.org/10.1016/j.scitotenv.2018.09.378

Mitik-Dineva N, Wang J, Truong VK et al (2009) Escherichia coli, Pseudomonas aeruginosa, and Staphylococcus aureus attachment patterns on glass surfaces with nanoscale roughness. Curr Microbiol 58:268–273. https://doi.org/10.1007/s00284-008-9320-8

Möller JN, Löder MG, Laforsch C (2020) Finding microplastics in soils: a review of analytical methods. Environ Sci Technol 54:2078–2090. [CrossRef]

Monkul MM, Ozhan HO (2021) Microplastic contamination in soils: a review from geotechnical engineering view. Polymers 13:1–24. https://doi.org/10.3390/polym13234129

Monreal CM, Sultan Y, Schnitzer M (2010) Soil organic matter in nano-scale structures of a cultivated Black Chernozem. Geoderma 159:237–242. [CrossRef]

Mustafa S, Bhatti HN, Maqbool M, Iqbal M (2021) Microalgae biosorption, bioaccumulation and biodegradation efficiency for the remediation of wastewater and carbon dioxide mitigation: Prospects, challenges and opportunities. J Water Process Eng 41:1–15. https://doi.org/10.1016/j.jwpe.2021.102009

Nakajima-Kambe T et al (1997) Determination of the polyester polyurethane breakdown products and distribution of the polyurethane degrading enzyme of Comamonas acidovorans strain TB-35. J Fermentation and Bioengineering 83(5):456–460

Nanda S, Sahu SS (2010) Biodegradability of polyethylene by Brevibacillus, Pseudomonas, and Rhodococcus spp. NY Sci J 3:95–98

Nauendorf A, Krause S, Bigalke NK et al (2016) Microbial colonization and degradation of polyethylene and biodegradable plastic bags in temperate fine-grained organic-rich marine sediments. Mar Pollut Bull 103:168–178. https://doi.org/10.1016/j.marpolbul.2015.12.024

Ng EL, Lin SY, Dungan AM et al (2021) Microplastic pollution alters forest soil microbiome. J Hazard Mater 409:1–10. https://doi.org/10.1016/j.jhazmat.2020.124606

Ng E-L, Lwanga EH, Eldridge SM, Johnston P, Hu H-W, Geissen V, Chen D (2018) An overview of microplastic and nanoplastic pollution in agroecosystems. Sci Total Environ 627:1377–1388. [CrossRef]

Nizzetto L, Futter M, Langaas S (2016) Are agricultural soils dumps for microplastics of urban origin?, 10777–10779

Oberbeckmann S, Kreikemeyer B, Labrenz M (2017) Environmental factors support the formation of specific bacterial assemblages on microplastics. Front Microbiol 8:1–12. https://doi.org/10.3389/fmicb.2017.02709

Ogonowski M, Motiei A, Ininbergs K et al (2018) Evidence for selective bacterial community structuring on microplastics. Environ Microbiol 20(8):2796–2808. https://doi.org/10.1111/1462-2920.14120

Omidoyin KC, Femi-Ola TO (2020) Isolation and screening of Streptomyces spp from soil samples of Ekiti State University Nigeria for antibacterial activity. Int J Res Innov Entrepreneurship 1:48–54. https://doi.org/10.15294/ijrie.v1i1

Ong H, Ling SW (2019) The growth and laccase activity of edible mushrooms involved in plastics degradation. Curr Top Toxicol 15:57–62

Paço A, Duarte K, da Costa JP et al (2017) Biodegradation of polyethylene microplastics by the marine fungus Zalerion maritimum. Sci Total Environ 586:10–15. https://doi.org/10.1016/j.scitotenv.2017.02.017

Petar P, Beaman B, Ravalison M, Boiron P (2015) Rodriguez-NavaV. Nocardia and Actinomyces. In: Yi-Wei T, Andrew S (eds) Molecular medical microbiology. 2nd ed. Amsterdam: Academic Press, pp 731–752

Pickett JE (2018) Weathering of plastics. In: Kutz M (ed) Handbook of environmental degradation of materials, 3rd edn. William Andrew Publishing, Oxford, pp 163–184

Piehl S, Leibner A, Löder MG, Dris R, Bogner C, Laforsch C (2018) Identification and quantification of macro-and microplas-tics on an agricultural farmland. Sci Rep 8:17950. [CrossRef]

Pinto M, Langer TM, Hüffer T, Hofmann T, Herndl GJ (2019) The composition of bacterial communities associated with plastic biofilms differs between different polymers and stages of biofilm succession. PLoS One 14:1–20. https://doi.org/10.1371/journal.pone.0217165

PlasticsEurope E (2019) Plastics—The Facts 2019. An analysis of European plastics production, demand and waste data. Available online: https://www.plasticseurope.org/en/resources/publications/1804-plastics-facts-2019. Accessed 15 August 2022

Pompilio A, Piccolomini R, Picciani C, D'Antonio D, Savini V, Di Bonaventura G (2008) Factors associated with adherence to and biofilm formation on polystyrene by Stenotrophomonas maltophilia: the role of cell surface hydrophobicity and motility. FEMS Microbiol Lett 287:41–47. https://doi.org/10.1111/j.1574-6968.2008.01292.x

Qi R, Jones DL, Li Z, Liu Q, Yan C (2020) Behavior of microplastics and plastic film residues in the soil environment: a critical review. Sci Total Environ 703:134722. [CrossRef] [PubMed]

Qian Z, Zhang H, Yang Z, Yuan L, Yong X, ChuanCheng F, Chen T, YongMing L (2016) Separation of microplastics from a coastal soil and their surface microscopic features. Chin Sci Bull 61:1604–1611

Ragusa A, Svelato A, Santacroce C, Catalano P, Notarstefano V, Carnevali O, Papa F, Rongioletti MCA, Baiocco F, Draghi S (2021a) Plasticenta: first evidence of microplastics in human placenta. Environ Int 146:106274

Ragusa A, Svelato A, Santacroce C, Catalano P, Notarstefano V, Carnevali O, Papa F, Rongioletti MCA, Baiocco F, Draghi S, D'Amore E, Rinaldo D, Matta M, Giorgini E (2021b) Plasticenta: first evidence of microplastics in human placenta. Environ Int 146:106274. https://doi.org/10.1016/j.envint.2020.106274

Ragusa A, Svelato A, Santacroce C, Catalano P, Notarstefano V, Carnevali O, Papa F, Rongioletti MCA, Baiocco F, Draghi S (2021) Plasticenta: first evidence of microplastics in human placenta. Environ Int 146:106274. [CrossRef] [PubMed]

Rajendran S, Kannan V, Natarajan K et al (2015) The role of microbes in plastic degradation. In: Chandra R (ed) Environmental waste management, 1st edn. CRC Press, Florida, pp 341–370

Rajik M, Pathak SP, Biswas SK, Naresh PR (2011) Effect of organic amendment on soil microflora and soil borne diseases of potato. Indian Phytopathol 64(3):280

Rillig M, Ryo M, Lehmann A et al (2019) The role of multiple global change factors in driving soil functions and microbial biodiversity. Science 366:886–890. https://doi.org/10.1126/science.aay2832

Rillig MC, Microplastic in terrestrial ecosystems and the soil?.

Rochman CM (2018a) Microplastics research—from sink to source. Science 360:28–29

Rochman CM (2018) Microplastics research—from sink to source. Science 360:28–29. [CrossRef]

Rong L, Zhao L, Zhao L et al (2021) LDPE microplastics affect soil microbial communities and nitrogen cycling. Sci Total Environ 773:1–11. https://doi.org/10.1016/j.scitotenv.2021.145640

S, K, Sahu, Alok Prasad D (2024) Microplastics degradation mechanism by aquatic plastiomes: a review. Environ Sci Eng, 213–226. https://doi.org/10.1007/978-3-031-50840-0_12

Sarkhel R et al (2020) Comparative biodegradation study of polymer from plastic bottle waste using novel isolated bacteria and fungi from marine source. J Polymer Res 27:1–8

Sarmah P, Rout J (2018) Algal colonization on polythene carry bags in a domestic solid waste dumping site of Silchar town in Assam. Phykos 48:67–77

Scheurer M, Bigalke M (2018) Microplastics in Swiss floodplain soils. Environ Sci Technol 52:3591–3598. [CrossRef]

Shah A, Hasan F, Akhter J, Hameed A, Ahmed S (2008) Degradation of polyurethane by novel bacterial consortium isolated from soil. Ann Microbiol 58:381–386. https://doi.org/10.1007/BF03175532

Shanmugam S, Hari A, Kumar D et al (2021) Recent developments and strategies in genome engineering and integrated fermentation approaches for biobutanol production from microalgae. Fuel 285:1–9. https://doi.org/10.1016/j.fuel.2020.119052

Sharma M, Dangi P, Choudhary M (2014) Actinomycetes: source, identification, and their applications. Int J Curr Microbiol Appl Sci 3:801–832

Shen M, Xiong W, Song B, Zhou C, Almatrafi E, Zeng G, Zhang Y (2022a) Microplastics in landfill and leachate: occurrence, environmental behavior and removal strategies. Chemosphere 305:135325

Shen M, Xiong W, Song B, Zhou C, Almatrafi E, Zeng G, Zhang Y (2022) Microplastics in landfill and leachate: Occurrence, environmental behavior and removal strategies. Chemosphere 305:135325. [CrossRef]

Shimpi N et al (2012) Biodegradation of polystyrene (PS)-poly (lactic acid)(PLA) nanocomposites using Pseudomonas aeruginosa. Macromolecular Res 20:181–187

Siira-Pietikäinen A et al (2001) Short-term responses of soil decomposer communities to forest management: clear felling versus alternative forest harvesting methods. Canadian J Forest Res 31(1):88–99

Sivan A, Szanto M, Pavlov V (2006) Biofilm development of the polyethylene-degrading bacterium Rhodococcus ruber. Appl Microbiol Biotechnol 72:346–352

Sudhakar M et al (2007) Marine bacteria mediated degradation of nylon 66 and 6. Int Biodeterioration & Biodegradation 60(3):144–151

Sun Y, Duan C, Cao N et al (2022) Effects of microplastics on soil microbiome: The impacts of polymer type, shape, and concentration. Sci Total Environ 806:1–11. https://doi.org/10.1016/j.scitotenv.2021.15051

Tareen A et al (2022) Biodeterioration of microplastics: a promising step towards plastics waste management. Polymers 14(11):2275

Temporiti MEE et al (2022) The analysis of the mycobiota in plastic polluted soil reveals a reduction in metabolic ability. J Fungi 8(12):1247

Thomas D, Schütze B, Heinze WM (2020) Steinmetz, Z. Sample preparation techniques for the analysis of microplastics in soil—a review. Sustainability 12:9074. [CrossRef]

Thompson RC, Olsen Y, Mitchell RP, Davis A, Rowland SJ, John AW, McGonigle D, Russell AE (2004) Lost at sea: where is all the plastic? Science 304:838. https://doi.org/10.1126/science.1094559

Tu C, Chen T, Zhou Q et al (2020) Biofilm formation and its influences on the properties of microplastics as affected by exposure time and depth in the seawater. Sci Total Environ 734:1–9. https://doi.org/10.1016/j.scitotenv.2020.139237

Ventura E, Marín A, Gámez-Pérez J, Cabedo L (2024) Recent advances in the relationships between biofilms and microplastics in natural environments. World J Microbiol Biotechnol 40(7):220

Verma R, Shankarappa V, Papireddy M, Gowda ANS (2016) Toxic pollutants from plastic waste—a review. Procedia Environ Sci 35:701–708. https://doi.org/10.1016/j.proenv.2016.07.069

Vethaak AD, Legler J (2021) Microplastics and human health. Science 371:672–674. [CrossRef]

Vethaak AD, Legler J (2021a) Microplastics and human health. Science 371:672–674

Volke T, Saucedo-Castañeda G, Gutiérrez-Rojas M, Manzur A, Favela-Torres E (2001) Thermally treated low density polyethylene biodegradation by Penicillium pinophilum and Aspergillus niger. J Appl Polym Sci 83:305–314. https://doi.org/10.1002/app.2245

Wang J, Huang M, Wang Q, Sun Y, Zhao Y, Huang Y (2020b) LDPE microplastics significantly alter the temporal turnover of soil microbial communities. Sci Total Environ 726:1–8. https://doi.org/10.1016/j.scitotenv.2020.138682

Wang F, Zhang X, Zhang S, Zhang S, Sun Y (2020c) Interactions of microplastics and cadmium on plant growth and arbuscular mycorrhizal fungal communities in an agricultural soil. Chemosphere 254:1–10. https://doi.org/10.1016/j.chemosphere.2020.126791

Wang J, Liu X, Li Y, Powell T, Wang X, Wang G, Zhang P (2019) Microplastics as contaminants in the soil environment: a mini-review. Sci Total Environ 691:848–857. [CrossRef]

Wang W, Ge J, Yu X, Li H (2020) Environmental fate and impacts of microplastics in soil ecosystems: progress and perspective. Sci Total Environ 708:134841. [CrossRef] [PubMed]

Wang C, Zhao J, Xing B (2021) Environmental source, fate, and toxicity of microplastics. J Hazard Mater 407:124357. [CrossRef]

Wang F, Wang Q, Adams CA, Sun Y, Zhang S (2022) Effects of microplastics on soil properties: Current knowledge and future perspectives. J Hazard Mater 424:127531. [CrossRef]

Wang T et al (2020) Adsorption-degradation of polycyclic aromatic hydrocarbons in soil by immobilized mixed bacteria and its effect on microbial communities. J Agric Food Chem 68(50):14907–14916

Warren MW, Zou X (2002) Soil macrofauna and litter nutrients in three tropical tree plantations on a disturbed site in Puerto Rico. For Ecol Manage 170(1–3):161–171

Webe, CJ, Opp C (2020) Spatial patterns of mesoplastics and coarse microplastics in floodplain soils as resulting from land use and fluvial processes. Environ Pollut 267:115390. [CrossRef] [PubMed]

Wei R, Zimmermann W (2017) Microbial enzymes for the recycling of recalcitrant petroleum-based plastics: how far are we? Microb Biotechnol 10:1308–1322. https://doi.org/10.1111/1751-7915.12710

Weithmann N, Möller JN, Löder MGJ, Piehl S, Laforsch C, Freitag R (2018) Organic fertilizer as a vehicle for the entry of microplastic into the environment. Sci Adv 4:eaap8060. https://doi.org/10.1126/sciadv.aap8060

Weithmann N, Möller JN, Löder MG, Piehl S, Laforsch C, Freitag R (2018) Organic fertilizer as a vehicle for the entry of microplastic into the environment. Sci Adv 4:eaap8060. [CrossRef]

World Steel Association (WSA), "Steel Statistical Yearbooks 1978 to 2016 [Internet]. Available from: www.worldsteel.org/steel-by-topic/statistics/steel-statistical-yearbook-.html

Xi B, Wang B, Chen M, Lee X, Zhang X, Wang S, Yu Z, Wu P (2022) Environmental behaviors and degradation methods of microplastics in different environmental media. Chemosphere 299:134354. [CrossRef]

Xiang Y, Jiang L, Zhou Y, Luo Z, Zhi D, Yang J, Lam SS (2022) Microplastics and environmental pollutants: key interaction and toxicology in aquatic and soil environments. J Hazard Mater 422:126843. https://doi.org/10.1016/j.jhazmat.2021.126843

Xu X, Wang S, Gao F et al (2019) Marine microplastic-associated bacterial community succession in response to geography, exposure time, and plastic type in China's coastal seawaters. Mar Pollut Bull 145:278–286. https://doi.org/10.1016/j.marpolbul.2019.05.036

Xu B, Liu F, Cryder Z, Huang D, Lu Z, He Y, Wang H, Lu Z, Brookes PC, Tang C (2020) Microplastics in the soil environment: occurrence, risks, interactions and fate–a review. Crit Rev Environ Sci Technol 50:2175–2222. [CrossRef]

Xu M, Du W, Ai F et al (2021) Polystyrene microplastics alleviate the effects of sulfamethazine on soil microbial communities at different CO_2 concentrations. J Hazard Mater 413: 1–11. https://doi.org/10.1016/j.jhazmat.2021.125286

Yan Y, Chen Z, Zhu F, Zhu C, Wang C, Gu C (2021) Effect of polyvinyl chloride microplastics on bacterial community and nutrient status in two agricultural soils. Bull Environ Contam Toxicol 107:602–609. https://doi.org/10.1007/s00128-020-02900-2

Yang Y, Liu W, Zhang Z, Grossart HP, Gadd GM (2020) Microplastics provide new microbial niches in aquatic environments. Appl Microbiol Biotechnol 104:6501–6511

Yang L, Zhang Y, Kang S, Wang Z, Wu C (2021a) Microplastics in soil: a review on methods, occurrence, sources, and potential risk. Sci Total Environ 780:146546

Yang L, Zhang Y, Kang S, Wang Z, Wu C (2021) Microplastics in soil: a review on methods, occurrence, sources, and potential risk. Sci Total Environ 780:146546. [CrossRef] [PubMed]

Yanto DHY et al (2019) Biodegradation of styrofoam waste by ligninolytic fungi and bacteria. IOP Conference series: earth and environmental science. Vol 308. No 1. IOP Publishing

Yi M, Zhou S, Zhang L, Ding S (2021) The effects of three different microplastics on enzyme activities and microbial communities in soil. Water Environ Res 93:24–32. https://doi.org/10.1002/wer.1327

Yoshida S, Hiraga K, Takehana T, Taniguchi I, Yamaji H, Maeda Y, Toyohara K, Miyamoto K, Kimura Y, Oda K (2016) A bacterium that degrades and assimilates poly (ethylene terephthalate). Science 351:1196–1199. [CrossRef]

Yu M, Van Der Ploeg M, Lwanga EH, Yang X, Zhang S, Ma X, Ritsema CJ, Geissen V (2019) Leaching of microplastics by preferential flow in earthworm (Lumbricus terrestris) burrows. Environ Chem 16:31–40. [CrossRef]

Yu L, Zhang J, Liu Y, Chen L, Tao S, Liu W (2021) Distribution characteristics of microplastics in agricultural soils from the largest vegetable production base in China. Sci Total Environ 756:143860. [CrossRef]

Yu H, Zhang Y, Tan W (2021) The "neighbor avoidance effect" of microplastics on bacterial and fungal diversity and communities in different soil horizons. Environ Sci Ecotechnol 8:1–9. https://doi.org/10.1016/j.ese.2021.100121

Yuan J, Ma J, Sun Y, Zhou T, Zhao Y, Yu F (2020) Microbial degradation and other environmental aspects of microplastics/ plastics. Sci Total Environ 715:1–9. https://doi.org/10.1016/j.scitotenv.2020.136968

Yurtsever M (2019) Glitters as a source of primary microplastics: An approach to environmental responsibility and ethics. J Agric Environ Ethics 32:459–478. [CrossRef]

Zettler ER, Mincer TJ, Amaral-Zettler LA (2013) Life in the "plastisphere": Microbial communities on plastic marine debris. Environ Sci Technol 47:7137–7146. https://doi.org/10.1021/es401288x

Zhang K, Xiong X, Hu H et al (2017) Occurrence and characteristics of microplastic pollution in Xiangxi Bay of Three Gorges Reservoir. China. Environ Sci Technol 51:3794–3801. https://doi.org/10.1021/acs.est.7b00369

Zhang M, Zhao Y, Qin X, Jia W, Chai L, Huang M, Huang Y (2019a) Microplastics from mulching film is a distinct habitat for bacteria in farmland soil. Sci Total Environ 688:470–478

Zhang L, Xie Y, Liu J, Zhong S, Qian Y, Gao P (2020a) An overlooked entry pathway of microplastics into agricultural soils from application of sludge-based fertilizers. Environ Sci Technol 54:4248–4255

Zhang J, Gao D, Li Q et al (2020b) Biodegradation of polyethylene microplastic particles by the fungus Aspergillus flavus from the guts of wax moth Galleria mellonella. Sci Total Environ 704:1–8. https://doi.org/10.1016/j.scitotenv.2019.135931

Zhang G, Liu Y (2018) The distribution of microplastics in soil aggregate fractions in southwestern China. Sci Total Environ 642:12–20. [CrossRef] [PubMed]

Zhang M, Zhao Y, Qin X, Jia W, Chai L, Huang M, Huang Y (2019) Microplastics from mulching film is a distinct habitat for bacteria in farmland soil. Sci Total Environ 688:470–478. [CrossRef]

Zhang L, Xie Y, Liu J, Zhong S, Qian Y, Gao P (2020) An overlooked entry pathway of microplastics into agricultural soils from application of sludge-based fertilizers. Environ Sci Technol 54:4248–4255. [CrossRef]

Zhang B, Yang X, Chen L, Chao J, Teng J, Wang Q (2020) Microplastics in soils: A review of possible sources, analytical methods and ecological impacts. J Chem Technol Biotechnol 95:2052–2068. [CrossRef]

Zhang X, Li Y., Ouyang D, Lei J, Tan Q, Xie L, Li Z, Liu T, Xiao Y, Farooq TH (2021) Systematical review of interactions between microplastics and microorganisms in the soil environment. J Hazard Mater 418:126288. [CrossRef] [PubMed]

Zhang Y, Zhang X, Li X, He D (2022) Interaction of microplastics and soil animals in agricultural ecosystems. Curr Opin Environ Sci Health 52:100327. [CrossRef]

Zhang J et al (2020) Biodegradation of polyethylene microplastic particles by the fungus Aspergillus flavus from the guts of wax moth Galleria mellonella. Sci Total Environ 704:135931

Zhou Y, Wang J, Zou M, Jia Z, Zhou S, Li Y (2020a) Microplastics in soils: a review of methods, occurrence, fate, transport, ecological and environmental risks. Sci Total Environ 748:141368

Zhou Q, Zhang H, Fu C, Zhou Y, Dai Z, Li Y, Tu C, Luo Y (2018) The distribution and morphology of microplastics in coastal soils adjacent to the Bohai Sea and the Yellow Sea. Geoderma 322:201–208. [CrossRef]

Zhou Y, Liu X, Wang J (2019) Characterization of microplastics and the association of heavy metals with microplastics in suburban soil of central China. Sci Total Environ 694:133798. [CrossRef]

Zhou Y, Wang J, Zou M, Jia Z, Zhou S, Li Y (2020) Microplastics in soils: a review of methods, occurrence, fate, transport, ecological and environmental risks. Sci Total Environ 748:141368. [CrossRef]

Zhou B, Wang J, Zhang H, Shi H, Fei Y, Huang S, Tong Y, Wen D, Luo Y, Barceló D (2020. Microplastics in agricultural soils on the coastal plain of Hangzhou Bay, east China: multiple sources other than plastic mulching film. J Hazard Mater 388:121814. [CrossRef]

Zhu F, Zhu C, Wang C, Gu C (2019a) Occurrence and ecological impacts of microplastics in soil systems: a review. Bull Environ Contam Toxicol 102:741–749. https://doi.org/10.1007/s00128-019-02623-z

Zhu D, Bi Q-F, Xiang Q, Chen Q-L, Christie P, Ke X, Wu L-H, Zhu Y-G (2018) Trophic predator-prey relationships promote transport of microplastics compared with the single Hypoaspis aculeifer and Folsomia candida. Environ Pollut 235:150–154. [CrossRef]

Zhu F, Zhu C, Wang C, Gu C (2019) Occurrence and ecological impacts of microplastics in soil systems: a review. Bull Environ Contam Toxicol 102:741–749. [CrossRef] [PubMed]

Precision Futures: Leveraging Modern Technology for Intelligent On-Farm Planning and Production Efficiency

Tayba Hasan, Gaurav Mudgal, Vasantharekha Ramasamy, Anil Kumar, and Donald Rugira Kugonza

Abstract Agriculture today is at a pivotal moment, navigating a landscape shaped by unpredictable climates, shrinking resources, and shifting socio-economic realities. As global food systems face mounting pressure to be more productive, resilient, and sustainable, traditional farming models—often fragmented, resource-intensive, and reactive—are struggling to keep pace. The call to rethink how we plan and manage agricultural systems is no longer optional; it's an urgent necessity grounded in real-world challenges. Meeting this moment requires a bold transformation—one that reimagines farming as a precision-driven, intelligence-enabled ecosystem. This isn't just about adopting new tools; it's about weaving together data, digital technologies, and inclusive decision-making frameworks that empower farmers to act with foresight and agility. Central to this shift are capabilities like spatial mapping of field variability, predictive modeling of crop responses, real-time monitoring of inputs and outputs, and user-friendly platforms that turn complex analytics into actionable strategies. Equally vital is the socio-technical backbone that supports this evolution: digital literacy, cooperative networks, and policy environments that ensure fair access and meaningful adoption across diverse farming communities. This chapter explores how advanced technologies are converging to reshape on-farm planning and operational efficiency. From satellite-based remote sensing and GIS-integrated land mapping to IoT sensor networks and AI-powered decision support systems, the landscape of agricultural innovation is rapidly expanding. We delve into emerging solutions like smart irrigation, autonomous machinery, digital twins, and blockchain-enabled traceability, alongside participatory tools such as mobile dashboards and multi-criteria decision analysis (MCDA). Case studies from varied agro-ecological contexts showcase

T. Hasan · G. Mudgal · A. Kumar
University Institute of Biotechnology, Chandigarh University, Mohali, Punjab, India

V. Ramasamy
Faculty of Engineering and Technology, Department of Biotechnology, SRM Institute of Science and Technology, Chennai, Tamil Nadu, India

D. R. Kugonza (✉)
Department of Animal and Range Sciences (DARS), School of Agricultural Sciences (SAS) Building, College of Agricultural and Environmental Sciences (CAES), Makerere University, Kampala, Uganda
e-mail: rugira.kugonza@mak.ac.ug

K. K. Kesari et al. (eds.), *Smart Sustainable Materials Advancing Re-manufacturing and Technology*, World Sustainability Series, https://doi.org/10.1007/978-3-032-17907-4_9

successful implementations, while evaluation frameworks—including Life Cycle Assessment (LCA), Data Envelopment Analysis (DEA), and Participatory Rural Appraisal (PRA)—offer multidimensional insights into ecological, economic, and social outcomes. Embedding intelligence into every layer of farm operations positions technology not as a substitute for farmer wisdom, but as a powerful catalyst for regenerative, inclusive, transformative, and future-ready agriculture.

1 Introduction

Global agriculture is facing unprecedented challenges, driven by a confluence of ecological degradation, climate volatility, demographic pressure, and technological disparity. The dominant conventional farming paradigm—centered on monoculture, synthetic inputs, mechanized tillage, and irrigation-intensive regimes—has indeed boosted productivity over decades but has simultaneously eroded natural resources and weakened agroecosystem resilience. For example, over 70% of global freshwater is consumed by agriculture (Kay et al. 2022; Ingrao et al. 2023), yet inefficient irrigation systems often lead to losses exceeding 50% due to seepage, evaporation, and runoff (Shangguan et al. 2002). Meanwhile, global soils have experienced a drastic decline in fertility, with the UK's arable lands losing up to 60% of organic carbon over the last century (Poulton et al. 2018). In India, over 29.77% of land is classified as degraded, affecting crop yields and farmer livelihoods alike (Naik et al. 2025; Mandal 2024). The conventional toolkits intended to remedy these challenges—including synthetic fertilizers, deep tillage, pesticide cycles, and centralized irrigation—are increasingly proving to be ecologically and economically unsustainable. Fertilizer imbalances, such as the skewed NPK ratios in India (often around 7.7:3.1:1 versus the ideal 4:2:1), are disrupting soil chemistry and suppressing microbial life essential for nutrient cycling (Singh and Kumari 2024). Mechanized tillage, while facilitating sowing, leads to compaction and erosion, thereby reducing arability and water infiltration. Pesticide overuse has decimated pollinator populations and induced pesticide resistance in several crops, while irrigation based on static schedules—rather than dynamic soil data—has accelerated salinity and aquifer depletion (Shah et al. 2025; Georgina Menon et al. 2025; Barwant et al. 2025).

Compounding these issues are external climate and socioeconomic factors. Seasonal variation, erratic rainfall, shifting monsoons, and heatwave frequency have created highly uncertain growing conditions (Fan et al. 2022; Yadav and Yadav 2024; Huggi et al. 2024). In India alone, climate modeling predicts up to a 22% reduction in rice yields by 2050 under certain emissions scenarios (Scott et al. 2019; Kumar et al. 2018). Moreover, the lack of diagnostics for soil health—such as soil texture, depth, microbial diversity, and water retention—prevents farmers from proactively managing fertility and input decisions. The Soil Health Card Scheme has created awareness, but farm-level precision data remains scarce in many regions, limiting its utility (Reddy 2019). Irrigation, in particular, presents a paradox: while critical to crop survival, most systems are calibrated without considering real-time soil moisture dynamics or weather variability. Traditional flood irrigation practices waste water and nutrients, while drip or sprinkler systems are often underutilized due to cost or maintenance issues. Without feedback from soil or plant sensors, these systems operate blindly. Likewise, arability and microbiota remain understudied in conventional agronomy frameworks, even though microbial diversity can profoundly influence nutrient availability, pathogen resistance, and soil structure.

In response to this complex landscape, a host of modern, digital, and ecological solutions have begun to reshape how farms plan and produce (Pace et al. 2025; Mansoor et al. 2025; Gamage et al. 2024; Naveed et al. 2024; Sishodia et al. 2020). At the forefront is precision agriculture, which uses satellite imagery, GIS maps, and soil sensors to measure spatial variability across fields. These tools enable variable-rate technologies (VRTs) (Ahmad and Mahdi 2018; He 2022) that tailor irrigation, seeding, and fertilization to microzones within a plot, thereby enhancing efficiency and productivity. For instance, NDVI (Normalized Difference Vegetation Index) analysis via drones allows farmers to detect early signs of stress or nutrient deficiency across crops and respond precisely (Meivel and Maheswari 2021; Miller et al. 2024; Raj et al. 2020). Equally impactful is the integration of Internet of Things (IoT) devices in irrigation systems (Fig. 1). Smart irrigation controllers linked to soil moisture probes and weather APIs adjust watering schedules dynamically, saving up to 50% of water usage. In addition, AI-powered decision support systems synthesize multispectral data, farm history, and weather forecasts to help farmers choose ideal crop varieties, planting windows, and input regimes (Mishra 2025; Gupta and Kumar Pal 2025). These tools are particularly relevant in the face of climate uncertainty, where traditional heuristics no longer suffice.

Meanwhile, AI and machine learning platforms are driving real-time pest and disease forecasting using image recognition, spectral modeling, and predictive analytics. Mobile apps now allow farmers to photograph a diseased leaf, receive instant diagnoses, and apply targeted remedies, drastically reducing pesticide use and improving crop health. Soil microbiota mapping is another emerging frontier, with startups offering portable sensors and genomics-based kits to detect microbial richness and activity, informing decisions on organic amendments, crop rotations, and bioinoculants. Beyond field-level interventions, digital twins—virtual simulations of farms—are enabling predictive modeling of crop scenarios, yield estimates, and input flows. These models use real-time and historical data to guide financial planning, logistics, and sustainability assessments (Lee et al. 2025; Ali et al. 2025; Omrany et al. 2025; Escribà-Gelonch et al. 2024; Qoseem et al. 2024). Complementary tools like farm management software (FMS) (Gupta and Kumar Pal 2025) help farmers consolidate information on inventory, labor, subsidies, and compliance, offering dashboards that support strategic planning. Biological inputs, such as biofertilizers, biostimulants, and plant-growth-promoting rhizobacteria (PGPRs) are gaining momentum as ecologically sound alternatives to synthetic inputs. These products enhance nutrient uptake, bolster plant immunity, and improve soil structure, while reducing runoff and dependence on chemicals (Singh et al. 2025; Dhar et al. 2024; Parashar et al. 2023; Kaur et al. 2022, 2023). CRISPR-based genetic

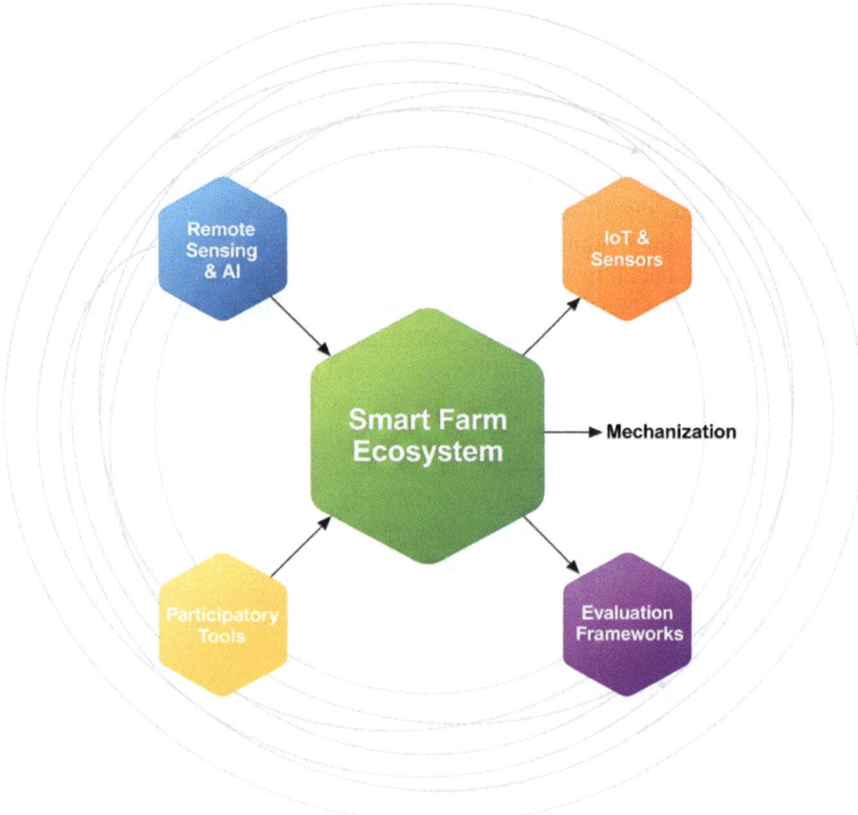

Fig. 1 Synergizing Technologies for Smart Agriculture. This systems diagram showcases the integrated technological landscape that underpins modern smart farming. From data acquisition via Remote Sensing and AI and IoT and Sensors, through automated Mechanization, to farmer engagement with Participatory Tools and performance assessment through Evaluation Frameworks, all components feed into and are optimized by the central Smart Farm Ecosystem, enhancing decision analytics, crop identification, environmental intelligence, and socio-economic adoption

technologies have also introduced high-yielding, stress-resistant crop varieties that can withstand drought, pests, and diseases without requiring heavy chemical support (Siddique et al. 2024; Abdi et al. 2024; Ahmad et al. 2021).

Energy efficiency is another domain seeing rapid innovation. Autonomous electric tractors and solar-powered machinery offer sustainable mechanization for small and medium farms. These machines reduce diesel dependence, lower emissions, and can be remotely controlled for precision operations. Sensor-equipped harvesters and robotic weeders further automate labor-intensive tasks while minimizing collateral crop damage. At the social and institutional level, technology adoption studies reveal that factors such as education, digital literacy, trust in innovation, and cooperative membership significantly influence uptake. Initiatives that offer bundled solutions—combining training, credit access, and tech deployment—have shown superior adoption rates and impact. Participatory planning tools and mobile advisory platforms ensure that innovations are context-sensitive and culturally appropriate.

Given this spectrum of challenges and solutions, this chapter proceeds to analyze emerging technologies through the lens of planning efficacy, productivity enhancement, and socio-economic integration. It will examine crop identification systems, environmental modeling, energy optimization tools, and adoption dynamics, supported by regional and global case studies. Evaluation frameworks—including multi-criteria analysis, life cycle assessments, and farmer-centered surveys—will be discussed to gauge the actual impact of technologies across ecological and human dimensions.

2 Smart Crop Identification and Classification

Accurate crop identification is foundational to modern agricultural planning, enabling farmers, researchers, and policymakers to make informed decisions about resource allocation, yield forecasting, and environmental impact mitigation. Traditionally, crop identification relied on manual field surveys, farmer declarations, and visual inspections—methods that are labor-intensive, time-consuming, and prone to error. In regions with fragmented landholdings and diverse cropping systems, such as India, these limitations are particularly acute. For example, studies have shown that farmer-reported crop types often diverge significantly from actual field conditions due to misclassification, local naming conventions, or memory bias (Wang et al. n.d.).

The advent of remote sensing technologies has revolutionized crop identification by enabling large-scale, repeatable, and objective monitoring of agricultural landscapes. Satellite platforms such as Sentinel-2, Landsat-8, and MODIS provide multispectral and temporal imagery that can be used to distinguish crop types based on phenological patterns, spectral signatures, and canopy structure. For instance, Sentinel-2's high spatial resolution (10–20 m) and frequent revisit cycle (5 days) make it ideal for tracking crop growth stages and differentiating between cereals, legumes, and oilseeds (Mai et al. 2025).

In a recent study from Iran, researchers demonstrated the power of remote sensing by using Sentinel-2 time series data, spectral indices, and ensemble classifiers to identify crops like wheat, barley, alfalfa, and rapeseed with over 80% accuracy—two months before harvest (Rahmati et al. 2022). This kind of early detection is a game-changer for agricultural planning. Building on this momentum, machine learning and deep learning models have significantly boosted the precision and scalability of crop classification. Algorithms such as Random Forest (RF), Support Vector Machines (SVM), and Convolutional Neural Networks (CNNs) are now widely applied to analyze spectral and temporal data, learning complex patterns from labeled datasets and generalizing across different regions and growing seasons. A standout example is the Cropformer model—a deep learning architecture designed for multi-scenario crop classification. It combines self-supervised pretraining with fine-tuned supervised learning to deliver high accuracy, even when labeled data is limited (Wang et al. 2023).This makes it especially valuable in regions where collecting ground truth data is costly or logistically difficult.

Transfer learning has emerged as a powerful strategy to overcome sample scarcity by leveraging knowledge from well-annotated source domains. In the Hexi Corridor of China, researchers used high-confidence pixels from the U.S. Cropland Data Layer (CDL) and transferred classification models to local datasets using algorithms like TrAdaBoost and XGBoost. The study achieved over 88% accuracy for maize and alfalfa, demonstrating the feasibility of cross-regional crop mapping without extensive local data collection (Mai et al. 2025). Such methods are increasingly relevant for smallholder-dominated landscapes, where heterogeneity and limited infrastructure pose significant barriers to conventional mapping.

In India, the Ministry of Agriculture has launched CROPIC (Collection of Real-Time Observations and Photos of Crops), an AI-powered mobile app that allows farmers to upload images of standing crops throughout the season. These images are analyzed using computer vision algorithms to detect crop type, growth stage, and stress indicators. The system supports automated insurance claim validation under the Pradhan Mantri Fasal Bima Yojana (PMFBY), reducing human dependency and improving transparency. CROPIC exemplifies how participatory data collection and AI analytics can democratize crop monitoring and empower farmers with actionable insights.

Beyond satellite and mobile-based platforms, unmanned aerial vehicles (UAVs) or drones are increasingly used for high-resolution crop identification. UAVs equipped with multispectral or hyperspectral cameras can capture centimeter-level imagery, enabling fine-grained analysis of plant health, canopy structure, and inter-row variability. Deep learning models trained on UAV imagery have achieved classification accuracies exceeding 96%, with explainable outputs that highlight key features influencing predictions (Teixeira et al. 2023). These capabilities are particularly useful for precision agriculture applications, where localized interventions can significantly improve input efficiency and yield.

Multimodal data fusion—combining satellite, UAV, ground sensor, and farmer-reported data—is another frontier in crop identification. Research continues to show that combining spectral, spatial, and temporal features from diverse sources significantly improves the robustness and adaptability of crop classification models. Datasets like EuroCropsML, which fuse Sentinel-2 imagery with field-level annotations across Europe, enable few-shot learning and cross-regional transferability—making it easier to generalize across varied agro-climatic zones. Similarly, the Sen4AgriNet benchmark dataset supports both segmentation and classification tasks using multi-year, multi-country data, providing a strong foundation for comparative research and algorithm development.

Yet, despite these advancements, key challenges persist. Models often grapple with class imbalance, spectral overlap between crop types, and temporal variability driven by weather conditions or farming practices. The absence of standardized ontologies and harmonized datasets further complicates interoperability and scalability. Overcoming these hurdles calls for collective action in data sharing, model benchmarking, and capacity building. Initiatives like the Radiant Earth Foundation's Crop Type Detection Challenge and the IGARSS benchmark datasets represent important steps toward more inclusive, transparent, and scalable solutions in agricultural AI.

3 Environmental Intelligence and Impact Modeling

Environmental intelligence in agriculture is reshaping how farms respond to their surroundings, by integrating real-time data on soil moisture, temperature, humidity, and weather forecasts directly into decision-making processes that boost productivity and sustainability. Unlike traditional farming systems, which often lack detailed awareness of environmental variability and rely on reactive, input-heavy practices, these intelligent systems offer a proactive approach. The absence of granular insights has historically led to inefficient resource use, increased vulnerability to climate stress, and missed opportunities for optimization (Singh et al. 2025; Parashar et al. 2023; Kaur et al. 2023, 2025; Mudgal et al. 2025; Seetharaman et al. 2025; Debnath et al. 2023; Verma et al. 2022; Sharma et al. 2021).

In contrast, environmental intelligence systems create dynamic feedback loops between the farm and its ecosystem by combining sensor networks, geospatial analytics, and predictive modeling. One of the most transformative tools in this space is the deployment of IoT-enabled sensors for soil and water tracking. These devices monitor key parameters—such as pH, salinity, nitrogen levels, and volumetric water content—and transmit the data to cloud-based platforms for visualization and analysis. A notable example is Doktar's Filiz sensor station (Doktar 2025), which delivers real-time insights into soil moisture, temperature, wind speed, and humidity to support precision agriculture. These insights help farmers fine-tune irrigation schedules, improving water-use efficiency and promoting consistent crop health and yield. Filiz also features a wind meter and Growing Degree Days (GDD) calculator, which support strategic decisions around pollination timing and seed selection—aligning dusting periods with favorable wind conditions to enhance reproductive success and yield potential (AgTech. 2021). While Filiz doesn't directly measure nutrient levels, its integration with agronomic advisory platforms enables informed fertilization planning.

These capabilities empower farmers to conserve resources, reduce input costs, and make data-driven decisions throughout the growing season. GIS-based soil mapping adds another layer of precision, helping farmers identify spatial variability in soil properties and implement site-specific management strategies that reduce waste and environmental impact (Xie et al. 2012; Assimakopoulos et al. 2003).

Remote sensing technologies—such as satellite imagery and UAVs—also play a vital role in environmental monitoring (Padhiary et al. 2025). These platforms capture multispectral and thermal data to assess vegetation health, detect water stress, and monitor land degradation. For example, NDVI (Normalized Difference Vegetation Index) derived from satellite data can reveal crop stress before symptoms are visible to the naked eye, allowing for timely intervention (Meivel and Maheswari 2021; Miller et al. 2024; Padhiary et al. 2025). When combined with GIS analytics, remote sensing enables the creation of high-resolution environmental maps that guide decisions on crop rotation, irrigation zoning, and pest control. Impact modeling tools further elevate environmental intelligence by simulating the outcomes of various farming scenarios under changing climatic and socio-economic conditions—helping farmers anticipate challenges and adapt with confidence. Models such as IMPACT (International Model for Policy Analysis of Agricultural Commodities and Trade) and IMPACTLite allow researchers and policymakers to evaluate the long-term effects of agricultural interventions on yield, water use, greenhouse gas emissions, and food security (Rosegrant 2012; Behrman 2010). These models integrate data from climate simulations, crop growth models, and economic forecasts to provide holistic assessments of sustainability and resilience. For example, IMPACT has been used to simulate the effects of irrigation expansion, fertilizer subsidies, and climate adaptation strategies across multiple regions and commodities. In India, GIS and remote sensing have been instrumental in managing water resources and soil health. The Soil Health Card Scheme, while a step forward, has been complemented by sensor-based platforms that offer continuous monitoring and diagnostics (Reddy 2019). RS-GIS integration has enabled drought forecasting, flood risk mapping, and groundwater potential zone identification, supporting climate-smart agriculture and disaster preparedness (Roy et al. 2022). These technologies are particularly valuable in regions facing erratic monsoons and groundwater depletion, where adaptive planning is essential.

Despite their promise, environmental intelligence systems face challenges related to data interoperability, infrastructure costs, and digital literacy. Many smallholder farmers lack access to high-resolution data or the tools to interpret it. Moreover, sensor networks require reliable connectivity and maintenance, which can be difficult in remote areas. Addressing these barriers will require inclusive technology design, capacity building, and supportive policy frameworks that promote open data standards and equitable access.

4 Energy Efficiency Through Precision Mechanization

Energy efficiency in agriculture is increasingly recognized as a cornerstone of sustainable farm management, particularly as mechanization becomes more widespread across diverse agroecological contexts. Traditional mechanized systems—dominated by diesel-powered tractors, harvesters, and irrigation pumps—have historically improved labor productivity and operational speed. However, they also account for a significant share of greenhouse gas emissions and energy costs. In fact, fossil fuels represent up to 75% of total energy consumption on industrial-scale farms, with diesel alone comprising the bulk of operational expenses (Jensen et al. 2025). This reliance on non-renewable energy sources not only contributes to climate change but also exposes farmers to volatile fuel markets and supply chain disruptions.

Recent innovations in electric and hybrid tractors are reshaping the mechanization landscape by offering cleaner, quieter, and more efficient alternatives. Companies like AutoNxt in India have developed electric tractors with operating costs one-fifth that of diesel models, enabling farmers to break even within two years of adoption (Chowdhary and Ranjan 2024). These tractors are equipped with smart steering, hydraulic automation, and crop health analytics, allowing for precision operations with minimal energy waste. Similarly, Monarch Tractor (Shamshiri 2024) in the U.S. has introduced autonomous electric tractors that reduce fuel consumption by 2,100 gallons annually, while cutting CO_2 emissions by 54 metric tons per unit. These machines integrate AI-based fleet management and remote diagnostics, enhancing both energy efficiency and labor productivity.

Mechanization is also advancing in the realm of precision tillage and planting, where GPS-guided seed drills and variable-depth trench planters optimize fuel use and soil engagement. For example, tractor-mounted multi-row seed-cum-fertilizer drills developed by ICAR place seeds and fertilizers at precise depths, improving nutrient uptake and reducing energy spent on rework.

Energy-efficient mechanization is rapidly redefining agricultural operations, offering both ecological and economic benefits. Field-tested systems have shown up to 82.4% operational efficiency, with notable reductions in fertilizer and fuel use. Conservation agriculture tools like strip-till and no-till equipment are central to this shift—minimizing soil disturbance, lowering tractor power demands, and improving soil health by reducing diesel consumption.

Solar-powered mechanization is emerging as a game-changer, especially in sun-rich regions with limited access to conventional power grids. Technologies such as solar irrigation pumps, cold storage units, and electric sprayers are replacing diesel-based systems, delivering up to 60% savings in energy costs while shrinking carbon footprints. Companies like Bullwork Mobility are leading the charge with solar-powered tractors and loaders capable of 5–8 h of continuous operation. These machines feature hybrid charging options to extend runtime and reduce range anxiety, and they're compatible with standard implements—offering torque levels on par with traditional diesel models, making them viable for widespread adoption.

Beyond the hardware, smart energy management platforms are empowering farmers to make informed decisions. Tools like WingspanAI and Farm Management Systems (FMS) provide real-time analytics on fuel consumption, machinery performance, and field coverage. These dashboards help optimize equipment deployment and schedule maintenance proactively—especially useful in cooperative farming setups where machinery is shared across multiple users and plots.

Despite these promising developments, adoption barriers remain. High initial costs, limited access to financing, and gaps in technical training often deter smallholders from embracing energy-efficient solutions. However, policy interventions and incentive programs are beginning to close these gaps. India's FAME scheme and the U.S.-based CORE 2025 initiative, for example, offer substantial support for electric tractor purchases and solar equipment installations (Monarch Tractor;

Farmonaut), paving the way for a more inclusive and sustainable mechanization future.

5 Technology Adoption and Socio-economic Dynamics

The journey toward adopting agricultural technologies is anything but linear—it's shaped by a nuanced blend of psychological, socio-demographic, economic, and institutional dynamics. While innovations like precision farming, smart irrigation, and AI-driven diagnostics offer clear benefits, their uptake among farmers—especially smallholders—remains uneven across geographies and income levels. Research shows that even when the value of sustainable technologies is widely acknowledged, adoption rates often lag behind expectations set by global frameworks such as the UN Sustainable Development Goals (SDGs), hindered by factors like perceived complexity, low trust in digital systems, and limited access to infrastructure (Rizzo et al. 2024).

Among the many variables influencing this adoption landscape, education stands out as one of the most powerful enablers. It not only shapes a farmer's ability to understand and evaluate new technologies but also influences openness to change, risk tolerance, and engagement with advisory networks. In essence, education acts as a gateway—transforming innovation from a distant possibility into a practical, empowering tool for agricultural transformation.

Farmers with higher levels of formal education are more likely to understand, evaluate, and implement new technologies. In India, educated farmers are up to 3.37% more likely to adopt modern techniques, with adoption rates rising further when supported by access to credit and extension services. However, education alone is not sufficient. Digital literacy, access to training programs, and participation in peer learning networks are powerful catalysts for agricultural technology adoption. They help demystify complex innovations, build user confidence, and foster a culture of experimentation and shared learning. When farmers are equipped with the skills to navigate digital platforms, exposed to hands-on demonstrations, and supported by trusted community networks, their perception of new technologies shifts—from skepticism to opportunity. These social and educational enablers not only reduce innovation aversion but also create ripple effects across farming communities, accelerating uptake and amplifying impact.

For example, farmers who participate in co-design initiatives and receive ongoing technical support are significantly more likely to sustain technology use over time (Addorisio et al. 2025).

Farm size and income level also influence adoption patterns. Larger farms tend to have better access to capital, infrastructure, and advisory services, making them more receptive to high-cost technologies like autonomous tractors or satellite-based crop monitoring. On the flip side, smallholder farmers frequently encounter structural and financial barriers that hinder their ability to adopt new technologies. Limited liquidity, fragmented landholdings, and a natural aversion to risk make it difficult to invest in unfamiliar tools—even when the long-term benefits are clear. A recent study from Kentucky, USA, underscores this dynamic: it found that larger farm size and greater farming experience were positively associated with the adoption of precision agriculture technologies, while age showed a negative correlation—indicating that younger farmers tend to be more receptive to innovation (Pandeya et al. 2025). This highlights the importance of tailoring outreach and support strategies to different demographic profiles within the farming community.

Gender dynamics add another layer of complexity to the agricultural technology adoption landscape. In regions like South Asia and Sub-Saharan Africa, women play a central role in agricultural labor yet often face systemic barriers—limited access to land, credit, and training—that restrict their ability to adopt new technologies. Empowering female farmers through targeted extension services and inclusive policy frameworks can significantly boost adoption rates and strengthen household resilience. Social capital also plays a pivotal role. Membership in cooperatives, farmer producer organizations (FPOs), and informal networks fosters knowledge exchange and collective bargaining, helping reduce transaction costs and improve access to innovations. From a policy standpoint, perceived economic benefits—such as profitability, return on investment, and input savings—remain the strongest motivators for adoption. Farmers are more inclined to embrace technologies that offer clear financial advantages, especially when supported by subsidies, tax incentives, or market linkages.

Yet, non-economic factors are equally influential. Environmental values, trust in institutions, and perceived ease of use shape adoption behavior in meaningful ways. The Technology Acceptance Model (TAM), widely applied in adoption studies, highlights that perceived usefulness and ease of use are critical drivers of behavioral intention. Evaluating farming technologies through a socio-economic lens requires moving beyond yield and profit metrics to include broader impacts on livelihoods, equity, and sustainability. A study across 32 districts in India found that technology adoption not only improved income and productivity but also enhanced social status, access to credit, and food security—particularly among small-scale farmers using drip irrigation and pest management tools. However, unintended consequences such as labor displacement, resource depletion, and rural–urban migration must also be factored into impact assessments. To capture these multidimensional outcomes, researchers increasingly rely on configurational approaches that analyze combinations of variables rather than isolated factors. This method recognizes that multiple pathways can lead to successful adoption depending on local context, enabling more nuanced policy design and targeted interventions. For instance, a farmer may choose to adopt a technology not just for its economic viability, but because it aligns with community norms or environmental values.

6 Participatory Planning and Decision Analytics

Participatory planning in agriculture marks a transformative move away from hierarchical, top-down approaches toward inclusive, data-informed collaboration among farmers, extension workers, researchers, and policymakers. As agricultural systems become increasingly complex and climate-related uncertainties grow, the demand for planning tools that are transparent, adaptive, and tailored to local realities has never been greater. At the heart of this shift are participatory decision analytics—digital platforms and multi-criteria frameworks that enable stakeholders to jointly design interventions, assess trade-offs, and allocate resources in alignment with shared objectives and real-time environmental and operational data. This approach not only democratizes decision-making but also strengthens resilience, accountability, and innovation across farming communities.

7 Use of Mobile Apps, Dashboards, and Collaborative Mapping Tools

Digital platforms are revolutionizing participatory planning in agriculture by making complex data accessible, actionable, and inclusive. With intuitive interfaces for collecting, visualizing, and interpreting information, these tools empower farmers and stakeholders to co-create strategies grounded in real-time insights. Mobile applications, in particular, have become game-changers—democratizing agronomic intelligence by allowing farmers to log field-level data, receive context-specific recommendations, and actively participate in collaborative decision-making. This shift not only enhances transparency and responsiveness but also fosters a sense of ownership and agency among farming communities, bridging the gap between grassroots knowledge and digital innovation.

- **Mobile Advisory Platforms**: Apps like *KrishiHub*, *AgriApp*, and *Plantix* provide farmers with localized weather forecasts, pest alerts, and crop management tips (Mahajan and Popma 2022; Tyagi and Jaryal 2024). These platforms often include feedback loops where farmers can report outcomes, enabling iterative refinement of recommendations.
- **Collaborative Dashboards**: Tools such as *Cropin's SmartFarm* and *Digital Green's Loop* allow extension agents and farmer groups to visualize farm-level data—such as crop health, input usage, and yield estimates—on shared dashboards (Halim-Lim et al. 2025; Rajeev et al. 2020). These dashboards support group decision-making, subsidy tracking, and compliance monitoring.
- **Participatory GIS Mapping**: Platforms like *GeoAgro* and *OpenStreetMap for Agriculture* enable farmers to co-create spatial maps of their fields, annotate soil types, water sources, and infrastructure, and plan interventions collaboratively

(Sánchez et al. 2024; Minghini and Frassinelli 2019). These maps are particularly useful for community irrigation planning, pest surveillance, and disaster preparedness.
- **Voice-Enabled Interfaces**: In regions with low literacy, voice-based apps such as *Gram Vaani* and *Kisan Call Centre* facilitate participatory engagement by allowing farmers to record queries, receive audio advisories, and participate in community polls (Rajput et al. n.d.).

These tools not only enhance transparency and accountability but also foster trust and ownership among users. By integrating farmer-generated data with expert analytics, participatory platforms bridge the gap between local knowledge and scientific insights, leading to more resilient and context-sensitive farming decisions.

8 Multi-criteria Decision Analysis (MCDA) in Crop and Input Selection

Multi-Criteria Decision Analysis (MCDA) is a structured methodology that supports complex decision-making by evaluating multiple, often conflicting criteria (Seif et al. 2024; Talukder 2021; Wotlolan et al. 2021). Digital platforms are revolutionizing participatory planning in agriculture by making complex data accessible, actionable, and inclusive. With intuitive interfaces for collecting, visualizing, and interpreting information, these tools empower farmers and stakeholders to co-create strategies grounded in real-time insights. Mobile applications, in particular, have become game-changers—democratizing agronomic intelligence by allowing farmers to log field-level data, receive context-specific recommendations, and actively participate in collaborative decision-making. This shift not only enhances transparency and responsiveness but also fosters a sense of ownership and agency among farming communities, bridging the gap between grassroots knowledge and digital innovation.

Multi-Criteria Decision Analysis (MCDA) is increasingly used to guide crop selection, input optimization, and resource allocation under uncertainty.

9 Frameworks and Techniques

Common MCDA methods include the Analytic Hierarchy Process (AHP), Technique for Order Preference by Similarity to Ideal Solution (TOPSIS), and Weighted Sum Model (WSM) (Singh and Pant 2021; Sałabun et al. 2020; Thakkar 2021). These techniques allow stakeholders to rank alternatives based on criteria such as yield potential, water requirements, market access, climate resilience, and input costs.

10 Participatory Weighting

MCDA frameworks often involve stakeholder workshops where farmers assign weights to decision criteria based on their priorities. For example, in drought-prone regions, water efficiency may be weighted more heavily than market price, whereas in peri-urban zones, shelf life and transportability may dominate.

11 Integration with Digital Tools

Platforms like AgroDSS, AgriDecide, and FarmSim embed MCDA algorithms into mobile apps and dashboards, enabling real-time scenario analysis. Users can simulate outcomes of different crop-input combinations, visualize trade-offs, and make informed choices aligned with their goals (Rupnik et al. 2019).

- Digital platforms are revolutionizing participatory planning in agriculture by making complex data accessible, actionable, and inclusive. With intuitive interfaces for collecting, visualizing, and interpreting information, these tools empower farmers and stakeholders to co-create strategies grounded in real-time insights. Mobile applications, in particular, have become game-changers—democratizing agronomic intelligence by allowing farmers to log field-level data, receive context-specific recommendations, and actively participate in collaborative decision-making. This shift not only enhances transparency and responsiveness but also fosters a sense of ownership and agency among farming communities, bridging the gap between grassroots knowledge and digital innovation.
- Multi-Criteria Decision Analysis (MCDA) is increasingly used to guide crop selection, input optimization, and resource allocation under uncertainty.

12 Frameworks and Techniques

- Common MCDA methods include the Analytic Hierarchy Process (AHP), Technique for Order Preference by Similarity to Ideal Solution (TOPSIS), and Weighted Sum Model (WSM). These techniques allow stakeholders to rank alternatives based on criteria such as yield potential, water requirements, market access, climate resilience, and input costs.
- **Participatory Weighting**
- MCDA frameworks often involve stakeholder workshops where farmers assign weights to decision criteria based on their priorities. For example, in drought-prone regions, water efficiency may be weighted more heavily than market price, whereas in peri-urban zones, shelf life and transportability may dominate.
- **Integration with Digital Tools**

- Platforms like AgroDSS, AgriDecide, and FarmSim embed MCDA algorithms into mobile apps and dashboards, enabling real-time scenario analysis. Users can simulate outcomes of different crop-input combinations, visualize trade-offs, and make informed choices aligned with their goals.

Case Example—India's PMKSY Program: Under the Pradhan Mantri Krishi Sinchayee Yojana (PMKSY), MCDA was used to prioritize irrigation investments across districts (Kumar and Konyak 2023). Criteria included soil type, rainfall variability, crop intensity, and socio-economic vulnerability. The participatory process ensured that marginalized regions received targeted support, improving equity and impact.

MCDA also supports adaptive planning in the face of climate uncertainty. By incorporating probabilistic forecasts and risk assessments, decision-makers can evaluate resilience metrics and choose strategies that balance productivity with long-term sustainability (Fig. 2).

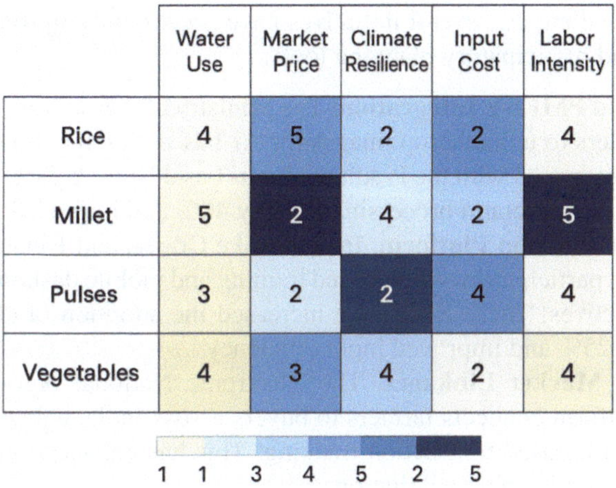

Fig. 2 Multi-Criteria Decision Analysis (MCDA) Matrix for Crop Selection in Smart Agriculture. This figure presents a weighted decision matrix used to evaluate four crop options—Rice, Millet, Pulses, and Vegetables—across five key criteria relevant to participatory planning in smart agriculture: Water Use, Market Price, Climate Resilience, Input Cost, and Labor Intensity. Each cell contains a score from 1 (least favorable) to 5 (most favorable), reflecting the relative performance of each crop under that criterion. Crop scores were derived using a multi-criteria framework informed by AHP and GIS-based suitability mapping (Wotlolan et al. 2021), empirical land evaluation parameters (Minh 2024), and participatory sustainability indicators (Talukder 2021), ensuring alignment with both scientific evidence and stakeholder priorities

13 Case Studies and Regional Applications

The real-world application of modern agricultural technologies varies widely across geographies, shaped by agroecological conditions, policy environments, infrastructure, and socio-cultural dynamics. Comparative case studies from India, Kenya, and Brazil offer valuable insights into how digital tools, precision systems, and participatory frameworks are being deployed to enhance farm planning and production efficiency.

14 India: Scaling Precision Agriculture in Smallholder Systems

India's agricultural landscape is characterized by fragmented landholdings, diverse cropping patterns, and high climate variability (Pankaj et al. 2020; Sati 2024). Despite these challenges, several initiatives have successfully introduced precision technologies and participatory planning tools:

- **CROPIC and PMFBY Integration**: The Ministry of Agriculture's CROPIC app enables farmers to upload crop images for AI-based classification and insurance validation. Integrated with the Pradhan Mantri Fasal Bima Yojana (PMFBY), this system has reduced claim processing time by 40% and improved transparency.
- **Digital Green's Loop Platform**: In states like Odisha and Bihar, Digital Green has deployed participatory video-based training and mobile dashboards to support farmer collectives. These tools have increased the adoption of climate-resilient practices by 25% and improved input efficiency.
- **eNAM and Market Linkages**: The Electronic National Agriculture Market (eNAM) platform connects farmers to buyers across India, using digital grading and blockchain-based transaction tracking. This has enhanced price realization and reduced middlemen exploitation.
- **Smart Irrigation in Maharashtra**: Solar-powered drip systems linked to soil moisture sensors have been deployed in drought-prone districts. Supported by Farmonaut and local NGOs, these systems have saved up to 50% water and increased yields by 18%.

India's experience highlights the importance of bundling technology with training, credit access, and institutional support to ensure equitable adoption and impact.

15 Kenya: Mobile-First Innovation for Climate-Smart Farming

Kenya has emerged as a hub for mobile-based agricultural innovation, leveraging high mobile penetration and vibrant tech ecosystems to support smallholder farmers (Mwikamba et al. 2024; Anning-Dorson 2025; Mapiye et al. 2023).

- **iShamba and M-Farm Platforms**: These mobile services offer weather alerts, market prices, and agronomic advice via SMS and voice. Over 1 million farmers use iShamba, with reported improvements in crop planning and input use.
- **Digital Soil Mapping**: The Kenya Agricultural and Livestock Research Organization (KALRO) has developed GIS-based soil maps integrated with mobile advisory tools. These maps guide fertilizer application and crop selection, improving productivity and reducing input waste.
- **Solar-Powered Cold Chains**: In regions like Meru and Kisumu, solar cold storage units linked to mobile dashboards help farmers preserve perishables and reduce post-harvest losses. These systems have increased income by 30% for horticulture producers.
- **Participatory Planning in Water Management**: Community mapping tools and mobile apps have been used to co-design irrigation schemes, improving water equity and reducing conflict in semi-arid zones.

Kenya's success underscores the role of mobile-first platforms and community engagement in scaling climate-smart technologies.

16 Brazil: Integrating Big Data and Agroecology

Farming landscape is a study in contrasts—where industrial-scale agriculture meets a rising wave of agroecological and regenerative practices. Innovation flows from both ends: corporate R&D labs and grassroots experiments are shaping the future together.

- **AgTech Valley in São Paulo**: This bustling innovation hub is home to startups working on AI-driven crop modeling, drone-based analytics, and digital twin technologies (Pivoto et al. 2018). Large farms tap into these tools to streamline logistics, cut emissions, and forecast yields with remarkable precision.
- **Embrapa's Digital Soil Platform**: Brazil's national agricultural research agency, Embrapa, has built a comprehensive soil database that integrates remote sensing and predictive modeling. It's a powerful tool for sustainable land use planning and carbon sequestration—helping farmers make smarter, greener decisions (Proença et al. 2025).
- **Participatory Agroforestry in Pará**: In the Amazon, smallholder cooperatives are using mobile mapping and multi-criteria decision frameworks to co-design agroforestry systems. These models aim to protect biodiversity while boosting

food production and household income—proving that tech can serve both planet and people.

- **Blockchain for Coffee Traceability**: In Minas Gerais, blockchain platforms track coffee from farm to cup, ensuring fair trade compliance and building consumer trust. These systems have improved market access and premium pricing for small producers.

Brazil's case illustrates how advanced analytics and participatory design can coexist to support both productivity and ecological stewardship (Giraldo and Rosset 2018; Souza et al. 2020).

17 Evaluation Frameworks and Impact Assessment

As agricultural technologies proliferate across diverse contexts (Table 1), robust evaluation frameworks are essential to assess their effectiveness, equity, and sustainability. Impact assessment in modern agriculture must go beyond yield metrics to encompass environmental outcomes, socio-economic benefits, and systemic resilience (Ibrahim et al. 2025; Movilla-Pateiro et al. 2021; Wijk et al. 2020; Mouratiadou et al. 2024). This section explores quantitative and participatory tools that help stakeholders measure the real-world performance of innovations and guide evidence-based policy and investment decisions.

18 Quantitative Tools for Evaluation

Several analytical frameworks have emerged to evaluate agricultural technologies across multiple dimensions:

- **Data Envelopment Analysis (DEA)**: DEA is a non-parametric method used to assess the relative efficiency of farms or technologies by comparing input–output ratios. It identifies best-performing units and benchmarks others against them. For example, DEA has been used to evaluate the energy efficiency of electric tractors versus diesel models across Indian farms, revealing up to 35% improvement in input utilization.
- **Life Cycle Assessment (LCA)**: LCA quantifies the environmental footprint of agricultural technologies across their entire lifecycle—from production and deployment to disposal. It measures indicators such as greenhouse gas emissions, water use, and land degradation. In Brazil, LCA has been applied to compare conventional soybean farming with regenerative practices, showing a 42% reduction in carbon emissions and improved soil health.
- **Cost–Benefit Analysis (CBA)**: This method weighs the costs of implementing a technology against the expected financial returns. It's especially useful for smallholder farmers deciding whether a new tool fits their budget and goals. In Kenya,

Table 1 Modern Farming Technologies. Here's a curated snapshot of cutting-edge farming technologies shaping agriculture in 2025—drawn from the latest industry insights and innovation platforms

Technology name	Core function	Key benefits	References
Precision agriculture *and* IoT	Real-time sensing, GPS mapping, and smart irrigation	Optimizes input use, reduces waste, and improves yield by 10–18%	Farmonaut
AI and machine learning	Predictive analytics for crop health, yield forecasting, and pest detection	Enhances decision-making, reduces chemical use, boosts productivity by 12–20%	Robovision, StartUs Insights
CRISPR and gene editing	Precise genetic modification of crops	Improves drought/pest resistance, enhances nutrition, and reduces chemical dependency	StartUs Insights, Future Food Cast
Autonomous farm equipment	Self-driving tractors, robotic harvesters	Reduces labor costs, increases precision, enables 24/7 operations	FJDynamics, MitraWeb
Vertical farming and CEA	Indoor stacked farming with climate control	Saves up to 95% water, enables urban agriculture, and year-round production	Farmonaut, Future Food Cast
Smart irrigation systems	IoT-based water delivery based on soil and weather data	Saves 30–50% water, improves crop hydration, lowers energy use	AgritechMind, KhetiGaadi
Blockchain in agriculture	Transparent supply chain tracking	Reduces fraud, builds consumer trust, streamlines logistics	StartUs Insights, FJDynamics
Satellite imaging and remote sensing	Crop health monitoring, disaster planning	Enables early detection, supports compliance, and improves yield by 7–14%	Farmonaut, StartUs Insights
Digital twins	Virtual farm simulations for predictive planning	Reduces risk, enhances decision-making, and optimizes resource use	FJDynamics, Cropin
Variable rate Technology (VRT)	Adjusts input application based on field variability	Reduces input costs, improves soil health, and increases yield	KhetiGaadi, MitraWeb
Robotic weeders	AI-guided weed removal	Minimizes herbicide use, reduces labor, and protects crops	StartUs Insights, KhetiGaadi
Smart greenhouses	Controlled environment with IoT sensors	Optimizes temperature, humidity, and light; improves productivity	Future Food Cast
Biodegradable packaging from agri-waste	Converts crop residues into eco-packaging	Reduces plastic use, adds value to waste, supports the circular economy	MitraWeb

(continued)

Table 1 (continued)

Technology name	Core function	Key benefits	References
Farm management software (FMS)	Digital planning, inventory, and subsidy tracking	Streamlines operations, improves financial planning, and supports compliance	FJDynamics, Cropin
Agricultural drones	Aerial crop monitoring, spraying, and mapping	Enables precision input application, early stress detection, and reduces labor	AgritechMind, Farmonaut
Soil sensors	Real-time monitoring of moisture, pH, salinity, and nutrients	Enables targeted irrigation and fertilization, improves soil health	KhetiGaadi
Smart harvesters	AI-enabled crop picking and sorting	Increases harvest efficiency, reduces waste, and improves quality	KhetiGaadi, StartUs Insights
Agroforestry tech platforms	Integrates trees with crops and livestock	Enhances biodiversity, improves soil, and diversifies income	CoirMedia, Farmonaut
Regenerative agriculture tools	Supports soil restoration, carbon sequestration	Improves fertility, enhances ecosystem services, and supports climate resilience	Future Food Cast, StartUs Insights
Biofertilizers and biostimulants	Microbial and organic inputs for soil and plant health	Reduces chemical dependency, improves nutrient uptake, boosts resilience	StartUs Insights, Agricells
AI-powered pest control apps	Image-based pest and disease identification	Reduces pesticide use, improves response time, and supports smallholders	AgritechMind

for example, solar irrigation systems showed a benefit–cost ratio of 2.3—meaning the financial gains more than doubled the investment, especially when subsidies were in place.

- **Multi-Criteria Decision Analysis (MCDA)**: MCDA helps stakeholders evaluate technologies across multiple dimensions—economic, environmental, and social. It's often used in participatory settings, allowing farmers, researchers, and policymakers to align on what matters most and choose solutions that reflect shared priorities.

Increasingly, these evaluation tools are built into digital dashboards and farm management platforms, making it easier to monitor progress, adjust plans, and make informed decisions in real time.

19 Socio-Economic Surveys and Participatory Appraisal

Numbers alone can't tell the full story. To truly understand how technologies impact farming communities, we need to listen—to lived experiences, local knowledge, and the rhythms of everyday life. That's where qualitative and participatory approaches come in, complementing quantitative models with depth and nuance.

- **Farmer-Centered Surveys**: These structured tools go beyond yield data to explore shifts in income, labor, food security, and social standing. In India, surveys across 32 districts showed that smart irrigation and pest control didn't just boost productivity—they also helped farmers gain access to credit and earn greater recognition in their communities.
- **Participatory Rural Appraisal (PRA)**: PRA techniques—like focus groups, seasonal calendars, and resource mapping—invite farmers to share their insights, evaluate what's working, and help design better solutions. It's a process that builds trust, fosters ownership, and ensures cultural relevance.
- **Social Impact Metrics**: Equity matters. Indicators such as gender inclusion, youth engagement, and cooperative participation are now central to evaluating tech adoption. In Brazil, blockchain-based traceability systems helped women gain access to premium markets by validating their contributions to production.
- **Resilience and Adaptation Indices**: These composite metrics look at how well farms can weather shocks, adapt to change, and evolve sustainably. They factor in things like crop diversity, access to knowledge, and institutional support—painting a fuller picture of long-term viability.

20 Integration and Visualization

Modern evaluation frameworks often combine multiple tools into integrated platforms, such a those outlined below:

- **SmartFarm Analytics (Cropin)**: Combines DEA, LCA, and socio-economic data into a unified dashboard for farm-level impact tracking.
- **AgImpact Explorer (IFPRI)**: Visualizes the outcomes of policy scenarios using spatial data, crop models, and economic forecasts.
- **Open Data Platforms**: Initiatives like Radiant Earth and FAO's Hand-in-Hand geospatial platform provide open-access datasets for benchmarking and comparative analysis.

When evaluation is built into every stage of a technology's journey—from early design to final rollout—it helps ensure that the innovation truly works for people. It becomes easier to measure impact, adapt to real-world needs, and make sure the benefits reach everyone.

21 Future Directions and Inclusive Innovation Pathways

Agriculture is stepping into a new chapter—one shaped by digital breakthroughs, climate pressures, and shifting social landscapes. In this evolving terrain, the future of farm planning and productivity will depend not just on smart technologies, but on how inclusive, ethical, and adaptable they are. Tools must be more than technically impressive; they need to respond to real-world challenges, regenerate ecosystems, and work across the full spectrum of farming contexts—from smallholders to large-scale operations. This section explores bold, forward-looking strategies that blend artificial intelligence, hands-on farmer training, ethical design principles, and supportive policy frameworks to build agroecosystems that are resilient, equitable, and ready for the future.

22 AI-Augmented Advisory Services and Farmer Training

Artificial intelligence is no longer just a behind-the-scenes tool—it's becoming a trusted partner in how farmers make decisions. By weaving together satellite images, sensor data, weather updates, and historical trends, AI-powered platforms now offer real-time, hyper-local advice that's tailored to each farm's unique conditions.

- **Personalized Crop Advisory**: Tools like *Plantix AI* and *AgNext* use machine learning to craft customized crop plans based on soil type, climate zone, and farmer preferences. This helps reduce waste, optimize inputs, and boost yield reliability.
- **Voice and Vernacular Interfaces**: Chatbots and voice assistants—such as *KisanGPT* and *Bhoomi AI*—speak the farmer's language, literally. By offering guidance in local dialects, they make digital support accessible to users with limited literacy, covering everything from pest alerts to irrigation tips and market prices.
- **Farmer Training Simulators**: Virtual reality and gamified learning are opening new doors for hands-on training. In India, *VR Kheti* lets farmers simulate crop scenarios, learn machinery operation, and explore regenerative practices in an interactive, risk-free environment.
- **AI for Extension Agents**: Field officers now have AI-powered dashboards that help them plan visits, diagnose problems, and monitor adoption rates. This means better service delivery and smarter use of resources.

Of course, technology alone isn't enough. These innovations must be backed by strong training programs, peer learning networks, and institutional support—so that every farmer, regardless of background, can benefit from the digital revolution.

23 Ethical Tech Design, Data Governance, and Policy Integration

As digital tools become part of everyday farming, we must ask: who owns the data, who benefits from the algorithms, and how do we ensure fairness? Ethical innovation in agriculture isn't just about technology—it's about trust, inclusion, and justice.

- **Data Sovereignty**: Farmers deserve to own and control their data. That means clear consent protocols, options to anonymize sensitive information, and governance models that put farmers first—like Digital Green's *Data Dignity Framework*, which champions respectful and transparent data practices.
- **Bias in AI Models**: Algorithms can unintentionally mirror societal biases—excluding certain regions, genders, or economic groups. Training AI on diverse datasets and using explainable tools like SHAP and Grad-CAM helps make these systems more transparent and trustworthy.
- **Open Standards and Interoperability**: To avoid vendor lock-in and foster collaboration, digital agriculture must embrace open data formats and APIs. This ensures that tools can work across different platforms and geographies, empowering farmers with choice and flexibility.
- **Policy Integration**: Governments play a key role in shaping ethical digital ecosystems. By aligning digital agriculture with rural development—through subsidies for smart tools, digital literacy programs, and robust data protection laws—they can ensure that innovation uplifts rather than excludes.

India's *AgriStack* initiative offers a compelling example: a unified digital infrastructure that connects land records, input usage, and market access under a secure, interoperable framework. When designed with care, such systems can amplify farmer agency and unlock inclusive growth.

24 Toward Resilient, Tech-Enabled Agroecosystems

The convergence of digital technologies, ecological intelligence, and participatory planning presents a transformative pathway toward resilient agroecosystems. Future innovations must be context-aware, reflecting local agroecological conditions, cultural practices, and resource constraints. They should be scalable and modular, enabling incremental adoption and capacity building across diverse farming systems. Collaborative design—engaging farmers, researchers, and policymakers—is essential to ensure relevance, usability, and long-term impact. Equally important is the regenerative orientation of these technologies, which must support soil restoration, biodiversity enhancement, and carbon sequestration. By embedding these principles into the innovation lifecycle, agriculture can transition from extractive to regenerative, from reactive to predictive, and from exclusive to inclusive—advancing sustainability through systems-level integration.

25 Conclusion: Toward Resilient, Tech-Enabled Agroecosystems

Agriculture is no longer waiting on the sidelines of technological progress—it's stepping into a new era where innovation is actively reshaping how farms are imagined, managed, and sustained. What once felt like a distant possibility is now unfolding across fields and communities, as digital tools begin to influence everything from crop planning to climate resilience. This transformation isn't just about machines or data—it's about rethinking how decisions are made, who gets to make them, and how those choices affect both people and the planet. In this evolving landscape, technologies like AI-driven advisory platforms, sensor-enabled irrigation systems, and collaborative mapping apps are doing more than streamlining operations—they're helping farmers respond to their environments with greater precision, flexibility, and confidence. These tools reflect a shift toward systems thinking, where ecological signals, local knowledge, and real-time analytics come together to guide sustainable action. But for technology to truly serve agriculture, it must be grounded in the realities of diverse agroecological zones and the lived experiences of farmers themselves. That's where context matters. Tools like variable-rate applicators, spectral imaging, and decision-support frameworks allow for site-specific interventions that reduce waste and optimize resources. Yet precision alone doesn't guarantee adoption. Farmers embrace innovation when it feels relevant, trustworthy, and accessible. Voice-enabled apps, participatory design processes, and inclusive platforms play a vital role in ensuring that women, youth, and marginalized communities are not just users, but co-creators of the agricultural future. Beyond boosting yields, technology must also help farms regenerate—building resilience against climate shocks and restoring ecological balance. Innovations like soil microbiome mapping, low-energy machinery, and regenerative inputs support long-term health of agroecosystems, allowing farms to thrive even under stress. And to know whether these interventions are truly working, we need to measure impact holistically, across economic, environmental, and social dimensions. Evaluation frameworks that blend data analytics with community insights, like DEA, LCA, and PRA, help us understand not just what works, but why it works, for whom, and under what conditions. Looking ahead, the future of agriculture depends on building ecosystems of innovation that are collaborative, ethical, and regenerative. Governments must provide enabling environments, through smart subsidies, infrastructure, and data governance. Farmers and extension agents need ongoing support, peer learning, and access to digital tools. Research institutions, startups, and cooperatives must work together to create solutions that are interoperable, scalable, and sensitive to local needs. Above all, technology must be designed with fairness, transparency, and ecological integrity at its heart. If we embrace these principles, agriculture can evolve into a resilient, tech-enabled system that nourishes both people and the planet—delivering productivity without compromise, sustainability without exclusion, and innovation without inequity.

References

Abdi G et al (2024) In Advances in genomics: methods and applications (ed Vijai Singh) 265–306 (Springer Nature Singapore)

Addorisio R, Spadoni R, Maesano G (2025) Adoption of innovative technologies for sustainable agriculture: a scoping review of the system domain. Sustainability 17:4224. https://doi.org/10.3390/su17094224

AgTech (2021) Using sensor technology to better monitor and predict pollination efficiency. https://agreads.com/2021/06/24/using-sensor-technology-to-better-monitor-and-predict-pollination-efficiency/

Ahmad S et al (2021) CRISPR-based crop improvements: a way forward to achieve zero hunger. J Agric Food Chem 69:8307–8323. https://doi.org/10.1021/acs.jafc.1c02653

Ahmad L, Mahdi SS (2018) In Satellite farming: an information and technology based agriculture 67–80 (Springer International Publishing)

Ali ZA et al (2025) Digital twins: cornerstone to circular economy and sustainability goals. Environ Dev Sustain. https://doi.org/10.1007/s10668-025-06221-4

Anning-Dorson T (2025) In Digital business transformation in Africa, Volume I: Business Model Innovation 189–211 (Springer Nature Switzerland)

Assimakopoulos JH, Kalivas DP, Kollias VJ (2003) A GIS-based fuzzy classification for mapping the agricultural soils for N-fertilizers use. Sci Total Environ 309:19–33. https://doi.org/10.1016/S0048-9697(03)00055-X

Barwant MM, Ogidi OI, Singh B, Patil VS, Sharma SAB (2025) In the interplay of pesticides and climate change: environmental dynamics and challenges (eds Babafemi Raphael Babaniyi & Ebunoluwa Elizabeth Babaniyi) 149–171 (Springer Nature Switzerland)

Behrman JR (2010) The international food policy research institute (IFPRI) and the Mexican PROGRESA anti-poverty and human resource investment conditional cash. World Dev 38:1473–1485. https://doi.org/10.1016/j.worlddev.2010.06.007

Chowdhary V, Ranjan V (2024) Energy evolution of India's agrarian economy. J Resour Energy Dev 21:31–40

Debnath S et al (2023) The enhanced affinity of WRKY reinforces drought tolerance in solanum lycopersicum L.: an innovative bioinformatics study. Plants 12:762

Dhar SK et al (2024) Novel Bacillus and Prestia isolates from Dwarf century plant enhance crop yield and salinity tolerance. Sci Rep 14:14645. https://doi.org/10.1038/s41598-024-65632-x

Doktar F (2025) Filiz: agricultural sensor station. https://www.doktar.com/en/products/agricultural-sensor-station-soil-temperature/

Escribà-Gelonch M et al (2024) Digital twins in agriculture: orchestration and applications. J Agric Food Chem 72:10737–10752. https://doi.org/10.1021/acs.jafc.4c01934

Fan X et al (2022) Impacts of extreme temperature and precipitation on crops during the growing season in South Asia. Remote Sensing 14:6093

Gamage A et al (2024) Advancing sustainability: the impact of emerging technologies in agriculture. Current Plant Biol 40:100420. https://doi.org/10.1016/j.cpb.2024.100420

Georgina Menon K, Venkateshwar Reddy K, Boje Gowd BH, Paul Vijay P, Jhansi R (2025) In the interplay of pesticides and climate change: environmental dynamics and challenges (eds Babafemi Raphael Babaniyi & Ebunoluwa Elizabeth Babaniyi) 129–147 (Springer Nature Switzerland)

Giraldo OF, Rosset PM (2018) Agroecology as a territory in dispute: between institutionality and social movements. J Peasant Studies 45:545–564. https://doi.org/10.1080/03066150.2017.1353496

Gupta G, Kumar Pal S (2025) Applications of AI in precision agriculture. Discover Agric 3:61. https://doi.org/10.1007/s44279-025-00220-9

Halim-Lim SA, Jamaludin AA, Islam A, Weerabahu S, Priyono A (2025) Unlocking potential for a circular bioeconomy transition through digital innovation, lean manufacturing and green

practices: a review. Manag Environ Qual Int J 36:130–154. https://doi.org/10.1108/MEQ-11-2023-0386
He L (2022) In Encyclopedia of smart agriculture technologies (ed Qin Zhang) 1–9 (Springer International Publishing)
Huggi L, Thimmegowda MN, Sridhara S, Manjunatha MH, Das B (2024) In climate change impacts on soil-plant-atmosphere continuum (eds Himanshu Pathak, Dibyendu Chatterjee, Saurav Saha, & Bappa Das) 713–758 (Springer Nature Singapore)
Ibrahim M, Krishna BKC, Fraser DG, E (2025) Bridging the gap in sustainability measurement and reporting for agroecosystems: overview and development of an adaptive sustainability assessment and monitoring framework. Ecol Indicators 170:113091. https://doi.org/10.1016/j.ecolind.2025.113091
Ingrao C, Strippoli R, Lagioia G, Huisingh D (2023) Water scarcity in agriculture: an overview of causes, impacts and approaches for reducing the risks. Heliyon 9(8):e18507. https://doi.org/10.1016/j.heliyon.2023.e18507
Jensen TA, Antille DL, Tullberg JN (2025) Improving on-farm energy use efficiency by optimizing machinery operations and management: a review. Agric Res 14:15–33. https://doi.org/10.1007/s40003-024-00824-5
Kaur J et al (2022) An exopolysaccharide-producing novel Agrobacterium pusense strain JAS1 isolated from snake plant enhances plant growth and soil water retention. Sci Rep 12:21330. https://doi.org/10.1038/s41598-022-25225-y
Kaur J et al (2023) Reactive Black-5, congo red and methyl orange: chemical degradation of Azo-Dyes by agrobacterium. Water 15:1664
Kaur S, Thakur H, Singh A, Ramasamy V, Mudgal G (2025) In Sustainable development goals towards environmental toxicity and green chemistry: environment and sustainability (eds Chander Prakash, Kavindra Kumar Kesari, & Arvind Negi) 167–200 (Springer Nature Switzerland)
Kay M, Bunning S, Burke J, Boerger V, Bojic D, Bosc P-M, Clark M, Dale D, England M, Hoogeveen J (2022) The state of the world's land and water resources for food and agriculture 2021. Systems at breaking point. FAO
Kumar JS, Konyak MWC (2023) A descriptive study on agriculture system in Nagaland: schemes and implementations. Int J Res Innov Soc Sci 7:644–658
Kumar A et al (2018) Possible future climate for rice growing regions in India: visualising 2050 and pest-related impact thereof. Indian J Agric Sci 88:197–211. https://doi.org/10.56093/ijas.v88i2.79169
Lee T, Ramp D, Bless A, Unlocking digital twin planning for grazing industries with farmer centred design. Agric Human Values.https://doi.org/10.1007/s10460-025-10752-x
Mahajan N, Popma F (2022) In Integrating new technologies in international business 187–203 (Apple Academic Press)
Mai J et al (2025) Enhancing crop type mapping in data-scarce regions through transfer learning: a case study of the Hexi corridor. Remote Sensing 17:1494
Mandal D, Roy T (2024) In climate change impacts on soil-plant-atmosphere continuum (eds Himanshu Pathak, Dibyendu Chatterjee, Saurav Saha, & Bappa Das) 139–161 (Springer Nature Singapore)
Mansoor S et al (2025) Integration of smart sensors and IOT in precision agriculture: trends, challenges and future prospectives. Frontiers Plant Sci 16. https://doi.org/10.3389/fpls.2025.1587869
Mapiye O, Makombe G, Molotsi A, Dzama K, Mapiye C (2023) Information and communication technologies (ICTs): the potential for enhancing the dissemination of agricultural information and services to smallholder farmers in sub-Saharan Africa. Inf Dev 39:638–658. https://doi.org/10.1177/02666669211064847
Meivel S, Maheswari S (2021) Remote sensing analysis of agricultural drone. J Indian Soc Remote Sens 49:689–701. https://doi.org/10.1007/s12524-020-01244-y

Miller JO, Mondal P, Sarupria M (2024) Sensor-based measurements of NDVI in small grain and corn fields by tractor, drone, and satellite platforms. Crop Environ 3:33–42. https://doi.org/10.1016/j.crope.2023.11.001

Minghini M, Frassinelli F (2019) OpenStreetMap history for intrinsic quality assessment: Is OSM up-to-date? Open Geospatial Data, Software and Standards 4:9. https://doi.org/10.1186/s40965-019-0067-x

Minh DT (2024) Assessment of crop-land suitability by the multi-criteria evaluation approach and geographic information system: a scoping review. CIGR J 26:1–22

Mishra V, Rethinking urban future: towards sustainable cities and communities. Anthropocene Sci.https://doi.org/10.1007/s44177-025-00088-w

Mouratiadou I et al (2024) The socio-economic performance of agroecology. a review. Agron Sustain Dev 44:19. https://doi.org/10.1007/s13593-024-00945-9

Movilla-Pateiro L, Mahou-Lago XM, Doval MI, Simal-Gandara J (2021) Toward a sustainable metric and indicators for the goal of sustainability in agricultural and food production. Crit Rev Food Sci Nutr 61:1108–1129. https://doi.org/10.1080/10408398.2020.1754161

Mudgal G et al (2025) In Sustainable remediation for pollution and climate resilience (eds Arafat Abdel Hamed Abdel Latef, Ehab M. Zayed, & Ahmad Alsayed Omar) 139–206 (Springer Nature Singapore)

Mwikamba JN, Otieno DJ, Oluoch-Kosura W (2024) Effect of using a mobile phone on technical efficiency and productivity of climate-smart horticulture farmers in Taita-Taveta county, Kenya. Heliyon 10, https://doi.org/10.1016/j.heliyon.2024.e36917

Naik SK, Shinde R, Mali SS, Sarkar PK, Das A (2025) In ecological solutions to agricultural land degradation (eds Subhash Babu et al.) 1–27 (Springer Nature Singapore)

Naveed M et al (2024) In Microbial data intelligence and computational techniques for sustainable computing (eds Aditya Khamparia, Babita Pandey, Devendra Kumar Pandey, & Deepak Gupta) 235–258 (Springer Nature Singapore)

Omrany H, Mehdipour A, Oteng D, Al-Obaidi KM (2025) The uptake of urban digital twins in the built environment: a pathway to resilient and sustainable cities. Comput Urban Sci 5:20. https://doi.org/10.1007/s43762-025-00177-x

Pace R et al (2025) Artificial intelligence in soil microbiome analysis: a potential application in predicting and enhancing soil health—a review. Discover Appl Sci 7:85. https://doi.org/10.1007/s42452-024-06381-4

Padhiary M, Kumar A, Sethi LN (2025) Emerging technologies for smart and sustainable precision agriculture. Discover Robot 1:6. https://doi.org/10.1007/s44430-025-00006-0

Pandeya S, Gyawali BR, Upadhaya S (2025) Factors influencing precision agriculture technology adoption among small-scale farmers in Kentucky and their implications for policy and practice. Agriculture 15:177. https://doi.org/10.3390/agriculture15020177

Pankaj PK et al (2020) In Food security and land use change under conditions of climatic variability: a multidimensional perspective (eds Victor R. Squires & Mahesh K. Gaur) 281–309 (Springer International Publishing)

Parashar M et al (2023) Two novel plant-growth-promoting Lelliottia Amnigena isolates from euphorbia Prostrata Aiton enhance the overall productivity of wheat and tomato. Plants 12:3081

Pivoto D et al (2018) Scientific development of smart farming technologies and their application in Brazil. Inf Process Agric 5:21–32. https://doi.org/10.1016/j.inpa.2017.12.002

Poulton P, Johnston J, Macdonald A, White R, Powlson D (2018) Major limitations to achieving "4 per 1000" increases in soil organic carbon stock in temperate regions: Evidence from long-term experiments at Rothamsted Research. United Kingdom. Global Change Biology 24:2563–2584. https://doi.org/10.1111/gcb.14066

Proença LC, Neira J, Cardoso VA, Farley J (2025) Post-growth for the global south: reframing agricultural policies in Brazil. Land 14:904. https://doi.org/10.3390/land14040904

Qoseem IO et al (2024) Unlocking the potentials of digital twins for optimal healthcare delivery in Africa. Oxford Open Digital Health 2, https://doi.org/10.1093/oodh/oqae039 (2024).

Rahmati A, Zoej MJV, Dehkordi AT (2022) Early identification of crop types using Sentinel-2 satellite images and an incremental multi-feature ensemble method (Case study: Shahriar, Iran). Adv Space Res 70:907–922. https://doi.org/10.1016/j.asr.2022.05.038

Raj R, Kar S, Nandan R, Jagarlapudi A (2020) In Unmanned aerial vehicle: applications in agriculture and environment (eds Ram Avtar & Teiji Watanabe) 7–23 (Springer International Publishing)

Rajeev PN, Joy S (2020) In Socio-tech innovation: harnessing technology for social good (eds Latha Poonamallee, Joanne Scillitoe, & Simy Joy) 289–304 (Springer International Publishing)

Rajput NK, Gandhi N. & Grover, B. A. in Innovations in bio-inspired computing and applications-IBICA2020. 429–440 (Springer International Publishing).

Reddy AA (2019) The soil health card scheme in India: lessons learned and challenges for replication in other developing countries. J Nat Resour Policy Res 9:124–156. https://doi.org/10.5325/naturesopolirese.9.2.0124

Rizzo G, Migliore G, Schifani G, Vecchio R (2024) Key factors influencing farmers' adoption of sustainable innovations: a systematic literature review and research agenda. Org Agric 14:57–84. https://doi.org/10.1007/s13165-023-00440-7

Rosegrant MW et al (2012) Development and testing of a global hydrological model for integrated assessment modeling. (International Food Policy Research Institute Washington, DC)

Roy S et al (2022) Evaluating urban environment quality (UEQ) for Class-I Indian city: an integrated RS-GIS based exploratory spatial analysis. Geocarto Int 38:2153932. https://doi.org/10.1080/10106049.2022.2153932

Rupnik R et al (2019) AgroDSS: a decision support system for agriculture and farming. Comput Electron Agric 161:260–271. https://doi.org/10.1016/j.compag.2018.04.001

Sałabun W, Wątróbski J, Shekhovtsov A (2020) Are MCDA methods benchmarkable? A comparative study of TOPSIS, VIKOR, COPRAS, and PROMETHEE II methods. Symmetry 12:1549. https://doi.org/10.3390/sym12091549

Sánchez MAS, Moreno-Calles AI, Hernández López JDJ, Casas A (2024) Collaborative, situated, and critical methodologies in transdisciplinary agroecologies for life sustainability. Discover Sustain 5:256, https://doi.org/10.1007/s43621-024-00479-w

Sati VP (2024) In Farming systems and sustainable agriculture in the Himalaya 63–78 (Springer Nature Switzerland)

Scott GJ, Petsakos A, Juarez H (2019) Climate change, food security, and future scenarios for potato production in India to 2030. Food Secur 11:43–56. https://doi.org/10.1007/s12571-019-00897-z

Seetharaman B et al (2025) In Sustainable development goals towards environmental toxicity and green chemistry: environment and sustainability (eds Chander Prakash, Kavindra Kumar Kesari, & Arvind Negi) 291–313 (Springer Nature Switzerland)

Seif AK et al (2024) Identifying managed aquifer recharge and rain water harvesting sites and structures for storing non-conventional water using GIS-based multi-criteria decision analysis approach. Appl Water Sci 14:181. https://doi.org/10.1007/s13201-024-02246-8

Shah SS et al (2025) Impact of irrigation, fertilizer, and pesticide management practices on groundwater and soil health in the rice–wheat cropping system—a comparison of conventional, resource conservation technologies and conservation agriculture. Environ Sci Pollut Res 32:533–558. https://doi.org/10.1007/s11356-024-35661-0

Shamshiri RR (2024) In Mobile robots for digital farming 89–106 (CRC Press)

Shangguan ZP, Shao MA, Lei TW, Fan TL (2002) Runoff water management technologies for dryland agriculture on the Loess. Int J Sust Dev World 9:341–350. https://doi.org/10.1080/13504500209470129

Sharma A, Chand K, Singh GB, Mudgal G (2021) In Free radical biology and environmental toxicity (eds Kavindra Kumar Kesari & Niraj Kumar Jha) 305–325 (Springer International Publishing)

Siddique AB, Ahmed T, Khan F (2024) In Molecular dynamics of plant stress and its management (eds Muhammad Shahid & Rajarshi Gaur) 429–446 (Springer Nature Singapore)

Singh GB, Dey S, Mudgal G (2025) In Microbial metabolomics: recent developments, challenges and future opportunities (eds Sukhminderjit Kaur, Sunny Dhiman, & Manikant Tripathi) 419–439 (Springer Nature Singapore).
Singh J, Kumari M (2024) Socio-economic impacts of agricultural technologies on farmers' livelihood. Nat Acad Sci Lett 47:227–231. https://doi.org/10.1007/s40009-023-01351-7
Singh M, Pant M (2021) A review of selected weighing methods in MCDM with a case study. Int J Syst Assur Eng Manag 12:126–144. https://doi.org/10.1007/s13198-020-01033-3
Sishodia RP, Ray RL, Singh SK (2020) Applications of remote sensing in precision agriculture: a review. Remote Sensing 12:3136. https://doi.org/10.3390/rs12193136
Souza GDSE, Gomes EG, Alves ERDA, Gasques JG (2020) Technological progress in the Brazilian agriculture. Socio-Econ Planning Sci 72:100879. https://doi.org/10.1016/j.seps.2020.100879
Talukder B (2021) Multi-criteria decision analysis (MCDA) for agricultural sustainability assessment. https://www.academia.edu/64503107/Multi_Criteria_Decision_Analysis_MCDA_for_Agricultural_Sustainability_Assessment
Teixeira I, Morais R, Sousa JJ, Cunha A (2023) Deep learning models for the classification of crops in aerial imagery: a review. Agriculture 13:965. https://doi.org/10.3390/agriculture13050965
Thakkar JJ (2021) Multi-criteria decision making. Vol. 336 390 (Springer)
Tyagi S, Jaryal RD (2024) Role of mobile apps for promotion of implementation of post-harvest technology. AgriGate Int Multidiscip e-Mag 4:461–468
Verma S, Negi NP, Pareek S, Mudgal G, Kumar D (2022) Auxin response factors in plant adaptation to drought and salinity stress. Physiol Plant 174:e13714. https://doi.org/10.1111/ppl.13714
Wang H et al (2023) Cropformer: a new generalized deep learning classification approach for multi-scenario crop classification. Frontiers Plant Sci 14. https://doi.org/10.3389/fpls.2023.1130659
Wang L et al, 571–585 (Springer International Publishing)
Wijk MT van, Merbold L, Hammond J, Butterbach-Bahl K (2020) Improving assessments of the three pillars of climate smart agriculture: current achievements and ideas for the future. Frontiers Sustain Food Syst 4. https://doi.org/10.3389/fsufs.2020.558483
Wotlolan DL, Lowry JH, Wales NA, Glencross K (2021) Land suitability evaluation for multiple crop agroforestry planning using GIS and multi-criteria decision analysis: a case study in Fiji. Agrofor Syst 95:1519–1532. https://doi.org/10.1007/s10457-021-00661-3
Xie YW et al (2012) A GIS-based fertilizer decision support system for farmers in Northeast China: a case study at Tong-le village. Nutr Cycl Agroecosyst 93:323–336. https://doi.org/10.1007/s10705-012-9519-8
Yadav V, Yadav N (2024) Beyond sustainability, toward resilience, and regeneration: an integrative framework for archetypes of regenerative innovation. Glob J Flex Syst Manag 25:849–879. https://doi.org/10.1007/s40171-024-00418-8

Technologies for Sustainable Agriculture Development

Ashmitha J. Mishma and R. Emilin Renitta

Abstract Global agriculture system is facing several challenges such as population growth, climate changes, resource scarcity, and environmental degradation. To overcome these challengers, the adoption of sustainable practices and advanced technologies are necessary to assure long-term food security. Sustainable agriculture development includes integrated farming system, comprising conservation practices, ecological farming systems, and precision-based interventions. This chapter examines the role of several sustainable agricultural practices in enhancing soil fertility, biodiversity, and carbon sequestration. Also, the chapter highlights the role of new technologies such as Internet of Things (IoT), unmanned aerial vehicles (UAVs), robotics, smart sensors, etc. for optimizing soil, water, and nutrient management, and also, in yield monitoring and crop protection. To protect the resource scarcity, emerging systems such as vertical farming, hydroponics, and controlled-environment agriculture have been invented. This chapter highlighted the importance of communication networks, smartphone-based applications, and advanced machinery in promoting efficiency and reducing labour dependency. Even tough challenges remain with climate resilience, limited resources, and poor connectivity, integrated farming, renewable energy usage and better policies will help in enhancing resilience, productivity, and sustainability of global food systems.

Keywords Sustainable agriculture · Integrated farming systems · Internet of Things (IoT) · Unmanned Aerial Vehicles (UAVs) · Vertical farming · Hydroponics · Communication networks · Smartphone-based applications

A. J. Mishma · R. Emilin Renitta (✉)
Division of Food Processing Technology, Karunya Institute of Technology and Sciences, Coimbatore, India
e-mail: emilinrenitta@karunya.edu

K. K. Kesari et al. (eds.), *Smart Sustainable Materials Advancing Re-manufacturing and Technology*, World Sustainability Series, https://doi.org/10.1007/978-3-032-17907-4_10

1 Introduction

As the global population increases, more people moving to well-developed area resulting in scattering of agricultural land. FAO, 2017 found that that global agriculture production should be increased by 70%, to meet the growing demand by 2050. To enhance global food security, productivity and nutritional standards of food production should be increased for low-income countries to reduce their vulnerability. To empower smallholder farmers, the adoption of sustainable agricultural technologies plays an important role by reducing labour demand, reducing delay in operations, increasing resource efficiency, and strengthening production systems against climate extremes (Viana et al. 2022).

Agriculture, nowadays faces a set of interlinked challenges, including effects of climate change such as droughts, floods, and heatwaves, that could alter crop cycles and enhance pest and disease spreads (Rashami et al. 2025). Another important challenge is water scarcity and declining soil moisture, which have decreased over 2,000 Gt over recent decades and have threatened the sustainability of agriculture and ecosystem integrity (Scholl et al. 2025). Approximately 40% of global soils are degrading due to excessive use of agrochemicals, erosion and monocropping that has resulted in reduced soil quality and biodiversity, thereby diminishing ecosystem services necessary for sustainable agriculture (Pimentel and Burgess 2013). Additionally, labour shortages due to urban migration and aging workforce along with economic pressures like volatile input cost and limited access to credit and technology, reduce the resilience of smallholder agriculture. Also, pollution and greenhouse gas emissions from intensive farming also results in climate change and environmental decline (Suchithkumar et al. 2024).

An alternative way to solve the issues of food security, environmental, and socioeconomic challenges, is by the introduction of sustainable agriculture. In earlier days, traditional agriculture practices have led to soil degradation, water depletion, biodiversity loss, and emitted high levels of greenhouse gases, have threatened ecosystem services essential for food production and human well-being (Boix-Fayos and de Vente 2023). While the introduction of sustainable farming systems have helped in restoring soil health, improved water retention, cycle nutrients, capture carbon, and foster biodiversity, enabling vital ecosystem services like pollination and natural pest control (Țopa et al. 2025). In livestock operations, techniques like rotational grazing, silvopasture, optimized feeding, has improved livestock welfare and yield, also helped in cutting emission and boost carbon sequestration (Wang et al. 2021).

The use of IoT sensors, UAVs, and field robotics have been developed to increase efficiency and resilience in agriculture, that enable real-time crop and environment monitoring with unprecedented precision (Duguma and Bai 2024). As per United Nations Department of Economic and Social Affairs USA, 2018 Several agricultural industries have collaborated with satellite networks and cellular networks to enable automated irrigation, fertilization, pest control, and data-driven decision-making through cloud-based platforms (Van Dunsen and Nunavut 2018). These systems allow farmers to reduce their time spent on crop monitoring and reduced labour demand. This chapter focus on the recent trends in advanced technology for agricultural development.

2 Types of Sustainable Agriculture Practices

Sustainable agriculture plays an important role for a healthier future by providing environmental, economic and social benefits to the entire population, by initiating sustainable agricultural practices such as conservation tillage, crop rotation, and alternative food options. By applying these techniques boots both our long-term health and also improve the standard the standard of living. Organic farming is one of several forms of sustainable agriculture practices that they help in reducing or avoiding synthetic input on the soil and focus on improving soil fertility, biodiversity and climate resilience through natural methods like crop rotation and composting (Reganold and Wachter 2016). Permaculture, support ecological diversity by providing minimal resource input, results in gaining long-term sustainability and regenerative systems (Krebs and Bach 2018). Agroforestry, which integrates trees into agricultural systems, increases soil fertility, stores carbon, and provides increased income to farmers (Jose 2009). Conservation agriculture, which controls erosion, conserves water, and restores soil health through soil cover, minimal tillage, and crop rotation (Rodríguez et al. 2022). Integrated crop-livestock, which combines crops and animals to improve soil nutrient cycling, income, and productivity (Pires et al. 2022). Precision agriculture uses technologies like GPS, sensors, and drones to optimize input use, detect problems early, reduce labor costs, and improve profitability and sustainability (Padhiary et al. 2024).

3 Major Applications, Services and Sensors for Sustainable Agriculture Development

Sustainable agriculture utilizes practices and technologies to ensure long-term food production while minimizing environmental impact and promoting social equity. Currently, various technologies such as wireless sensors and IoTs and unmanned Aerial vehicle (UAVs) have been implemented which leads to many innovations in crop improvement. For advanced agricultural applications, several applications for monitoring, sensors and services hierarchy have been mentioned below.

3.1 Air Monitoring

Air quality monitoring is important in sustainable agriculture development for protecting crop health and, optimizing resource use. They help the farmers in identifying and mitigating threats to their crops, such as air pollution and extreme weather events, and also helps in minimising environmental damage (Srivastava et al. 2024). Bhatnagar et al. (2020) proposed a real-time monitoring system of micro climate based on wireless sensor networks (WSN), helped in sensing temperature and relative humidity (SHT15) powered by solar panels and supported by ZigBee communication technology.

In 2020, Kim proposed an IoT solution, Geostationary Environment Monitoring Spectrometer (GEMS), which is based on GPRS technology for monitoring atmospheric constituents such as ozone, nitrogen dioxide carbon monoxide, sulphur dioxide as well as cloud and UV indices, also helps in monitoring variables such as relative humidity, temperature, and radiation by testing over five different regions of China over a 2-year period by.

3.2 Soil Monitoring

Monitoring the quality of the soil is essential because it has a significant impact on plant growth. Measuring the amount of nutrients in the soil is the primary goal of soil monitoring. Every year in the spring, soil quality is checked; nevertheless, it can vary based on meteorological and environmental factors (Martínez-Fernández et al. 2016).

In order to estimate the amount of fertilizer needed, recent research has produced tools, technologies, and devices to monitor soil health, including salinity, pH, soil organic carbon, electrical conductivity, nitrogen, potassium, and phosphorous, as well as soil features, including moisture, water-holding capacity, and physical, chemical, and water-holding characteristics. In 2017, Agro Cares has launched a laboratory in box instrument and scanner to gather data on soil conditions, such as soil fertility monitoring, foliar nutrient analysis to support precision farming decisions (Talavera et al. 2017).

Moisture remote sensors like the Advanced Microwave Scanning Radiometer Earth Observing System (AMSR-E), the Soil Moisture Ocean Salinity system (SMOS), and the Soil Moisture Active Passive system (SMAP) (29) can be utilized for monitoring the moisture in the soil levels. Numerous vision-based and mobile sensors such as FarmBots and Agribots were created for identifying agricultural aspects such soil depth, seeding, weeding, fertilizing, and irrigating. Agribot is a multipurpose robot that helps growers with planting, fertilizer application, field irrigation, and health of the soil assessment. Whereas FarmBot is a free and open-source computerized precision agricultural farming device that uses software and the internet to plant, water, and monitor crops (Devaux et al. 2014).

3.3 Irrigation

Nearly 97% of the Earth's water is from the oceans, of which only 3% constitute to freshwater, and among that freshwater supply is limited and 2–3% is glaciers, and approximately half is available as surface water, such as lakes, rivers, and reservoirs, and the remain exist as groundwater. The global agriculture uses nearly 75% of the freshwater. The United Nations Convention to Combat Desertification (UNCCD), has reported that by 2030, about 168 countries face considerable water shortages for irrigation.

Several new irrigation techniques have been invented due to the increased demand for water, and to minimize water loss and address shortages include sprinkler, surface drip irrigation, subsurface drip irrigation, subsurface capillary irrigation (Ali et al. 2025). These advanced methods reduce the use of water by considering factors like soil type, moisture levels, and climatic conditions. Also, the use of the Internet of Things (IoT) plays a significant role in enhancing crop quality and productivity, for example, Crop Water Stress Index (CWSI), is a wireless sensor-based monitoring system used to assess moisture levels and estimate irrigation requirements effectively (Ayaz et al. 2018).

3.4 Fertilization Monitoring

Fertilization is the important factor in agriculture, because the essential nutrients in fertilizers help in plant growth, development, and reproduction. But the imbalanced application of fertilizers may affect the plant health and leads to economic loss, but it also impacts the quality of air, water and soil. Introduction of new technologies such as the Internet of Things (IoT) are useful by accurately estimating the necessary rates of fertilizer application in smart agriculture and also helps to reduce negative effects on environmental conditions (Ayaz et al. 2018).

Innovative technologies are now used to assess the spatial and temporal application of nutrients for effective fertilization, making the process less labour-intensive while maximizing efficiency. Generally, the Normalized Difference Vegetation Index (NDVI) is employed to monitor vegetation health, which aids in predicting nutrient requirements in the soil (Amani and Moghadam 2023). The nutrients can be effectively applied by using modern methods such as geographic information systems (GIS), variable rate technology, and global positioning systems (GPS), that can estimate nutrient levels in the fields and reduce the negative environmental impact (Khan et al. 2021).

3.5 Crop Harvesting Monitoring

The yield monitoring system is used to track and assess moisture levels, grain flow, crop yields, and the number of grains produced per crop. Regular crop monitoring during harvesting period is a critical process. Measuring crop yield is not only important at harvest time but also during the crop's developmental stages and prior to harvest. Several factors must be considered when monitoring crop yield, including achieving a high level of pollination, particularly in changing environmental conditions, which necessitates predicting seed outcomes (Torbick et al. 2017).

Crop forecasting is used to estimate yields before harvest, and helps farmers to make short-term plans and decisions, also, they can analyse crop maturity and quality, and ideal time for harvesting of crops. Effective crop disease management strategies and the overall development of crops can be improved through accurate crop predictions. New technologies, such as advanced yield monitors, can be integrated with machines used for harvesting. A smart phone application called FarmRTX, a web-based program that helps to provide precise harvesting data. Furthermore, it allows users to create detailed mapping tools that can be shared with other experts and growers, facilitating the export of information to additional agricultural tools for monitoring crop yield (Khan et al. 2021).

4 Advanced Agricultural Methods for Sustainable Development

To improve crop development and yield, several advanced methods such as greenhouses, hydroponics, and vertical farming have been implemented by farmers to address the challenges faced in agriculture and they help in increasing crop production.

Green house farming is one of the sustainable agriculture methods which gives year-round production in controlled environments with optimum temperature, CO_2 levels, and light. These farming reduces the water usage, fertilizers and pesticide runoff, soil erosion and helps in increased crop yield and resource efficiency (Benke and Tomkins 2017). In 2024, Lee introduced an AI-powered greenhouse environmental control system (AI-GECS), which is an automated solution which manages greenhouse environments by integrating weather forecasts, microclimate predictions, and crop physiology. Another method is hydroponics and aeroponics, where soilless cultivation is increased by 90–95% that results in reduced water usage and eliminates soil pathogens (Niu and Masabni 2022). Farmers achieved about to 70 times higher productivity in vertical farming and they conserves about 97% of water and land, and supports pesticide-free, high-density crop production (Mishra et al. 2024). Finally, usage of drones in phenotyping gives high-throughput imagining, and machine learning that enhance plant breeding and crop management, and improves several selections of traits such as drought tolerance, disease resistance, and yield through precise genetic-phenotypic linkage (Yang et al. 2020).

5 Advanced Technologies for Sustainable Agriculture Development

Agriculture has recently transformed from small and medium farms to extensive large-scale industries by the use of advanced agricultural technologies. In olden days, the traditional agricultural mainly relay on large machinery, massive robots, and tractors that utilize various remote sensing technologies. But the modern farming, the use of several technologies such as Geographic Information Systems (GIS) and Global Positioning Systems (GPS) has been implemented with crops for irrigation, sowing, fertilization, and harvesting. Additionally, the advanced technologies involve the usage of smartphone sensors, applications and technologies including cellular networks, Bluetooth, and Zigbee and sensor devices.

5.1 Smartphone Sensors and it's Applications

In modern farming techniques, use of smartphone sensors and mobile applications plays an important role for improving the agriculture development. To measure leaf area (Brandoli et al. 2021), evaluate fruit ripeness, harvest timing, disease presence, and soil erosion, image sensors or cameras are used by capturing the photographs (Baumüller 2018). In order to get the precise geographic data for land measurement, crop mapping, and farm management, GPS are used, while microphones detect machine sounds for maintenance, monitor pest activity, and enable voice commands (Khan et al. 2021).

In addition to smartphone sensors, various mobile applications have been developed to support agricultural activities. PETAFA, a GIS-based application, provides NDVI data across different crop growth stages and offers soil analysis linked to specific locations (Ayaz et al. 2019). L and PKS, is also a GIS-based application. That's aids in assessing the soil quality by analysing texture, depth and landscape, and offer recommendations for sustainable land management and climate change (Herrick et al. 2016). AMACA helps estimate machinery and operational costs using a QFD-based design tailored to user requirements (Orlando et al. 2016). AgriMaps is another GIS-based application that provides visual, site-specific land management recommendations using spatial data analysis (Jordan et al. 2016). SWApp improves irrigation efficiency in arid regions by monitoring soil moisture levels and incorporating historical weather data (Zhao et al. 2024).

Table 1 Communication technologies and their roles in modern agriculture

Technology	Role in agriculture
Cellular networks (2G–5G)	Reliable network access is essential in rural areas; advanced generations (like 5G) provide increased bandwidth. Satellite connections are another option, yet they can be excessively costly for small and medium-sized farms (Masek et al. 2016)
Low power wide area network (LPWAN)	Tailored for low-bandwidth, extensive-range, and energy-saving communication, LPWAN offers extended battery life and is cost-effective. Its application includes managing crops and pastures (Zulkifli and Noor 2017)
Bluetooth (BLE/Mesh)	Facilitates close-range connections among smartphones, gateways, and sensors. Ideal for setting up devices, collecting local data in small greenhouses, and facilitating interactions between workers and devices (Hong and Hsieh 2016)
Zigbee	It is designed to be a substitute for traditional equipment. I t is well-suited for IoT applications in greenhouse environments, to enable the monitoring of humidity and soil moisture as well as to manage irrigation or fertigation systems (Borrero and Zabalo 2020)

5.2 Agricultural Communications

The prompt communication and reporting of information is regarded as a crucial element of modern agriculture. A strong connection and involvement of various factors are essential to achieve effective and meaningful objectives. Telecom providers play a vital role in ensuring reliable communication for agricultural development. The Table 1 illustrates the different communication technologies used in contemporary agriculture.

5.3 Sensor Devices

Among the various tools employed in modern agriculture, sensor devices are essential and play a vital role in gathering data on crop conditions and other relevant information. A wide range of sensors has been integrated into modern agriculture to improve precision, sustainability, and operational efficiency. To collect data from remote areas telematics sensors monitor the operational status of agricultural machinery, and assist with location tracking and route optimization (Yeong et al. 2024). To get the acquisition and storage of geospatial, environmental, and climatic data, remote sensing technologies are used and supports applications such as forecasting, crop monitoring, yield assessment, land degradation analysis, and pest management (Vikas and Rajan 2024). Remote sensing technologies includes satellites, Light Detection and Ranging (LiDAR), and unmanned aerial vehicles (UAVs) (Yalew et al. 2016). The use of optical sensors, used to evaluate parameters such as soil moisture, organic matter content, mineral composition, and plant health status by utilizing

Table 2 Advanced agricultural machines: functions and benefits

Machine	Function and advantage
Automatic in row weeder	This machine helps in locating and removing weeds, that are 8 mm from plant stems, and also helps in removing the demand for chemicals (Lahre and Satpathy 2024; Scholz 2018)
Olive harvester	This machine is used to harvest olives from trees without damaging the entire plant and also this machine is not time consuming and usage of labours are also reduced (Messina et al. 2025)
Small-scale potato harvester	Mechanical harvesting of potatoes by eliminating the intensive labour. But this method is not suitable for large-scale farmers. For large-scale farming, tractor-towed harvesters are used (Johnson and Auat Cheein 2023)
Robotic lettuce harvester	The lettuce is harvested using artificial intelligence (AI) and robotic arms eliminates the need for labours (Mail et al. 2023)

light reflectance properties (Yew et al. 2014). In order to assess the tank level monitoring, spray distance measurement, and crop canopy assessment ultrasonic sensors are used (Zheng et al. 2024). In 2023, electromagnetic sensors are invented, and they can assess soil permeability, texture, and moisture levels, facilitating variable-rate input applications (Mensah et al. 2023). Recently in 2024, electrochemical sensors are introduced for the evaluation of soil pH and nutrient concentrations, and they provide faster alternative to conventional laboratory analyses (Hossain et al. 2024).

6 Advanced Machinery for Agriculture Development

Introduction of advanced agricultural machinery helps the farmers to increase their production and efficiency, and they automate processes that required human or animal labour. The main aim of this machinery is to produce affordable, user-friendly equipment where they can ease harvesters and also reduce labour requirements, improve the efficiency of agricultural fields. Several advanced agricultural machines are listed in Table 2.

7 Current Challenges and Future Prospects

Considering initiatives to increase food security and environmental sustainability farming still facing significant worldwide issues. Over 800 million individuals still go hungry, based on recent UN and WHO figures, underscoring the pressing need to boost food production in order to feed the world's expanding population, which is predicted to approach 10 billion people by 2050.The lack of resources is one of the main problems; as a result of soil erosion, urbanization, and climate change, there is

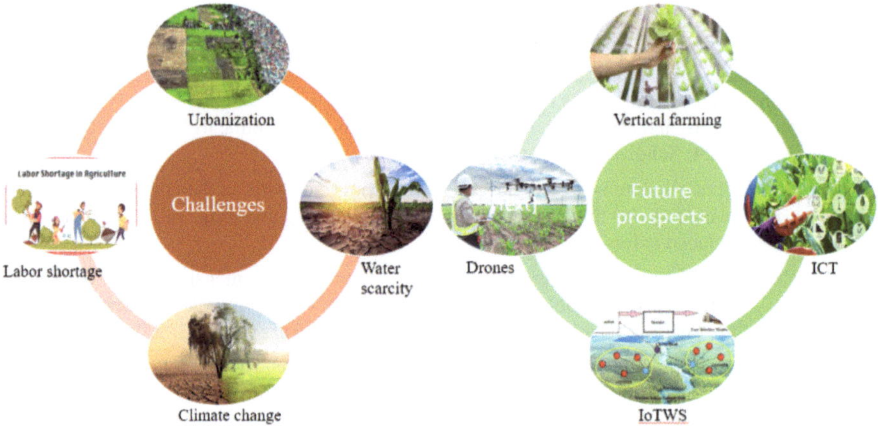

Fig. 1 Challenges and future prospects in sustainable agriculture development

less arable land. Also, water scarcity is an important challenge in agriculture development. Moreover, the agricultural development is still getting worse by extreme weather events, warming temperatures, and shifting rainfall patterns (Wyche and Steinfield 2016) (Fig. 1).

Agriculture must adopt smart technologies in the future to solve these problems. By adopting technologies such as Artificial intelligence (AI), big data, robotics, cloud computing, and the Internet of Things, helps in enhancing farm management, and reduces waste production and increases the resource utilization. They also help the farmers to manage environmental consequences, anticipate yields, automate processes, and monitor crops. In future, farming will be more resilient, and effective by including proper integrated systems that link data, technology, and human skill and also by providing proper assistance such as funding, education, and policy reform (Javaid et al. 2023).

When IoT-based wireless systems (IoTWS) are deployed properly throughout agricultural fields, they can give farmers timely updates and real-time insights, facilitate well-informed decision-making and increase crop output. The Internet of Things (IoT) is used for optimizing the farming operations for sustainable agriculture, by include keeping an eye on agricultural transportation, getting maintenance warnings for equipment, using water and power efficiently, and for obtaining current market pricing. They also contribute for automation and simplicity of the equipment and machinery and gives farmers more authority over their land, tools, and resources, causing traditional farming methods to change to more accurate and data-driven methods. Moreover,, immediate connectivity and data availability will be greatly enhanced by upcoming developments in sensor networks that are wireless and the introduction of 5G communication technologies. Recent trends indicate that by 2021, there will be more than 75 million IoT devices operating in agriculture, and by 2050, the average farm may produce almost 4.1 million points of data every day (Wyche and Steinfield 2016).

In agriculture, robust and pervasive connectivity is essential to the real success of IoT. From a telecom's perspective, every step of the agricultural value chain can be significantly impacted by increased connectivity and the emergence of value-added services. Even while connectivity is provided by the majority of telecom companies worldwide, these services currently only account for a small portion of the market for smart agriculture as a whole. Given how important connectivity is, particularly in rural or isolated places, telecom providers must offer more than just basic services. They ought to provide complete, end-to-end solutions that are suited to farmers' requirements. Services must be simple to use and comprehend because many farmers may lack technical expertise. This strategy would expand the marketplace reach of telecommunications and mobile businesses in addition to assisting farmers in implementing digital tools (Meola 2016; FAO 2018).

Monitoring agricultural data and finding trends requires the use of artificial intelligence (AI) and machine learning (ML). More specifically, these technologies assist evaluate the genes that are best suited for producing high-yield crop types. This gives farmers all across the world access to better seeds that are suited to particular temperatures and geographical areas. Furthermore, by determining which agricultural items have great demand and which are scarce, ML and AI systems can forecast market trends. With this knowledge, farmers may decide what to grow in the future with more knowledge (Khan et al. 2021).

Although agricultural technology is facing a number of obstacles, the general use of IoT in farming uses advanced agricultural methods particularly GPS, generalized package radio service (GPRS), and long-term sensor operations limit the farmers in rural area to implement high-tech farming solutions because they frequently rely on costly renewable energy sources that must be acquired separately. To increase the efficiency of farming from rural areas, more sustainable and planned approach to power management by the introduction of smart grids and microgrids, renewable energy options such as rechargeable battery packs, and distributed energy resources (DERs). They help in providing energy to the places that lacks steady power supply, especially in rural areas. However funding and public awareness are need to effectively implement these systems (Enescu et al. 2020).

In addition to implementing contemporary technologies, new farming methods are necessary to solve the problems caused by scarce resources and land. Urbanization is increasing as a result of the steady decline of arable land; roughly 3 million people are relocating to cities globally, further taxing already scarce urban resources. In future, the only use of traditional farming may not satisfy the food demand, so the use of modern techniques are also important. For example, vertical farming helps in water management and land scarcity and is also a promising solution for feeding the expanding city population because it is particularly well-suited to urban and peri-urban settings (Touliatos et al. 2016). Another option is the using the hydroponics which also uses less water and land. Farming operations are now more resource-efficient using these techniques and advancements in technology (Elsheshtawy 2008).

8 Conclusion

The demand for food is increasing and agriculture must need to incorporate cutting-edge and sustainable technologies in response to the world's expanding population and urbanization increases. To reduce the gap between farmers, suppliers, and customers, modern farming including various technologies like the Internet of Things (IoT), unmanned aerial vehicles (UAVs), satellite imaging, intelligent sensors, cloud computing, and mechanization have been introduced. It is not an alternative to conventional agricultural models, but may help in increase crop yield and management, improve resource management that helps in promoting environmental sustainability and food security. The use of sustainable agriculture is need which not only provides food but also helps in preserving natural ecosystem and improves the life of farmers for the generation.

References

Ali A, Hussain T, Zahid A (2025) Smart irrigation technologies and prospects for enhancing water use efficiency for sustainable agriculture. AgriEngineering 7(4):106. https://doi.org/10.3390/agriengineering7040106

Amani S, Shafizadeh-Moghadam H (2023) A review of machine learning models and influential factors for estimating evapotranspiration using remote sensing and ground-based data. Agric Water Manag 284:108324. https://doi.org/10.1016/j.agwat.2023.108324

Ayaz M, Ammad-Uddin M, Sharif Z, Mansour A, Aggoune E-HM (2019) Internet-of-Things (IoT)-based smart agriculture: toward making the fields talk. IEEE Access 7:129551–129583. https://doi.org/10.1109/ACCESS.2019.2932609

Ayaz M, Ammad-uddin M, Baig I (2018) Wireless sensor's civil applications, prototypes, and future integration possibilities: a review. IEEE Sens J 18(1):4–30. https://doi.org/10.1109/JSEN.2017.2766364

Baumüller H (2018) The little we know: an exploratory literature review on the utility of mobile phone-enabled services for smallholder farmers. J Int Dev 30(1):134–154. https://doi.org/10.1002/jid.3314

Benke K, Tomkins B (2017) Future food-production systems: vertical farming and controlled-environment agriculture. Sustainability: science, practice and policy 13(1):13–26. https://doi.org/10.1080/15487733.2017.1394054

Bhatnagar V, Singh G, Kumar G, Gupta R (2020) Internet of things in smart agriculture: applications and open challenges. Int J Stud Res Technol Manage 8(1):11–17. https://doi.org/10.18510/ijsrtm.2020.812

Boix-Fayos C, de Vente J (2023) Challenges and potential pathways towards sustainable agriculture within the European Green Deal. Agric Syst 207:103634. https://doi.org/10.1016/j.agsy.2023.103634

Borrero JD, Zabalo A (2020) An autonomous wireless device for real-time monitoring of water needs. Sensors 20(7):2078. https://doi.org/10.3390/s20072078

Brandoli B, Spadon G, Esau T, Hennessy P, Carvalho ACPL, Amer-Yahia S, Rodrigues-Jr JF (2021) DropLeaf: a precision farming smartphone tool for real-time quantification of pesticide application coverage. Comput Electron Agric 180:105906. https://doi.org/10.1016/j.compag.2020.105906

Cárceles Rodríguez B, Durán-Zuazo VH, Soriano Rodríguez M, García-Tejero IF, Gálvez Ruiz B, Cuadros S (2022) Conservation agriculture as a sustainable system for soil health: a review. Soil Syst 6(4):87. https://doi.org/10.3390/soilsystems6040087

Devaux A, Kromann P, Ortiz O (2014) Potatoes for sustainable global food security. Potato Res 57(3–4):185–199. https://doi.org/10.1007/s11540-014-9265-1

Duguma AL, Bai X (2024) How the internet of things technology improves agricultural efficiency. Artif Intell Rev 58(2):63. https://doi.org/10.1007/s10462-024-11046-0

Van Dusen J, Nunavut NWT (2018) United Nations department of economic and social affairs. Indigenous peoples' rights to lands, territories and resources at the centre of UN annual forum

Elsheshtawy Y (ed) (2008) The evolving Arab city. Routledge. https://doi.org/10.4324/9780203696798

Enescu D, Chicco G, Porumb R, Seritan G (2020) Thermal energy storage for grid applications: current status and emerging trends. Energies 13(2):340. https://doi.org/10.3390/en13020340

FAO, FAOSTAT (2018) Food and Agriculture Organization of the United Nations, Rome, 403. https://faostat.fao.org

Herrick JE, Beh A, Barrios E, Bouvier I, Coetzee M, Dent D, Elias E, Hengl T, Karl JW, Liniger H, Matuszak J, Neff JC, Ndungu LW, Obersteiner M, Shepherd KD, Urama KC, Bosch R, Webb NP (2016) The land-potential knowledge system (landpks): mobile apps and collaboration for optimizing climate change investments. Ecosyst Health Sustain 2(3). https://doi.org/10.1002/ehs2.1209

Hong G-Z, Hsieh C-L (2016) Application of integrated control strategy and bluetooth for irrigating romaine lettuce in greenhouse. IFAC-PapersOnLine 49(16):381–386. https://doi.org/10.1016/j.ifacol.2016.10.070

Hossain MI, Khaleque MA, Ali MR, Bacchu MS, Hossain MS, Shahed SMF, Saad Aly MA, Khan MdZH (2024) Development of electrochemical sensors for quick detection of environmental (soil and water) NPK ions. RSC Adv 14(13):9137–9158. https://doi.org/10.1039/D4RA00034J

Javaid M, Haleem A, Khan IH, Suman R (2023) Understanding the potential applications of artificial intelligence in agriculture sector. Adv Agrochem 2(1):15–30. https://doi.org/10.1016/j.aac.2022.10.001

Johnson CM, Auat F (2023) Machinery for potato harvesting: a state-of-the-art review. Front Plant Sci 14. https://doi.org/10.3389/fpls.2023.1156734

Jordan R, Eudoxie G, Maharaj K, Belfon R, Bernard M (2016) AgriMaps: improving site-specific land management through mobile maps. Comput Electron Agric 123:292–296. https://doi.org/10.1016/j.compag.2016.02.009

Jose S (2009) Agroforestry for ecosystem services and environmental benefits: an overview. Agrofor Syst 76(1):1–10. https://doi.org/10.1007/s10457-009-9229-7

Khan N, Ray RL, Sargani GR, Ihtisham M, Khayyam M, Ismail S (2021) Current progress and future prospects of agriculture technology: gateway to sustainable agriculture. Sustainability 13(9):4883. https://doi.org/10.3390/su13094883

Krebs J, Bach S (2018) Permaculture—scientific evidence of principles for the agroecological design of farming systems. Sustainability 10(9):3218. https://doi.org/10.3390/su10093218

Lahre J, Satpathy S (2024) Advancements in precision control in intra-row weeding: a comprehensive review. Int J Res Agron 7(4S):15–29. https://doi.org/10.33545/2618060X.2024.v7.i4Sa.509

Lee M-H, Yao M-H, Kow P-Y, Kuo B-J, Chang F-J (2024) An artificial intelligence-powered environmental control system for resilient and efficient greenhouse farming. Sustainability 16(24):10958. https://doi.org/10.3390/su162410958

Mail MF, Maja JM, Marshall M, Cutulle M, Miller G, Barnes E (2023) Agricultural harvesting robot concept design and system components: a review. AgriEngineering 5(2):777–800. https://doi.org/10.3390/agriengineering5020048

Martínez-Fernández J, González-Zamora A, Sánchez N, Gumuzzio A, Herrero-Jiménez CM (2016) Satellite soil moisture for agricultural drought monitoring: assessment of the SMOS derived Soil

Water Deficit Index. Remote Sens Environ 177:277–286. https://doi.org/10.1016/j.rse.2016.02.064
Masek P, Masek J, Frantik P, Fujdiak R, Ometov A, Hosek J, Andreev S, Mlynek P, Misurec J (2016) A harmonized perspective on transportation management in smart cities: the novel IoT-driven environment for road traffic modeling. Sensors 16(11):1872. https://doi.org/10.3390/s16111872
Mensah C, Katanda Y, Krishnapillai M, Cheema M, Galagedara L (2023) Estimation of soil water content using electromagnetic induction sensors under different land uses. Environ Res Commun 5(8):085002. https://doi.org/10.1088/2515-7620/acebbd
Meola A (2016) Why IoT, big data & smart farming are the future of agriculture. Bus Insider 20
Messina G, Sbaglia M, Bernardi B (2025) Mechanical harvesting of olive orchards: an overview on trunk shakers. AgriEngineering 7(3):52. https://doi.org/10.3390/agriengineering7030052
Mishra N, Hangshing L, Kadam DS, Tapang T, Shameena S (2024) Advances in vertical farming: opportunities and challenges. J Sci Res Rep 30(8):212–222. https://doi.org/10.9734/jsrr/2024/v30i82241
Niu G, Masabni J (2022) Hydroponics. In: Plant factory basics, applications and advances. Elsevier, pp 153–166. https://doi.org/10.1016/B978-0-323-85152-7.00023-9
Orlando F, Movedi E, Coduto D, Parisi S, Brancadoro L, Pagani V, Guarneri T, Confalonieri R (2016) Estimating Leaf Area Index (LAI) in vineyards using the PocketLAI smart-app. Sensors 16(12):2004. https://doi.org/10.3390/s16122004
Padhiary M, Saha D, Kumar R, Sethi LN, Kumar A (2024) Enhancing precision agriculture: a comprehensive review of machine learning and AI vision applications in all-terrain vehicle for farm automation. Smart Agric Technol 8:100483. https://doi.org/10.1016/j.atech.2024.100483
Pimentel D, Burgess M (2013) Soil erosion threatens food production. Agriculture 3(3):443–463. https://doi.org/10.3390/agriculture3030443
Pires GC, de Oliveira Denardin LG, Silva LS, de Freitas CM, Gonçalves EC, de Camargo TA, Bremm C, de Faccio PC, de Souza ED (2022) System fertilization increases soybean yield through soil quality improvements in integrated crop-livestock system in tropical soils. J Soil Sci Plant Nutr 22(4):4487–4495. https://doi.org/10.1007/s42729-022-01050-0
Rashami M, Yadav P, Chouhan R (2025) Climate change and its impact on agriculture. The PLANTA Res Book Ser 6(1):1794–1801
Reganold JP, Wachter JM (2016) Organic agriculture in the twenty-first century. Nat Plants 2(2):15221. https://doi.org/10.1038/nplants.2015.221
Scholl MA, McCabe GJ, Olson CG, Powlen KA (2025). Climate change and future water availability in the United States. https://doi.org/10.3133/pp1894E
Scholz M (2018) Enhancing adoption of integrated weed management—an Australian farmer's perspective. Outlooks Pest Manage 29(2):66–69. https://doi.org/10.1564/v29_apr_04
Srivastava RK, Purohit S, Alam E, Islam MK (2024) Advancements in soil management: optimizing crop production through interdisciplinary approaches. J Agric Food Res 18:101528. https://doi.org/10.1016/j.jafr.2024.101528
Suchithkumar C, Reddy S, Divyavani P, Divya S, Sravya M, Mani M, Manoharan G (2024) Modern challenges in agriculture, 040051. https://doi.org/10.1063/5.0195744
Talavera JM, Tobón LE, Gómez JA, Culman MA, Aranda JM, Parra DT, Quiroz LA, Hoyos A, Garreta LE (2017) Review of IoT applications in agro-industrial and environmental fields. Comput Electron Agric 142:283–297. https://doi.org/10.1016/j.compag.2017.09.015
Ţopa D-C, Căpşună S, Calistru A-E, Ailincăi C (2025) Sustainable practices for enhancing soil health and crop quality in modern agriculture: a review. Agriculture 15(9):998. https://doi.org/10.3390/agriculture15090998
Torbick N, Chowdhury D, Salas W, Qi J (2017) Monitoring rice agriculture across Myanmar using time series Sentinel-1 assisted by landsat-8 and PALSAR-2. Remote Sens 9(2):119. https://doi.org/10.3390/rs9020119
Touliatos D, Dodd IC, McAinsh M (2016) Vertical farming increases lettuce yield per unit area compared to conventional horizontal hydroponics. Food Energy Secur 5(3):184–191. https://doi.org/10.1002/fes3.83

Viana CM, Freire D, Abrantes P, Rocha J, Pereira P (2022) Agricultural land systems importance for supporting food security and sustainable development goals: a systematic review. Sci Total Environ 806:150718. https://doi.org/10.1016/j.scitotenv.2021.150718

Vikas, Ranjan R (2024) Agroecological approaches to sustainable development. Front Sustain Food Syst 8. https://doi.org/10.3389/fsufs.2024.1405409

Wang F, Harindintwali JD, Yuan Z, Wang M, Wang F, Li S, Yin Z, Huang L, Fu Y, Li L, Chang SX, Zhang L, Rinklebe J, Yuan Z, Zhu Q, Xiang L, Tsang DCW, Xu L, Jiang X, Chen JM (2021) Technologies and perspectives for achieving carbon neutrality. The Innovation 2(4):100180. https://doi.org/10.1016/j.xinn.2021.100180

Wyche S, Steinfield C (2016) Why don't farmers use cell phones to access market prices? technology affordances and barriers to market information services adoption in Rural Kenya. Inf Technol Dev 22(2):320–333. https://doi.org/10.1080/02681102.2015.1048184

Yalew SG, van Griensven A, Mul ML, van der Zaag P (2016) Land suitability analysis for agriculture in the Abbay basin using remote sensing, GIS and AHP techniques. Model Earth Syst Environ 2(2):101. https://doi.org/10.1007/s40808-016-0167-x

Yang W, Feng H, Zhang X, Zhang J, Doonan JH, Batchelor WD, Xiong L, Yan J (2020) Crop phenomics and high-throughput phenotyping: past decades, current challenges, and future perspectives. Mol Plant 13(2):187–214. https://doi.org/10.1016/j.molp.2020.01.008

Yeong DJ, Panduru K, Walsh J (2024) Smart agriculture: software platform for telematics monitoring in farm machinery. In: 2024 35th Irish signals and systems conference (ISSC), pp 1–6. https://doi.org/10.1109/ISSC61953.2024.10602840

Yew TK, Yusoff Y, Sieng LK, Lah HC, Majid H, Shelida N (2014) An electrochemical sensor ASIC for agriculture applications. In: 2014 37th international convention on information and communication technology, electronics and microelectronics (MIPRO), pp 85–90. https://doi.org/10.1109/MIPRO.2014.6859538

Zhao Y, Li G, Li S, Luo Y, Bai Y (2024) A review on the optimization of irrigation schedules for farmlands based on a simulation-optimization model. Water 16(17):2545. https://doi.org/10.3390/w16172545

Zheng Y, Hui X, Cai D, Shoukat MR, Wang Y, Wang Z, Ma F, Yan H (2024) Calibrating ultrasonic sensor measurements of crop canopy heights: a case study of maize and wheat. Front Plant Sci 15. https://doi.org/10.3389/fpls.2024.1354359

Zulkifli CZ, Noor NN (2017) Wireless Sensor Network and Internet of Things (IoT) Solution in Agriculture. Pertanika J Sci Technol 25:91–100

Aquatic Agritecture: Technology-Led Shifts from Terrestrial Farming to Ocean-Based Food Systems

Abhijeet Singh, Jaspreet Kaur, Sanjoy K. Dhar, Simrandeep Kaur, Shivam Kumar Lal, Harshita Thakur, Payal Ghosh, and Gaurav Mudgal

Abstract As farmland shrinks, freshwater grows scarce, and climate patterns become increasingly unpredictable, humanity is looking seaward for answers. The ocean—vast, resilient, and largely untapped—is emerging as a new frontier for food security. This chapter invites readers to explore how ocean-based agriculture is transforming from a bold idea into a practical solution, blending science, ethics, and innovation. Imagine oceans not just as wild expanses, but as living farms. Picture intelligent seaweed cultivation guided by artificial intelligence, vertical farms anchored offshore, and underwater greenhouses like *Nemo's Garden* that grow land crops beneath the waves using sunlight and stable microclimates. These aren't science fiction—they're real, and they're reshaping how we think about growing food. We dive into the world of polyculture systems, especially *Integrated Multi-Trophic Aquaculture (IMTA)*, where fish, shellfish, and seaweeds are farmed together in harmony. Each species plays a role—some feed, some filter, some absorb nutrients—creating a balanced ecosystem that boosts productivity and protects biodiversity. Through case studies—from Norway's high-tech salmon farms to Zanzibar's women-led seaweed cooperatives and the Faroe Islands' climate-smart kelp initiatives—we see how ocean

A. Singh · P. Ghosh
School of Bio Sciences and Technology (SBST), VIT University, Vellore, Tamil Nadu, India
e-mail: payel.ghosh@vit.ac.in

J. Kaur
India Kanya Maha Vidyalaya, Jalandhar, Punjab, India

S. K. Dhar
Banophool Adibashi Green Heart College, Mirpur, Dhaka, Bangladesh

S. Kaur · G. Mudgal (✉)
University Institute of Biotechnology, Chandigarh University, Mohali, Punjab, India
e-mail: gauravmdgl@gmail.com

S. K. Lal
University Institute of Agricultural Sciences, Chandigarh University, Mohali, Punjab, India

H. Thakur
Amity Institute of Biotechnology (AIB), Faculty of Science, Engineering and Technology, Amity University Haryana, Gurugram, India

K. K. Kesari et al. (eds.), *Smart Sustainable Materials Advancing Re-manufacturing and Technology*, World Sustainability Series, https://doi.org/10.1007/978-3-032-17907-4_11

farming is already changing lives and landscapes. These stories show what's possible when innovation meets local wisdom. Of course, ocean farming comes with challenges. We take a clear-eyed look at risks like pollution, disease spread, and habitat disruption. We also explore deeper questions: How do we ensure animal welfare? Who gets to benefit? How do we include communities in shaping this future? Policy plays a crucial role. We examine frameworks like India's Blue Economy Strategy and the EU's Ocean Multi-use Action Plan, which help guide sustainable investment and marine planning. On the tech horizon, we spotlight robotic aquaculture, solar-powered desalination, and multi-use ocean zones—all aligned with global goals like SDG 14 and climate resilience. This chapter ends with a call to action: to build ocean agriculture not just with cutting-edge tools, but with collaboration, fairness, and deep respect for marine life. The ocean can feed us—but only if we nurture it wisely.

Keywords Blue foods · Vertical aquaculture · Seaweed bioplastics · Submerged greenhouses · Ocean zoning · Offshore biospheres · 3D ocean farming · Marine cooperatives · Kelp carbon sequestration

1 Introduction—Redefining Agricultural Frontiers

The global food system is under unprecedented pressure. With the world population projected to reach nearly 10 billion by 2050, the demand for food—particularly protein-rich and nutrient-dense sources—is expected to surge by more than 70% compared to current levels (Mossler 2020). Historically, terrestrial agriculture has been the backbone of food production, but its limitations are becoming increasingly evident. Land-based farming accounts for over 70% of the global freshwater resources. It contributes more than 20% of greenhouse gas emissions, making it one of the largest drivers of deforestation, biodiversity loss, and nutrient pollution (Lovat et al. 2025; Patrick and Getty, 2025).

Despite technological advances, terrestrial farming continues to face diminishing returns due to soil degradation, climate variability, and land scarcity. A recent global clustering study revealed that 46% of land under cultivation and grazing is scattered and vulnerable to degradation, while only 16.2% is under intense food production (Jung et al. 2024). Moreover, the energy input required for crop production varies drastically, with high-input systems in Western Europe exceeding 30 GJ/ha, compared to less than 1 GJ/ha in low-input systems across Africa (Wang 2022). These disparities highlight the inefficiencies and ecological costs of conventional agriculture.

In light of these constraints, the ocean emerges as a vast and largely untapped frontier for food production. Covering over 70% of the Earth's surface, marine ecosystems offer immense potential for sustainable protein sources, including seaweed, bivalves, and finfish. The ocean currently provides 59 billion kilograms of food annually, but with improved management and technological innovation, this figure could rise to

80–103 billion kilograms by 2050—a 36–74% increase (Mossler 2020). This expansion is not merely theoretical; it is grounded in economic and ecological feasibility, as demonstrated by modeling studies that account for feed technologies, policy reforms, and consumer demand shifts. Seaweed farming, in particular, has gained attention for its low environmental footprint and high nutritional value. Seaweeds are rich in vitamins, minerals, and bioactive compounds, and their cultivation requires no freshwater, fertilizers, or arable land. In South Asia and Africa, seaweed farming is emerging as a livelihood strategy and a climate-resilient food source (Schmidt 2025). Bivalve aquaculture—think mussels, oysters, and clams—is quietly revolutionizing how we produce food from the sea. These filter-feeding organisms naturally clean the water as they grow, requiring minimal human intervention. Because they thrive without feed inputs or complex infrastructure, they represent one of the most sustainable ways to produce protein on the planet (Food from the Sea 2021).

Beyond the familiar promise of seaweed and shellfish, a new wave of innovation is diving deeper—literally. Submerged agriculture is redefining what it means to grow food, with technologies that allow us to cultivate land-based crops beneath the ocean's surface. From *aquapods* that drift like futuristic coral reefs to *air-dome habitats* and *submerged greenhouses* that mimic terrestrial farming conditions, these systems create controlled environments where tomatoes, herbs, and leafy greens can thrive underwater. It's a bold reimagining of agriculture—one that blends marine engineering with ecological design to unlock new possibilities for resilient food production. These systems leverage photosynthetic engineering, nutrient cycling, and pressure-adapted plant varieties to grow food in salty conditions. Although in experimental stages, such technologies represent a great change in agricultural thinking that embraces the ocean not just as a source of fish, but as a platform for diversified food systems (Tont and Delistraty 1977). The concept of "aquatic agritecture" is gaining traction as a framework for designing ocean-based food systems. These systems are envisioned as modular, scalable, and resilient, capable of operating in offshore zones, coastal margins, and even urban waterfronts. They combine renewable energy sources (e.g., solar buoys, wave-powered pumps) with smart monitoring technologies to optimize growth conditions and minimize ecological impact. Such innovations align with the goals of the blue economy, which seeks to balance economic development with ocean conservation.

However, the transition to ocean-based agriculture is not without challenges. Marine ecosystems are fragile and already under stress from overfishing, pollution, and climate change. According to the FAO, only 62.3% of marine fish stocks are biologically sustainable, and 23.8 million tons of aquatic foods are lost or wasted annually in the supply chain (Schmidt 2025). Moreover, the legal and regulatory frameworks governing ocean farming are fragmented and often ill-equipped to handle emerging technologies. Maritime zoning, tenure rights, and environmental impact assessments must evolve to support responsible innovation. Ethical considerations also come into play. The expansion of ocean farming must ensure equitable access for coastal and indigenous communities, protect biodiversity, and avoid replicating the extractive models of terrestrial agriculture. Social protections, gender inclusion, and participatory governance are essential to building inclusive aquatic food systems.

As noted by CGIAR, even modest increases in aquatic food consumption during early childhood can significantly reduce stunting and enhance cognitive development, underscoring the societal value of marine nutrition (Schmidt 2025).

In this chapter, we explore the technological, ecological, and socio-economic dimensions of shifting food regimes toward the ocean. We begin by examining the limitations of terrestrial farming and the drivers behind the search for alternative food systems. We then delve into the science and engineering of seaweed cultivation, bivalve aquaculture, and submerged agriculture, highlighting key innovations and case studies. The chapter also addresses the policy landscape, environmental risks, and ethical frameworks necessary for sustainable ocean farming. By reimagining crop farming other than on the lands, we attempt to open new possibilities for climate-resilient, nutrient-rich, and ecologically viable food systems. For centuries, we've looked to the ocean mainly as a source of fish. But today, that view is evolving. The ocean is being reimagined, not just as a place to harvest, but as a living platform for smart, sustainable agriculture. This transformation isn't just technological; it reflects a deeper philosophical shift. We're moving toward a mindset of planetary stewardship, where food production aligns with ecological balance, and where access to nutritious, ocean-grown resources becomes a matter of equity, not privilege.

2 Technological Advancements in Seaweed Cultivation

Seaweed farming has rapidly evolved from artisanal coastal practices into a cutting-edge, technology-driven sector that supports climate resilience, food security, and bioeconomic growth. Smart tools like autonomous underwater vehicles (AUVs), sensor networks, and artificial intelligence (AI) now enable precision aquaculture at unprecedented scales—boosting productivity, enhancing environmental monitoring, reducing labor costs, and optimizing biomass quality for applications ranging from food to biofuels. Among the most transformative innovations are AI-supported autonomous inspection systems, which use underwater drones with optical and acoustic sensors to monitor kelp growth, detect disease, and assess nutrient dynamics in real time; Bell et al. (2020) demonstrated the effectiveness of satellite and drone-based platforms for tracking canopy-forming kelps across offshore farms with high-frequency spatial coverage (Bella 2020), while Ackerman (2017) reported Marine BioEnergy's use of drone submarines to cycle kelp through nutrient-rich ocean layers, improving photosynthesis and yield (Ackerman 2017). These systems are further enhanced by machine learning algorithms that analyze underwater imagery to classify kelp health, predict harvest timing, and optimize farm layouts—exemplified by the BioDiMoBot project, which integrates biohybrid sensors and AI analytics to detect early signs of ecosystem stress and support biodiversity and carbon tracking (Robotics and kelp farming project aims to boost biodiversity and carbon capture 2025; Brandslet 2024). Offshore cultivation has also expanded beyond sheltered zones, with vertical ocean farms using the full water column to grow kelp, mussels, and oysters in stacked arrays that clean water, sequester carbon, and foster marine biodiversity; the GreenWave model in the U.S. showcases this approach by combining 3D farming with community aquaculture to restore habitats and remove excess nitrogen (Gertz 2017; Hassan 2024). Meanwhile, floating sea farms equipped with solar-powered desalination units and hydroponic platforms are enabling the growth of terrestrial crops on ocean surfaces—such as those developed by researchers at the University of South Australia, who created autonomous systems that convert seawater into freshwater for irrigation, proving the viability of integrated food-energy-water systems at sea (Owen 2023). These innovations reflect a broader shift toward decentralizing food production and reducing pressure on land-based agriculture.

Seaweed farming has rapidly evolved from artisanal coastal practices into a cutting-edge, technology-driven sector that supports climate resilience, food security, and bioeconomic growth. Smart tools like autonomous underwater vehicles (AUVs), sensor networks, and artificial intelligence (AI) now enable precision aquaculture at unprecedented scales—boosting productivity, enhancing environmental monitoring, reducing labor costs, and optimizing biomass quality for applications ranging from food to biofuels. Among the most transformative innovations are AI-supported autonomous inspection systems, which use underwater drones with optical and acoustic sensors to monitor kelp growth, detect disease, and assess nutrient dynamics in real time; Bell et al. (2020) demonstrated the effectiveness of satellite and drone-based platforms for tracking canopy-forming kelps across offshore farms with high-frequency spatial coverage (Bella et al. 2020, #82), while Ackerman (2017) reported Marine BioEnergy's use of drone submarines to cycle kelp through nutrient-rich ocean layers, improving photosynthesis and yield (Ackerman 2017). These systems are further enhanced by machine learning algorithms that analyze underwater imagery to classify kelp health, predict harvest timing, and optimize farm layouts—exemplified by the BioDiMoBot project, which integrates biohybrid sensors and AI analytics to detect early signs of ecosystem stress and support biodiversity and carbon tracking (Robotics and kelp farming project aims to boost biodiversity and carbon capture 2025; Brandslet 2024). Offshore cultivation has also expanded beyond sheltered zones, with vertical ocean farms using the full water column to grow kelp, mussels, and oysters in stacked arrays that clean water, sequester carbon, and foster marine biodiversity; the GreenWave model in the U.S. showcases this approach by combining 3D farming with community aquaculture to restore habitats and remove excess nitrogen (Gertz 2017; Hassan 2024). Meanwhile, floating sea farms equipped with solar-powered desalination units and hydroponic platforms are enabling the growth of terrestrial crops on ocean surfaces—such as those developed by researchers at the University of South Australia, who created autonomous systems that convert seawater into freshwater for irrigation, proving the viability of integrated food-energy-water systems at sea (Owen 2023). These innovations reflect a broader shift toward decentralizing food production and reducing pressure on land-based agriculture.

The climate benefits of seaweed farming are no longer just theoretical—they're increasingly backed by robust empirical evidence. Studies have shown that seaweed cultivation can sequester significant amounts of carbon, mitigate ocean acidification, and improve coastal resilience. As data accumulates from long-term monitoring and remote sensing platforms, the role of seaweed as a nature-based climate solution is gaining scientific and policy traction, reinforcing its potential within blue carbon strategies and sustainable ocean economies. Duarte et al. (2025) conducted a global study across 20 farms and found that carbon burial rates beneath seaweed beds can match those of natural blue carbon ecosystems like mangroves and seagrasses, with older farms storing up to 140 tonnes of carbon per hectare (Duarte 2025). These findings are corroborated by Chaudhari et al. (2024a), who emphasized the role of species selection, oceanic conditions, and farm design in maximizing carbon sequestration potential (Chaudhari et al. 2024b). To ground the discussion on species selection in seaweed farming, Table 1 presents a comparative analysis of prominent macroalgae cultivated across global marine environments. It synthesizes core attributes such as growth rate, optimal depth, biochemical composition, carbon sequestration capacity, and region-specific applicability. This data provides critical context for evaluating how different species contribute not only to yield and biodiversity but also to climate mitigation and market versatility (Chaudhari, et al. 2024a; Canvin et al. 2024; Ross et al. 2022). The table reinforces the argument that strategic species selection—guided by ecological performance and economic relevance—is a cornerstone of smart seaweed cultivation. In summary, the convergence of AI, robotics, remote sensing, and offshore engineering is redefining the possibilities of seaweed cultivation. These technologies not only enhance productivity and ecological performance but also position seaweed farming as a cornerstone of the emerging blue economy. As the sector matures, continued investment in scalable infrastructure, data-driven management, and interdisciplinary research will be essential to unlock its full potential. Figure 1 visually complements these narratives on AI-enabled underwater drones, remote sensing platforms, and smart monitoring innovations. It reinforces the section's emphasis on how data-driven technologies are revolutionizing ocean farming practices.

3 Submerged Agriculture—The Rise of Underwater Farming

Submerged agriculture represents a radical rethinking of food production, one that challenges the terrestrial bias of conventional farming systems. By cultivating crops beneath the ocean surface, researchers and entrepreneurs are exploring how controlled underwater environments can support plant growth while bypassing many of the limitations of land-based agriculture. These innovations are emerging in response to urgent global challenges: the overuse of pesticides, dwindling freshwater supplies, and escalating climate disruptions like soil salinization and extreme weather events. Underwater farming offers a compelling alternative—one that sidesteps many of the limitations of land-based agriculture. As highlighted by the World Economic Forum, this approach could potentially nourish the entire global population using

Table 1 Comparative attributes of selected seaweed species for cultivation and carbon sequestration potential

Seaweed species	Growth rate (cm/day)	Optimal depth (m)	Carbon sequestration potential (tons/ha/year)	Biochemical highlights	Primary applications	Cultivation regions	References
Saccharina latissima *(Sugar Kelp)*	**5–10**	**5–15**	**6–10**	**Rich in alginate, mannitol, iodine**	**Bioplastics, animal feed, biofuels**	**North Atlantic, Europe**	Canvin et al. (2024)
Macrocystis pyrifera *(Giant Kelp)*	10–15	10–30	10–15	High biomass yield, fucoidan, laminarin	Bioenergy, packaging, carbon credits	Pacific coasts, Chile, USA	Ross et al. (2022)
Gracilaria edulis	**4–7**	**1–3**	**3–6**	**Source of agar, antioxidants, minerals**	**Cosmetics, food additives, bioplastics**	**India, Southeast Asia**	Chaudhari, et al. (2024a)
Eucheuma spp.	3–6	1–3	2–5	Carrageenan-rich, high sulfate polysaccharides	Food stabilizers, pharmaceuticals	Philippines, Tanzania	Meinita et al. 2022)
Ulva spp. *(Sea Lettuce)*	**2–4**	**0.5–3**	**1–3**	**Ulvan polysaccharides, protein-rich**	**Animal feed, biofertilizers, nutraceuticals**	**Mediterranean, India**	Morais (2020)
Asparagopsis taxiformis	1–3	5–10	2–4	Bromoform content, methane-reducing properties	Livestock feed additive, climate mitigation	Australia, Hawaii	Mahrose and Michalak (2022)
Sargassum spp.	**3–5**	**Surface–5**	**2–4**	**Polyphenols, alginate, antioxidants**	**Bioplastics, fertilizers, feed additives**	**Gulf of Mexico, India**	Zaimis et al. (2021)
Kappaphycus alvarezii	4–6	1–3	3–5	Carrageenan, high growth efficiency	Food gels, bioplastics, cosmetics	India, Indonesia	Maguire (2024)

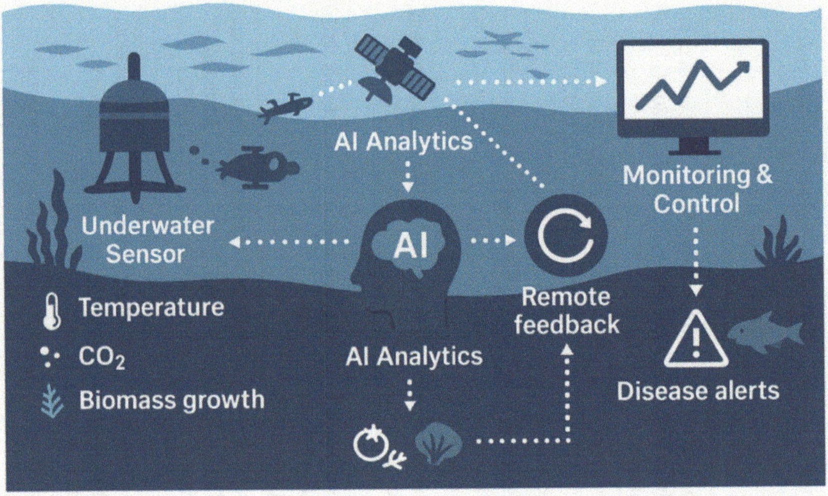

Fig. 1 Tech-enabled monitoring systems in ocean farming. A systems diagram depicting the feedback loop of smart ocean farming. It illustrates the flow from underwater sensors—embedded in autonomous underwater vehicles (AUVs), satellite arrays, and drone imaging platforms—into AI analytics for real-time diagnostics on temperature, CO_2 concentration, biomass growth rates, and disease alerts. Data is remotely accessed and interpreted by operators to adjust farming parameters, ensuring precision cultivation, ecosystem resilience, and scalable productivity. The figure highlights the integration of robotics, artificial intelligence, and remote sensing as central pillars of blue agritech

just 2% of the ocean's surface, all while eliminating pesticide dependence and dramatically cutting water usage (Marchant 2021a, b).

One of the most striking examples of submerged agriculture is *Nemo's Garden*, an experimental underwater farming initiative located off the coast of Noli, Italy. Developed by the Ocean Reef Group, this project features air-filled plastic biospheres anchored between 4.5 and 11 m below the sea surface. These domes act as miniature greenhouses, maintaining stable temperature and humidity levels while allowing sunlight to filter through the water. Inside, crops like basil, tomatoes, beans, and lettuce are grown hydroponically in nutrient-rich solutions. The system cleverly uses natural seawater evaporation and condensation to produce freshwater for irrigation, eliminating the need for external water inputs after the initial setup (Portalanza 2025). Each biosphere is equipped with sensors that monitor carbon dioxide, oxygen, humidity, temperature, and light levels—parameters that are remotely managed via solar-powered systems to optimize plant growth. The spherical design ensures consistent internal pressure and temperature, while tidal movements naturally ventilate the air. Early studies suggest that crops grown in these underwater greenhouses not only grow faster but also contain higher levels of antioxidants than their land-based counterparts, pointing to submerged agriculture as a promising path toward both sustainability and enhanced nutritional value (Underwater farming: exploring the new agricultural frontier and its environmental benefits 2025).

Beyond *Nemo's Garden*, submerged farming is branching into even more futuristic terrain. One emerging concept is vertical farming inside submarines—designed to produce food in isolated or extreme environments such as deep-sea habitats, polar stations, or disaster zones. These enclosed systems use stacked hydroponic or aquaponic layers, illuminated by energy-efficient LED grow lights. Sophisticated climate control and water recycling technologies maintain optimal growing conditions, allowing crops to flourish in places where traditional agriculture is impossible. This approach not only expands the spatial logic of farming but also aligns with long-term visions for resilient, decentralized food systems in a changing climate.

Submarine farms can produce leafy greens, herbs, and small fruits while simultaneously improving air quality through photosynthesis. The integration of aquaponics—where fish waste provides nutrients for plants and plants purify water for fish—creates a closed-loop ecosystem that is both efficient and sustainable (Khandakar et al. 2024; Abusin and Mandikiana 2020; Nair 2025).

The physiological basis for underwater crop cultivation lies in photosynthetic engineering and nutrient cycling. Submerged plants must adapt to reduced light penetration and slower gas diffusion in water. Research has shown that certain terrestrial crops, such as rice, can perform underwater photosynthesis by producing carbohydrates and oxygen, which support internal aeration and energy metabolism during submergence (Pedersen et al. 2013). These adaptations include thinner cuticles, hydrophobic leaf surfaces that retain gas films, and enhanced carbon-concentrating mechanisms. Such traits are critical for maintaining photosynthetic efficiency and plant survival in underwater environments.

Moreover, submerged agriculture leverages controlled microclimates to mitigate the effects of climate change. Flooding, droughts, and heatwaves are increasingly disrupting terrestrial farming, but underwater farms remain insulated from these extremes. The marine environment provides natural pest isolation, stable temperatures, and renewable water sources through evaporation–condensation cycles. These features make submerged agriculture particularly attractive for coastal regions facing soil salinization and freshwater scarcity (Cook 2022; Lizundia et al. 2022). Despite its promise, submerged agriculture faces technical and economic challenges. The initial cost of installing biospheres or submarine farms is high, and maintenance requires specialized equipment and monitoring to prevent algae growth or structural damage. Crop variety is also limited, as not all plants can adapt to underwater conditions. However, ongoing research into pressure-adapted plant varieties and photosynthetic optimization is expanding the range of cultivable species. Advances in biotechnology, such as gene editing and synthetic biology, may further enhance the viability of submerged crops in the near future (Croce et al. 2024; Pedersen and Colmer 2014).

To contextualize the diversity of submerged agriculture technologies, Table 2 provides a comparative overview of leading systems deployed or conceptualized across global marine settings. It details critical parameters, including depth range, crop compatibility, structural design, and environmental control mechanisms, to illustrate how each platform adapts to unique underwater conditions. From

Nemo's Garden's biospheric innovation to experimental vertical farms in pressurized capsules, these technologies reflect both the engineering ingenuity and ecological considerations driving underwater cultivation forward (Dini et al. 2016; Bersani et al. 2020). Submerged farming is far from a one-size-fits-all approach, it's a rapidly evolving ecosystem of innovation, shaped by technological precision and climate-conscious design. As the table illustrates, this field encompasses everything from biosphere-based greenhouses and vertical ocean farms to AI-driven kelp monitoring and floating desalination platforms. Together, these diverse models reflect a shared ambition: to reimagine agriculture in harmony with the marine environment. In conclusion, submerged agriculture stands as a bold and visionary frontier in sustainable food production. By leveraging the ocean's unique properties and integrating smart technologies, underwater farming offers a compelling pathway to climate-resilient, pesticide-free, and resource-efficient systems that could redefine how, and where, we grow our food. As terrestrial systems face mounting pressures, the ocean may well become the next fertile ground for feeding the planet.

4 Marine Food Systems and Ecosystem Integration

Marine food systems are undergoing a transformative shift toward ecological integration, driven by the urgent need to produce sustainable protein while preserving ocean health (Fig. 2). Traditional monoculture aquaculture—focused on single-species farming such as salmon or shrimp—has been criticized for its environmental footprint, including nutrient loading, habitat degradation, and disease proliferation. In response, researchers and practitioners are increasingly adopting polyculture and integrated multi-trophic aquaculture (IMTA) models that mimic natural ecosystems by cultivating complementary species across trophic levels. These systems not only enhance productivity but also offer biomitigation benefits, such as nutrient recycling and habitat restoration (Sukhdhane et al. 2018; Nguyen and Wang 2024). IMTA systems typically combine fed species (e.g., finfish or shrimp) with organic extractive species (e.g., bivalves like mussels and oysters) and inorganic extractive species (e.g., seaweeds). The waste generated by fed species—uneaten feed, feces, and excreted nutrients—is assimilated by filter feeders and macroalgae, thereby reducing eutrophication and improving water quality. For example, co-farming bivalves with seaweed has been shown to result in net reductions in CO_2 emissions and nutrient pollution, while also increasing biomass yield and economic diversification (Nguyen and Khuong 2022; Food Planet Prize 2021).

The GreenWave model of regenerative ocean farming exemplifies this approach. It employs a 3D polyculture system where sugar kelp, scallops, mussels, oysters, and clams are cultivated vertically throughout the water column. This spatially efficient design enables high-density food production without relying on fertilizers, freshwater, or feed inputs—making it one of the most resource-conscious agricultural models available. Beyond productivity, these systems offer vital ecosystem services: they help sequester carbon, remove excess nitrogen, and buffer coastlines against

Table 2 Comparative overview of submerged agriculture technologies and their functional features

Technology type	Depth range (m)	Supported crops	Environmental control features	Structural design & materials	Power source	Notable examples/ locations	Scientific reference
Nemo's garden biospheres	**5–10**	**Basil, Lettuce, Tomatoes, Beans**	**Passive humidity control, CO_2/O_2 sensors, solar-powered ventilation**	**Air-filled polymer domes anchored to the seabed**	**Solar + Passive**	**Noli, Italy**	Dini et al. (2016)
Submarine vertical farms	10–30	Leafy Greens, Herbs, Microgreens	LED grow lights, hydroponic nutrient dosing, and climate control	Multi-tiered hydroponic racks in pressurized vessels	Battery + Renewable	Conceptual R&D, Norway	Lovat et al. (2025)
Floating agropods	**Surface–5**	**Tomatoes, Spinach, Strawberries**	**Salinity sensors, moisture regulation, and solar desalination**	**Floating platforms with hydroponic beds**	**Solar Powered**	**Open Access Gov Prototypes**	Owen (2023)
Underwater aquadomes	5–15	Beans, Peppers, Leafy Greens	Temperature stabilization, nutrient cycling, and light diffusion membranes	Transparent geodesic domes with internal hydroponics	Hybrid (solar + grid)	Experimental setups, UAE & Japan	de Dieu Marcel Ufitikirez et al. (2024)

(continued)

Table 2 (continued)

Technology type	Depth range (m)	Supported crops	Environmental control features	Structural design & materials	Power source	Notable examples/ locations	Scientific reference
Submerged hydroponic capsules	**3–8**	**Lettuce, Kale, Herbs**	**Oxygenation pumps, pH and EC monitoring, UV sterilization**	**Cylindrical capsules with sealed hydroponic chambers**	**Solar + Backup Battery**	**Pilot trials, Mediterranean Basin**	Brandslet (2024)
Air-dome greenhouses	4–12	Spinach, Chard, Mint	Evaporative cooling, humidity sensors, and CO_2 enrichment	Inflatable domes with internal climate zones	Solar + Wind Hybrid	Conceptual models, South Korea	Fukao et al. (2019)

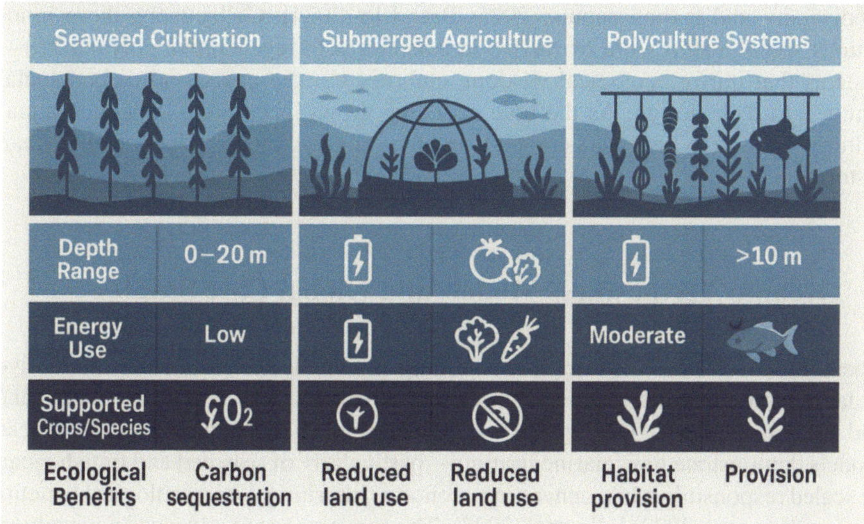

Fig. 2 Comparative overview of ocean farming modalities. A visual matrix showcasing three distinct modalities of ocean agriculture—smart seaweed cultivation, submerged agriculture, and polyculture systems. Each approach is illustrated using stylized iconography to highlight defining features such as operational depth, energy requirements, supported crops and species, and ecological contributions including carbon sequestration, biodiversity enhancement, and nutrient cycling. The figure underscores the diversity and design logic of marine farming systems, emphasizing their alignment with climate-smart principles and regenerative ocean stewardship

storm surges. Empirical studies have shown that kelp forests can reduce wave energy by 30–50% during storms, while shellfish act as natural filters, removing nutrients that would otherwise fuel harmful algal blooms and trigger hypoxic conditions (Carr 2021).

Integrated marine food systems are proving to be more than just ecologically beneficial—they're engines of economic resilience and inclusive growth. By cultivating a diverse mix of species and products, from edible seaweeds and shellfish to bioplastics, fertilizers, and cosmetics, ocean farmers can tap into multiple revenue streams and buffer themselves against market fluctuations. Initiatives like *GreenWave* are leading the charge, aiming to train 10,000 regenerative ocean farmers over the next decade with support in permitting, seed access, and buyer networks. Indigenous-led efforts such as Alaska's *Native Conservancy* are also embracing polyculture systems to restore culturally significant species like herring, while advancing food sovereignty and environmental justice (Carr 2021). On the technology front, smart sensors and AI-driven platforms are transforming ecosystem management—enabling real-time monitoring of water quality, species health, and environmental conditions. These tools support adaptive decision-making and early detection of stressors. The *BioDiMoBot* project, for instance, uses biohybrid sensors and machine learning to track biodiversity and carbon dynamics in kelp farms, offering a scalable model for ecosystem-integrated aquaculture (Robotics and kelp farming project aims to boost biodiversity and carbon capture 2025; Brandslet 2024). Collectively, these innovations signal a shift from extractive monocultures to regenerative polycultures—systems that mirror ecological wisdom, embrace climate goals, and embed social equity into their design. As the blue economy grows, ecosystem-integrated aquaculture is poised to become a cornerstone of sustainable food production and ocean restoration.

5 Oceanic Agribusiness Models and Global Case Studies

Ocean-based agribusiness is rapidly emerging as a viable and sustainable alternative to land-based food production, with countries like Norway, Tanzania (Zanzibar), and the Faroe Islands leading the way in innovation and implementation. These models demonstrate how marine farming—particularly of seaweed and finfish—can be scaled responsibly while generating economic, nutritional, and ecological benefits (Sugumaran et al. 2022; Jolly et al. 2023). The convergence of advanced aquaculture technologies, community-led entrepreneurship, and supportive policy frameworks is reshaping the global food landscape.

In Norway, the aquaculture industry has pioneered offshore salmon farming through innovations such as the Havfarm and SubFarm systems. The Havfarm, developed by Nordlaks, is a ship-shaped, semi-submersible platform capable of housing up to 10,000 tonnes of salmon. It is designed to withstand wave heights of up to 10 m and includes dynamic positioning systems that allow it to relocate during storms, minimizing environmental stress and maximizing fish welfare (Baker 2018; Ocean rainforest secures $4.5 m to ramp up Californian kelp production 2023). Meanwhile, SubFarm's submerged cage technology keeps salmon at optimal depths, away from sea lice and toxic algae, and is engineered for co-location with offshore wind farms—an approach that enhances spatial efficiency and reduces visual pollution (Submersible Aquaculture Cage System https://subfarm.no/). These systems represent a shift toward multi-use ocean zones and integrated aquaculture-energy platforms. Norway's salmon industry also emphasizes fish welfare and environmental compliance, with strict regulations on sea lice thresholds and waste discharge. Companies like SalMar and Lerøy have adopted cleaner fish strategies and advanced feed formulations to reduce parasitic loads and improve sustainability metrics (Dyrevern 2024; Norwegian salmon aquaculture sustainable solution or a source of pollution 2021). These efforts have positioned Norway as a global leader in responsible aquaculture, with over 1.5 million tonnes of fish produced annually and exported to more than 50 countries.

In Zanzibar, seaweed farming has become a cornerstone of coastal livelihoods and gender empowerment. Over 25,000 women—known locally as the Seaweed Mamas—cultivate Eucheuma species in shallow tidal zones, threading seedlings onto ropes anchored to the seabed. This practice, introduced in the late 1980s, has evolved into a thriving industry that now accounts for over 90% of Zanzibar's marine exports, surpassing traditional spice commodities like cloves and nutmeg (Onyango 2025; Kashyap 2025).

Women-led enterprises such as *Mwani Zanzibar* have transformed raw seaweed into high-value skincare products, including soaps, oils, and balms, using proprietary extraction methods that preserve marine actives. These products are handcrafted in micro-batches and sold to tourists and luxury markets, generating monthly incomes of $250–$300, nearly seven times the average for traditional seaweed farmers (Mwani 2025; Mwani Centre https://www.zanzibar.com/places/mwani-centre/). The Mwani Mamas also co-own land, contribute to household incomes, and participate in local governance, marking a profound shift in social dynamics and economic agency.

Despite climate challenges such as rising ocean temperatures and shorter growing seasons, Zanzibar's seaweed sector is adapting through innovation and infrastructure investment. The Zanzibar Seaweed Company (ZASCO) is constructing an $8 million processing facility—the largest in Africa—which will enable farmers to move up the value chain and access global markets for bioplastics, animal feed, and pharmaceuticals (Onyango 2025).

To illustrate the diversity and scale of ocean farming initiatives worldwide, Fig. 3 and Table 3 present a comparative analysis of notable case studies spanning different geographies, species, and operational models. It captures metrics such as dominant products, community frameworks, climate resilience strategies, and environmental footprints, highlighting how regions from Norway to Zanzibar are adapting aquaculture to local ecosystems and socioeconomic contexts. This synthesis underscores that ocean agribusiness is not monolithic but highly contextual, shaped by policy, innovation, and cultural dynamics (Onyango 2025; Froehlich et al. 2017; Jeeva et al. 2022). The table provides a strategic snapshot to support discussions on replicability and regional suitability in scaling sustainable marine food systems.

6 Environmental and Ethical Considerations in Ocean-Based Food Systems

As ocean-based food systems expand, they bring with them a complex web of environmental and ethical considerations. While marine aquaculture and submerged agriculture offer innovative responses to land scarcity and climate stress, they are not without ecological and ethical complexities. These ocean-based systems must be carefully managed to avoid unintended harm to fragile marine ecosystems and to ensure that benefits are distributed equitably. One of the most urgent environmental challenges is the impact of intensive aquaculture—especially of carnivorous species like salmon—on biodiversity and water quality. High-density fish farming has been associated with nutrient overload, eutrophication, and the proliferation of parasites such as sea lice. In Norway, for instance, salmon aquaculture has resulted in substantial organic waste buildup in fjords, disrupting benthic habitats and posing serious risks to wild fish populations (Norwegian salmon aquaculture sustainable solution or

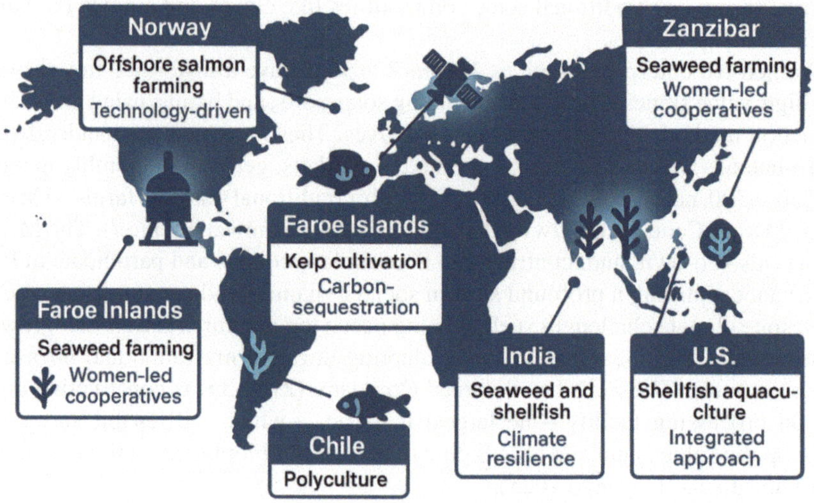

Fig. 3 Global distribution of ocean-based agribusiness models. A geospatial visualization showcasing diverse ocean farming enterprises across Norway, Zanzibar, the Faroe Islands, India, Chile, and the USA. Each regional pop-up highlights dominant species (e.g., Atlantic salmon, Eucheuma seaweed, sugar kelp), operational models (e.g., women-led cooperatives, SME-led farms, corporate aquaculture), climate resilience innovations (e.g., submerged cages, hurricane-proof anchors, seasonal adaptation), and key ecological footprints such as carbon sequestration and nitrogen removal. The figure underscores the localized adaptation and global scalability of marine food systems

a source of pollution 2021; Asche et al. 2013). Addressing these issues is critical if ocean farming is to evolve as a truly regenerative, inclusive, and ecologically sound practice.

Sea lice infestations not only compromise farmed fish welfare but also act as disease vectors for wild salmon, prompting regulatory thresholds and costly delousing procedures (Hernández-Cruz 2015).

To mitigate these impacts, innovations such as submersible cage systems and multi-trophic aquaculture are gaining traction. Norway's SubFarm technology, for instance, submerges salmon cages to depths that avoid sea lice and toxic algae, while co-locating with offshore wind farms to optimize spatial use and reduce visual pollution (Submersible Aquaculture Cage System, https://subfarm.no/). Integrated multi-trophic aquaculture (IMTA) systems combine finfish with seaweed and bivalves, creating symbiotic relationships that recycle nutrients and improve water quality. These systems have shown promise in reducing environmental footprints while enhancing productivity and resilience.

Ethical concerns also arise around animal welfare in aquaculture. High stocking densities, stressful handling procedures, and disease outbreaks contribute to elevated mortality rates. In 2023 alone, over 100 million farmed salmon died in Norwegian facilities, with causes ranging from gill disease to thermal delousing treatments

Table 3 Comparative metrics of global ocean farming case studies

Location	Dominant product	Community model	Climate resilience features	Environmental impact profile	Annual yield (approx.)	Innovation highlight	Scientific references
Norway	**Atlantic Salmon**	**Corporate-industrial**	**Submersible cages, AI monitoring, offshore siting**	**Regulated waste discharge, sea lice mitigation**	**~10,000 tons/farm**	**Co-location with offshore wind farms**	Froehlich et al. (2017)
Zanzibar (Tanzania)	Eucheuma Seaweed	Women-led cooperatives	Low-tech, climate-sensitive, seasonal adaptation	Minimal input farming, habitat restoration	~15–20 tons/farm	Skincare conversion, local equity empowerment	Onyango (2025)
Faroe Islands	**Sugar Kelp**	**SME-led**	**Cold-water kelp strains, carbon modeling**	**Positive carbon balance, biodiversity support**	**~30 tons/farm**	**Ocean Rainforest supply chains, EU integration**	Inletkeeper (2020)
USA (Connecticut)	Kelp, Mussels, Oysters	Non-profit regenerative farms	Hurricane-proof anchors, polyculture resilience	Nitrogen removal, carbon sequestration	~250,000 shellfish + 30 tons kelp/acre	3D ocean farming, open-source farmer training	Christou et al. (2013)
India (Andhra Pradesh)	**Indian Pompano, Seabass**	**Government-supported pilot clusters**	**Cage farming in estuaries and open sea zones**	**Moderate nutrient loading, monitored stocking**	**~1.5–2 tons/cage**	**Income doubling via cage aquaculture**	Jeeva et al. (2022)
Chile	Kelp, Mussels	Private–public partnerships	Integrated marine spatial planning	Moderate ecological footprint, export-oriented	~20–25 tons/farm	Sustainable Ocean Plan, blue carbon integration	Christou et al. (2013), Khan et al. (2024)

(continued)

Table 3 (continued)

Location	Dominant product	Community model	Climate resilience features	Environmental impact profile	Annual yield (approx.)	Innovation highlight	Scientific references
Australia (Cygnet Bay)	**Pearl Oysters**	**Indigenous-led enterprises**	**Longline spacing, biofouling management**	**High water filtration, low-density farming**	**~1400–1600 oysters/farm**	**Marine pearl farming with ecosystem services**	Marine Pearl Farming (2024)
Denmark	Mussels, Seaweed	Urban community farms	Educational platforms, modular platforms	Low-impact, high engagement	~1 ton mussels/ platform	Havhøst regenerative education model	Waycott (2022)

known to induce panic and pain (Dyrevern 2024). Cleaner fish, used to control parasites, often suffer near-total mortality and are treated as disposable tools, raising serious welfare questions. Addressing these issues requires stricter welfare standards, alternative parasite control methods, and investment in humane slaughter technologies.

On the ethical front, gender equity and community empowerment are central to the success of ocean-based food systems. In Zanzibar, over 25,000 women—known as the Seaweed Mamas—have transformed seaweed farming into a vehicle for financial independence, entrepreneurship, and social change. Through cooperatives like Mwani Zanzibar, women produce high-value skincare products, co-own land, and contribute to household incomes, challenging traditional gender roles and fostering inclusive development (Onyango 2025; Mawani 2025). However, climate change threatens this progress, with rising ocean temperatures shortening growing seasons and reducing yields. Investments in climate-resilient farming techniques and local processing infrastructure are critical to sustaining these gains. Transparency and traceability are also ethical imperatives. Consumers increasingly demand to know where their seafood comes from and how it was produced. Yet, studies show that most salmon consumers are unaware of the origin or sustainability of their food, highlighting a gap in labeling and public awareness (Norwegian salmon aquaculture sustainable solution or a source of pollution 2021). Certification schemes, blockchain-based traceability, and consumer education campaigns can help bridge this gap and promote responsible consumption. Therefore, the environmental and ethical dimensions of ocean-based food systems must be addressed with the same rigor as technological innovation. By embedding ecological intelligence, animal welfare, and social equity into the design and governance of marine agribusiness, we can ensure that the blue economy is not only productive but also principled.

7 Policy, Infrastructure, and the Blue Economy

The expansion of ocean-based food systems hinges not only on technological innovation and ecological integration but also on robust policy frameworks and enabling infrastructure. The concept of the blue economy—defined as the sustainable use of ocean resources for economic growth, improved livelihoods, and ecosystem health—has gained traction globally, with governments and multilateral institutions crafting strategies to align marine development with climate and biodiversity goals (Indian blue economy—a draft policy framework 2020). In India, the Draft Blue Economy Policy Framework envisions optimal utilization of maritime sectors including aquaculture, fisheries, ocean energy, and marine biotechnology. It emphasizes integrated coastal zone management, spatial planning, and ecosystem-based governance. The policy proposes the establishment of a National Blue Economy Advisory Council (BEAC) comprising secretaries of relevant ministries, coastal state representatives, and industry stakeholders to ensure cross-sectoral coordination and strategic oversight (Indian blue economy—a draft policy framework 2020; Ministry of Earth Science 2022).

Infrastructure development is a cornerstone of blue economy implementation. The Deep Ocean Mission, spearheaded by India's Ministry of Earth Sciences, aims to enhance ocean observation systems, deploy data buoys, and support deep-sea mining and marine biodiversity mapping. These initiatives are complemented by investments in marine clusters, aquaparks, and coastal logistics hubs, which facilitate the growth of aquaculture and mariculture industries while improving supply chain efficiency (Gopalakrishnan, et al. 2024). Globally, the EU Ocean Multi-use Action Plan provides a template for integrating multiple maritime activities—such as offshore wind, aquaculture, and tourism—within shared ocean zones. It recommends legal reforms, stakeholder engagement, and spatial planning tools to promote coexistence and minimize conflict among ocean users (Ocean Multi-use Action Plan, https://maritime-spatial-planning.ec.europa.eu/practices/ocean-multi-use-action-plan; Kenchington and Day 2011). Zoning, as demonstrated in the Great Barrier Reef Marine Park, is a fundamental cornerstone of effective marine spatial planning, enabling ecosystem-based management and sustainable multi-use governance (Kenchington and Day 2011).

Legal frameworks must evolve to support multi-use marine spatial planning (MSP). Countries like South Africa and Australia have enacted legislation that balances conservation with economic development, incorporating stakeholder rights, environmental impact assessments, and adaptive zoning regulations (Taljaard and Niekerk 2013). These models highlight the importance of participatory governance and legal clarity in advancing sustainable ocean farming. Investment strategies are also evolving. The Ocean Investment Protocol (de Vos, et al. 2025) outlines a roadmap for financing sustainable ocean sectors, including aquaculture, offshore energy, and marine conservation. It calls for public–private partnerships, blue bonds, and blended finance models to unlock capital for infrastructure, innovation, and ecosystem restoration. The protocol emphasizes the need for reliable ocean data, global standards, and inclusive planning to ensure that investments are equitable and climate-aligned. Hence, the success of ocean-based food systems depends on the convergence of policy coherence, infrastructure investment, and legal innovation. By embedding sustainability into governance and planning, nations can harness the full potential of the blue economy while safeguarding marine ecosystems and empowering coastal communities.

8 Future Roadmaps and Innovation Pathways

The future of ocean-based agriculture is unfolding at the intersection of visionary design, technological innovation, and policy integration—charting a course toward scalable, sustainable food systems that actively restore marine ecosystems. As land-based farming grapples with soil degradation, freshwater scarcity, and climate volatility, ocean farming is emerging as a climate-resilient, nutrient-dense, and spatially efficient alternative. The next wave of innovation will center on weaving together regenerative aquaculture, vertical ocean farming, and multi-use marine zones into unified frameworks that support biodiversity, community livelihoods, and long-term ecological health. These strategies are increasingly aligned with global sustainability targets, particularly the United Nations Sustainable Development Goal 14: *Life Below Water* (Pecl et al. 2022).

One of the most promising models is 3D ocean farming, pioneered by Bren Smith and the GreenWave initiative. This technique utilizes the entire water column to cultivate seaweed, shellfish, and other species in vertically stacked arrangements, creating polyculture systems that mimic natural ecosystems. These farms require no fertilizers, freshwater, or feed, making them among the most sustainable forms of food production. A single acre of 3D ocean farm can yield up to 30 tons of kelp and 250,000 shellfish annually, while also providing habitat for marine life and buffering coastlines against storm surges (Gertz 2017; Smith 2021). Scaling such models requires investment in infrastructure, training, and market access. GreenWave's open-source framework and farmer support programs offer a replicable blueprint for coastal communities worldwide. However, broader adoption will depend on reforms in permitting, financing, and supply chain logistics. The World Economic Forum's Global Sustainable Aquaculture Roadmap outlines four key pathways for systemic change: responsible production, better livelihoods, healthy consumption, and enabling environments. These pathways emphasize the need for inclusive governance, climate-smart technologies, and equitable access to blue food markets.

The vision for Aquatic Agritecture is rapidly transitioning from speculative concept to urgent reality—driven by mounting climate pressures, shrinking arable land, and nutrient runoff from conventional agriculture. Realizing this shift toward ocean-based food systems demands a multidisciplinary lens, precisely the kind cultivated across decades of academic and research legacies. Fields such as microbial bioprospecting, synthetic biology, plant biotechnology, sustainable materials science, toxicological assessment, and pollution mitigation collectively offer the scaffolding needed to operationalize Aquatic Agritecture. At its core lies the understanding of resilient biological systems (Dhar et al. 2024; Parashar 2023; Kaur 2023; Kaur et al. 2022), which informs the strategic bioprospecting of underutilized plants and their endophytes to enhance salinity tolerance and boost yields. These insights underpin the development of tailored biofertilizer consortia for saline marine cultivation—particularly for halophytic crops and seaweed—laying the groundwork for scalable, climate-adaptive agriculture beneath the waves.

The knowledge of microbial traits such as indole acetic acid production, biofilm formation, and inorganic phosphate solubilization offers a microbial toolkit to sustain productivity in marine farms where traditional fertilizers are ineffective or ecologically harmful (Kaur et al. 2022). Furthermore, works on PGPRs affirms the significance of symbiotic microbes in nutrient cycling (Singh et al. 2025). This is pivotal for aquatic systems where nitrogen management can be complex and energetically costly. Extrapolating these findings to marine analogs can enable rhizobial-like consortia for macroalgae or salt-loving crops grown in coastal aquaponics (Serazetdinova, et al. 2025; Fotina et al. 2024). The mechanistic insight into extracellular polymeric substances (EPS) described in microbial bionocculants showcases microbial strategies for enhancing structural integrity of substrates—critical for anchoring crops in floating beds or vertical ocean farms. EPS can also aid in water retention, buffering salinity fluctuations, and providing biological interfaces to mimic soil functionality in water-based environments. Synthetic biology and genomic tools for enhancing plant and microbial performance under environmental stress form another critical pillar (Abdi 2024a, b). Such studies present strategic examples of genomic fine-tuning, which could be extended to marine crops like kelp, spirulina, or duckweed (Abdi 2024c, d). Applying site-directed mutagenesis and CRISPR-Cas systems to ocean-grown food sources could usher in cultivars optimized for salt stress, rapid biomass accumulation, and targeted metabolite production, which are essential for making marine food systems economically viable (Abdi 2024a). Furthermore, understanding biodiversity through DNA barcoding and conservation biology, which informs species selection and ecological management in Aquatic Agritecture would also be highly sought strategy (Bhat, et al. 2024). Cataloging resilient marine organisms, evaluating biodiversity compatibility, and applying genetic barcoding to monitor contamination and invasive species together form a foundational framework for intelligent ocean farming. These practices enable precision in species selection, ecosystem integration, and biosecurity management. Parallel advances in material science, particularly in cellulose-based innovation (Khandakar et al. 2024; Das et al. 2025; Rao 2025; Mudgal 2025; Thakur et al. 2025), are reshaping marine infrastructure design. There is an urgent need to develop ocean-friendly alternatives to conventional plastics. Such materials could be engineered into biodegradable cultivation platforms for seaweed and mussels, edible coatings that extend the shelf life of seafood, or sustainable packaging solutions that help mitigate the maritime plastic footprint. This convergence of biotechnology and sustainable materials offers a pathway toward regenerative aquaculture systems that are both ecologically sound and industrially scalable. The design principles proposed by many also align with circular materials use in aquaculture gear, waste reuse, and antifouling surfaces (Singh 2025).

Pollution management is imperative in ocean-based farming systems (Kaur et al. 2025; Seetharaman 2025; Singh et al. 2024), which directly calls attention to the hazards of persistent toxins, plastics, and runoff entering aquatic food webs. Bacterial systems developed for dye degradation may be adapted to neutralize chemical waste from marine fisheries, boats, and floating processing units (Kaur 2023).

These approaches align seamlessly with Aquatic Agritecture's imperative for embedded bioremediation strategies. Behavioral and toxicological studies using zebrafish offer a powerful asset for aquatic systems monitoring, with numerous studies highlighting their role as biosensors for feed toxicity and pollutant burden (Balu 2024, #21). These models are readily adaptable to marine contexts, enabling the detection of bioaccumulating toxins in filter feeders and small fish, and facilitating real-time monitoring of aquafeed additives (Balu et al. 2024; Herath et al. 2025; Shahid et al. 2016; Pawar et al. 2016). Complementary research has shown that microbial consortia originally developed for degrading terrestrial dyes and converting waste into energy can be reengineered for oceanic waste management—particularly in processing fish waste, shellfish shells, and macroalgal biomass into value-added products or renewable energy within self-sustaining Aquatic Agritecture zones. Additionally, plant tissue culture and regeneration methodologies (Kaur and Mudgal 2021; Parashar et al. 2020; Kaur et al. 2023), are gaining relevance in aquatic contexts, offering replication techniques and shoot regeneration models that can be extrapolated to salt-tolerant aquatic plants. This enables rapid scaling of controlled ocean farms and supports biopharmaceutical production using sea-derived botanicals (Binh et al. 1992; Al-Khateeb et al. 2020). Finally, the application of existing food chain monitoring tools ensures consumer safety and regulatory compliance across marine systems (Vinayak et al. 2021), reinforcing the viability of Aquatic Agritecture as a holistic, resilient, and ethically grounded framework for ocean-based agriculture.

Other works inform how abiotic stress hormones and gene expression markers in terrestrial plants may guide selection and enhancement of stress-resilient marine crops, paving the way for deeper integration of hormonal engineering in sea-based agriculture (Saeed 2023; Verma et al. 2022).

Technological innovation will play a pivotal role in scaling ocean agriculture. Underwater drones, robotic fish, and AI-powered monitoring systems are already being deployed to assess fish health, optimize nutrient cycling, and automate harvesting. Companies like Aquaai and Aquabyte are integrating machine learning with sensor networks to create intelligent aquaculture platforms that reduce labor costs and environmental impact (Moore 2024). Meanwhile, floating sea farms equipped with solar-powered desalination units are enabling the cultivation of terrestrial crops on ocean surfaces, offering a hybrid model of food-energy-water systems that could revolutionize offshore agriculture (Owen 2023). Visionary models also include multi-use ocean zones, where aquaculture coexists with offshore wind, conservation areas, and tourism. The EU's Ocean Multi-use Action Plan and India's Blue Economy Policy Framework advocate for integrated marine spatial planning that balances ecological integrity with economic development (Indian blue economy—a draft policy framework 2020; Ocean Multi-use Action Plan, https://maritime-spatial-planning.ec.europa.eu/practices/ocean-multi-use-action-plan). These frameworks support the creation of marine clusters and aquaparks, which can serve as innovation hubs for ocean farming, processing, and distribution.

Looking ahead, regenerative ocean farming is poised to become a cornerstone of climate mitigation. Seaweed farms can sequester up to 173 million tons of carbon annually, while bivalves improve water quality and biodiversity. Integrating these systems with carbon credit schemes, blue bonds, and ecosystem service payments could unlock new financing mechanisms for sustainable ocean agriculture (Degnarain 2020). To build on this vision, the future of ocean agriculture will be shaped by the convergence of science, design, and policy—where interdisciplinary collaboration becomes the engine of innovation, technological foresight drives scalable solutions, and inclusive governance ensures equitable access and long-term stewardship. Scaling visionary models like 3D ocean farming, deploying smart infrastructure, and embedding these systems within global sustainability frameworks—particularly the

UN SDGs—can shift our relationship with the ocean from one of extraction to one of regeneration. In this paradigm, the ocean becomes not just a food source, but a living blueprint for climate resilience, ecological restoration, and socio-economic transformation.

9 Conclusion and Strategic Takeaways

Strategically, the way forward hinges on four key imperatives: **(1) Ecological integration**, where ocean farming systems are designed to regenerate biodiversity, sequester carbon, and harmonize with local ecosystems; **(2) Technological foresight**, ensuring that AI, robotics, and bioengineering are deployed ethically and inclusively to enhance productivity without compromising marine health; **(3) Inclusive governance**, embedding coastal communities, Indigenous knowledge systems, and gender equity into decision-making processes to foster ownership and resilience; and **(4) Policy alignment**, where marine zoning, climate adaptation strategies, and sustainability metrics are codified into national and global frameworks to guide responsible growth. Together, these imperatives form the scaffolding for a regenerative blue economy—one that transforms the ocean from a site of extraction into a living laboratory for sustainable innovation, food security, and planetary healing. More stringent ways forward hinge on four key imperatives:

- **Systemic Integration**: Align ocean agriculture with energy, water, and biodiversity goals through cross-sectoral planning and spatial design.
- **Inclusive Innovation**: Invest in training, access, and equity to ensure marginalized coastal communities benefit from ocean-based farming.
- **Data-Driven Monitoring**: Utilize smart sensors and open-source platforms for real-time tracking of environmental health, yields, and risks.
- **Global Collaboration**: Encourage international cooperation on standards, marine protection, and climate finance for scalable impact.

In reimagining our relationship with the ocean, we are also reimagining what food can be—nutritious, locally embedded, ecologically restorative, and economically inclusive. Ocean agriculture is not a replacement for terrestrial farming, but a complement to it, helping us diversify and future-proof our food systems. As we chart this course into the deep blue, we do not sail alone; we carry with us centuries of coastal wisdom, cutting-edge science, and a shared responsibility to nourish life both above and beneath the waves.

References

Abdi G et al (2024a) Genomic Precision: unveiling the transformative role of genome editing in advancing genomics research and applications. In: Singh V (eds) Advances in Genomics. Springer, Singapore. https://doi.org/10.1007/978-981-97-3169-5_13

Abdi G et al (2024b) DNA barcoding and its applications. In: Singh V (eds) Advances in Genomics. Springer, Singapore. https://doi.org/10.1007/978-981-97-3169-5_5

Abdi G et al (2024c) Engineering genomic landscapes: Synthetic biology approaches in genomic rearrangement. In: Singh V (eds) Advances in Genomics. Springer, Singapore. https://doi.org/10.1007/978-981-97-3169-5_12

Abdi G et al (2024d) Progress on synthetic genomics. In: Singh V (eds) Advances in Genomics. Springer, Singapore. https://doi.org/10.1007/978-981-97-3169-5_9

Abusin SAA, Mandikiana BW (2020) Towards sustainable food production systems in Qatar: assessment of the viability of aquaponics. Glob Food Sec 25:100349. https://doi.org/10.1016/j.gfs.2020.100349

Ackerman E (2017) Robotic kelp farms promise an ocean full of carbon-neutral, low-cost energy. https://spectrum.ieee.org/robotic-kelp-farms-promise-an-ocean-full-of-carbon-neutral-low-cost-energy

Al-Khateeb SA, Al-Khateeb AA, Sattar MN, Mohmand AS (2020) Induced in vitro adaptation for salt tolerance in date palm (Phoenix dactylifera L.) cultivar Khalas. Biol. Res. 53:37. https://doi.org/10.1186/s40659-020-00305-3

Asche F, Roll KH, Sandvold HN, Sørvig A, Zhang D (2013) Salmon aquaculture: larger companies and increased production. Aquac Econ Manag 17:322–339. https://doi.org/10.1080/13657305.2013.812156

Baker J (2018) Havfarm: a new salmon fishing revolution in Norway. https://www.ship-technology.com/features/havfarm-fish-farm-vessel/

Balu UR et al (2024) Linking EDC-laden food consumption and modern lifestyle habits with preeclampsia: a non-animal approach to identifying early diagnostic biomarkers through biochemical alterations. Food Chem Toxicol 194:115073. https://doi.org/10.1016/j.fct.2024.115073

Bella TW et al (2020) The utility of satellites and autonomous remote sensing platforms for monitoring offshore aquaculture farms: a case study for Canopy forming kelps. Front Marine Sci 7 (2020). https://doi.org/10.3389/fmars.2020.520223

Bersani C, Ouammi A, Sacile R, Zero E (2020) Model predictive control of smart greenhouses as the path towards near zero energy consumption. Energies 13:3647. https://doi.org/10.3390/en13143647

Bhat AR et al (2024) Curating and validating universally applicable primers for efficient ITS2-based DNA barcoding across plant taxa. Int J Health Allied Sci 13, Article 4

Binh DQ, Heszky LE, Gyulai G, Csillag A (1992) Plant regeneration of NaCl-pretreated cells from long-term suspension culture of rice (Oryza sativa L.) in high saline conditions. Plant Cell, Tissue Organ Cult 29:75–82. https://doi.org/10.1007/BF00033611

Brandslet S (2024) Artificial intelligence makes smarter use of seaweed and kelp. https://partner.sciencenorway.no/artificial-intelligence-biotechnology-nature-conservation/artificial-intelligence-makes-smarter-use-of-seaweed-and-kelp/2346636

Canvin MC, Moore PJ, Smale DA (2024) Quantifying growth, erosion and dislodgement rates of farmed kelp (Saccharina latissima) to examine the carbon sequestration potential of temperate seaweed farming. J Appl Phycol 36:3091–3102. https://doi.org/10.1007/s10811-024-03323-w

Carr G (2021) Regenerative ocean farming: how can polycultures help our coasts? https://smea.uw.edu/currents/regenerative-ocean-farming-how-can-polycultures-help-our-coasts/

Chaudhari V et al (2024b) The global potential of seaweed farming for carbon sequestration and removal. Int J Res Agron. https://doi.org/10.33545/2618060X.2024.v7.i9a.1425

Chaudhari V et al (2024a) The global potential of seaweed farming for carbon sequestration and removal. Int J Res Agron 7. https://doi.org/10.33545/2618060X.2024.v7.i9a.1425

Christou P, Savin R, Costa-Pierce B, Misztal I, Whitelaw B (2013) Sustainable food production, 1 edn. 1869. Springer, New York, NY
Cook M (2022) Adaptations of ocean plants. https://www.sciencing.com/adaptations-ocean-plants-6560351/
Croce R et al (2024) Perspectives on improving photosynthesis to increase crop yield. Plant Cell 36:3944–3973. https://doi.org/10.1093/plcell/koae132
Das D, Singh A, Adari R, Mudgal G, Radha P (2025) Biocompatible cellulose derivatives: green chemistry and its sustainable applications. In: Kesari KK, Prakash C, Khalid M, Negi A (eds) Functionalized Cellulose Materials. Engineering Materials. Springer, Cham.https://doi.org/10.1007/978-3-031-76953-5_7
de Vos K, Johnson S, Naas J-S, Nanninga R, Martínez Toral K (2025) Ocean investment protocol – 2025: a multi-stakeholder plan to enable funding for the Sustainable Ocean Economy. United Nations Global Compact; Ocean Fox Advisory; Blue Biome Consulting. https://unfccc.int/sites/default/files/resource/Informal_Summary_Report_Ocean_Dialogue_2025.pdf
de Dieu Marcel Ufitikirez J et al (2024) Agricultural waste valorization: exploring environmentally friendly approaches to bioenergy conversion. Sustainability 16:3617. https://www.mdpi.com/2071-1050/16/9/3617
Degnarain N (2020) Ocean crops: is this the next frontier for agriculture? https://www.forbes.com/sites/nishandegnarain/2020/07/29/ocean-crops-is-this-the-next-frontier-for-agriculture/
Dhar SK et al (2024) Novel Bacillus and Prestia isolates from Dwarf century plant enhance crop yield and salinity tolerance. Sci Rep 14:14645. https://doi.org/10.1038/s41598-024-65632-x
Dini G, Princi E, Gamberini S, Gamberini L, in Published in: OCEANS 2016 MTS/IEEE Monterey. IEEE, pp 1–6
Duarte C (2025) Seaweed farms dish up climate benefits. Nat Clim Chang 15:180–187. https://doi.org/10.1038/s41558-024-02238-1
Dyrevern (2024) Fish farming in Norway. https://dyrevern.no/dyrevern/fish-farming-in-norway/
Food from the Sea (2021).https://worldoceanreview.com/en/wor-7/food-from-the-sea/
Fotina NV et al (2024) Enhancement of wheat growth by plant growth-stimulating bacteria during phytopathogenic inhibition. Biocatal Agric Biotechnol 60:103294. https://doi.org/10.1016/j.bcab.2024.103294
Food Planet Prize (2021) GreenWave – Taking ocean farming to new depths. The curt bergfors food planet prize. https://foodplanetprize.org/initiatives/greenwave-taking-ocean-farming-to-new-depths/
Froehlich HE, Smith A, Gentry RR, Halpern BS (2017) Offshore aquaculture: I know it when I see it. Front Mar Sci 4. https://doi.org/10.3389/fmars.2017.00154
Fukao T, Barrera-Figueroa BE, Juntawong P, Peña-Castro JM (2019) Submergence and waterlogging stress in plants: a review highlighting research opportunities and understudied aspects. Front Plant Sci 10:340
Gertz E (2017) Vertical ocean farms that can feed us and help our seas. https://ideas.ted.com/vertical-ocean-farms-that-can-feed-us-and-help-our-seas/
Gopalakrishnan A et al (2024) Propelling India's blue economy: technological and governance perspectives in fisheries and aquaculture. Curr Sci 126. https://doi.org/10.18520/cs/v126/i2/169-176
Hassan T (2024) What is vertical ocean farming? https://agrinextcon.com/vertical-ocean-farming-sustainable-and-shellfish/
Herath HMLPB et al (2025) Innovative transgenic zebrafish biosensor for heavy metal detection. Environ Pollut 366:125547. https://doi.org/10.1016/j.envpol.2024.125547
Hernández-Cruz et al (2015) Growth performance and gene expression in gilthead sea bream (S parus aurata) fed microdiets with high docosahexaenoic acid and antioxidant levels. Aquac Nutr 21. https://doi.org/10.1111/anu.12213
Indian blue economy—a draft policy framework (2020)
Inletkeeper C (2020) GreenWave's regenerative ocean farming: a case study in generating ecosystem services. https://inletkeeper.org/regenerative-ocean-farming/

Jeeva C et al (2022) Success of cage farming of marine finfishes in doubling farmers' income: a techno-social impact analysis. Curr Sci 123:1031. https://doi.org/10.18520/cs/v123/i8/1031-1037

Jolly CM et al (2023) Dynamics of aquaculture governance. J World Aquac Soc 54:427–481. https://doi.org/10.1111/jwas.12967

Jung M et al (2024) A global clustering of terrestrial food production systems. PLoS One 19:e0296846. https://doi.org/10.1371/journal.pone.0296846

Kashyap T (2025) Meet the 'Seaweed Mamas' holding up Zanzibar's coastal economy. https://thenodmag.com/content/seaweed-mamas-zanzibar-coastal-economy

Kaur J et al (2022) An exopolysaccharide-producing novel Agrobacterium pusense strain JAS1 isolated from snake plant enhances plant growth and soil water retention. Sci Rep 12:21330. https://doi.org/10.1038/s41598-022-25225-y

Kaur J et al (2023) GC-MS validated phytochemical up-leveling with in vitro-raised Sansevieria trifasciata [Prain]: The Mother in Law's tongue gets more antibacterial. Curr Plant Biol 35–36:100308. https://doi.org/10.1016/j.cpb.2023.100308

Kaur J et al (2023) Reactive Black-5, Congo Red and Methyl Orange: chemical degradation of Azo-Dyes by agrobacterium. Water 15:1664

Kaur J, Mudgal G (2021) An efficient and quick protocol for in vitro multiplication of snake plant, Sansevieria trifasciata var. Laurentii [Prain]. In: Plant cell, tissue and organ culture (PCTOC), vol 147, pp 405–411. https://doi.org/10.1007/s11240-021-02132-0

Kaur S, Thakur H, Singh A, Ramasamy V, Mudgal G (2025) Poisoned Seas: chemical threats to marine life and human health. In: Prakash C, Kesari KK, Negi A (eds) Sustainable Development Goals Towards Environmental Toxicity and Green Chemistry. World Sustainability Series. Springer, Cham.https://doi.org/10.1007/978-3-031-77327-3_10

Kenchington RA, Day JC (2011) Zoning, a fundamental cornerstone of effective Marine Spatial Planning: lessons learnt from the Great Barrier Reef. Aust J Coast Conserv 15:271–278. https://doi.org/10.1007/s11852-011-0147-2

Khan N, Sudhakar K, Mamat R (2024) Macroalgae farming for sustainable future: Navigating opportunities and driving innovation. Heliyon 10:e28208. https://doi.org/10.1016/j.heliyon.2024.e28208

Khandakar A et al (2024) Smart aquaponics: an innovative machine learning framework for fish farming optimization. Comput Electr Eng 119:109590. https://doi.org/10.1016/j.compeleceng.2024.109590

Lizundia E, Luzi F, Puglia D (2022) Organic waste valorisation towards circular and sustainable biocomposites. Green Chem 24:5429–5459. https://doi.org/10.1039/D2GC01668K

Lovat SJ, Noor E, Milo R (2025) Vertical farming limitations and potential demonstrated by back-of-the-envelope calculations. Plant Physiol 198:kiaf056. https://doi.org/10.1093/plphys/kiaf056

Maguire J (2024) Seaweeds from sustainable aquaculture as feedstock for biodegradable bioplastics.https://cordis.europa.eu/project/id/606032/reporting

Mahrose KM, Michalak I (2022) Seaweeds for animal feed, current status, challenges, and opportunities. In: Sustainable global resources of seaweeds. Volume 1: Bioresources, cultivation, trade, and multifarious applications, pp 357–379. https://doi.org/10.1007/978-3-030-91955-9_19

Marchant N (2021a) What are underwater farms? And how do they work? https://www.weforum.org/stories/2021/07/underwater-farms-sustainable/

Marchant N (2021b) Underwater farms & how they work. https://www.agritecture.com/blog/2021/7/23/underwater-farms-amp-how-they-work

Marine Pearl Farming (2024). https://www.aquaculturescience.org/content/dam/tnc/nature/en/documents/Pearl_Case_Study_TNC_web1.pdf

Mwani M (2025) Farming in rhythm with the moon and tides of the Indian Ocean and artfully handcrafting skincare. https://mwanizanzibar.com/pages/the-mwani-zanzibar-mamas

Meinita MDN, Harwanto D, Choi J-S (2022) Seaweed exhibits therapeutic properties against chronic diseases: an overview. Appl Sci 12:2638. https://doi.org/10.3390/app12052638

Ministry of Earth Science (2022) Blue Economic Policy. https://www.pib.gov.in/PressReleasePage.aspx?PRID=1845257
Moore S (2024) Under the sea: importance of underwater robots and drones in sea farming. https://www.azolifesciences.com/article/Under-the-Sea3b-Importance-of-Underwater-Robots-and-Drones-in-Sea-Farming.aspx
Morais T et al (2020) Seaweed potential in the animal feed: a review. J Mar Sci Eng 8:559. https://doi.org/10.3390/jmse8080559
Mossler M (2020) The future of food from the sea, explained, https://sustainablefisheries-uw.org/food-from-the-ocean-2050/
Mudgal G et al (2025) Game Changer: how cellulose-based bioleather is transforming the market. In: Kesari KK, Prakash C, Khalid M, Negi A (eds) Functionalized Cellulose Materials. Engineering Materials. Springer, Cham. https://doi.org/10.1007/978-3-031-76953-5_12
MUSES Project Partners (2017) Ocean multi-use action plan. The European Maritime Spatial Planning Platform; SUBMARINER Network. https://maritime-spatial-planning.ec.europa.eu/practices/ocean-multi-use-action-plan
Mwani Centre.https://www.zanzibar.com/places/mwani-centre/
Nair CS et al (2025) Recent advancements in aquaponics with special emphasis on its sustainability, 56:e13116. https://doi.org/10.1111/jwas.13116
Nguyen HN, Khuong DA (2022) New Trends in Pyrolysis Methods: opportunities, limitations, and advantages. In: Ramola S, Mohan D, Masek O, Méndez A, Tsubota T (eds) Engineered Biochar. Springer, Singapore.https://doi.org/10.1007/978-981-19-2488-0_7
Nguyen HP, Wang CM (2024) Advances in offshore aquaculture and renewable energy production. J Mar Sci Eng 12:1679. https://doi.org/10.3390/jmse12091679
Norwegian salmon aquaculture sustainable solution or a source of pollution? (2021). https://storymaps.arcgis.com/stories/ea8428ce086e4088a9e9861efe2ae255
Ocean rainforest secures $4.5 m to ramp up Californian kelp production (2023). https://thefishsite.com/articles/ocean-rainforest-secures-4-5-m-to-ramp-up-californian-kelp-production
Onyango S (2025) Seaweed turns the financial tide for Zanzibari women. https://africasolutionsmediahub.org/2025/05/24/seaweed-turns-the-financial-tide-for-zanzibari-women/
Owen (2023) Floating sea farms: a vision to feed the world. https://www.openaccessgovernment.org/floating-sea-farms-a-vision-to-feed-the-world/166272/
Parashar M, Mudgal G, Jaspreet K (2020) In vitro tissue culture studies on Anagallis arvensis. Plant Cell Biotechnol Mol Biol 21:130–135
Parashar M et al (2023) Two novel plant-growth-promoting Lelliottia amnigena Isolates from Euphorbia prostrata Aiton enhance the overall productivity of wheat and tomato. Plants 12:3081
Patrick P, Getty (2025) Farming within earth's limits is still possible—but it will take a Herculean effort. https://theconversation.com/farming-within-earths-limits-is-still-possible-but-it-will-take-a-herculean-effort-259901
Pawar N et al (2016) Development of a fluorescent transgenic zebrafish biosensor for sensing aquatic heavy metal pollution. Transgenic Res 25:617–627. https://doi.org/10.1007/s11248-016-9959-z
Pecl GT et al (2022) Future Seas 2030: pathways to sustainability for the UN ocean decade and beyond. Rev Fish Biol Fish 32:1–7. https://doi.org/10.1007/s11160-022-09705-y
Pedersen O, Colmer TD, Sand-Jensen K (2013) Underwater photosynthesis of submerged plants–recent advances and methods. Front Plant Sci 4:140. https://doi.org/10.3389/fpls.2013.00140
Pedersen O, Colmer TD (2014) Underwater photosynthesis and internal aeration of submerged terrestrial wetland plants. low-oxygen stress in plants: oxygen sensing and adaptive responses to hypoxia, pp 315–327. https://doi.org/10.3389/fpls.2013.00140
Portalanza D (2025) Underwater gardens: how deep-sea farming can save the future. https://www.theweather.net/news/plants/underwater-gardens-how-deep-sea-farming-can-save-the-future.html
Rao PM, Singh AO, Sriniketha Narayanan K, Shankar R, Mudgal G, Radha P (2025) Cellulose-Based Packaging: an emerging sustainable packaging option for future use. In: Kesari KK, Prakash C, Khalid M, Negi A (eds) Functionalized Cellulose Materials. Engineering Materials. Springer Cham.https://doi.org/10.1007/978-3-031-76953-5_4

Robotics and kelp farming project aims to boost biodiversity and carbon capture (2025). https://envirotecmagazine.com/2025/05/14/robotics-and-kelp-farming-project-aims-to-boost-biodiversity-and-carbon-capture/
Ross F, Tarbuck P, Macreadie PI (2022) Seaweed afforestation at large-scales exclusively for carbon sequestration: critical assessment of risks, viability and the state of knowledge. Front Mar Sci 9:1015612. https://doi.org/10.3389/fmars.2022.1015612
Saeed T et al (2023) Exploring the effects of selenium and brassinosteroids on photosynthesis and protein expression patterns in tomato plants under low temperatures. Plants (Basel) 12:3351
Schmidt (2025) Panel underscores untapped potential of aquatic food systems to tackle multiple sustainable development goals, including ending hunger, protecting ocean. https://press.un.org/en/2025/sea2229.doc.htm
Seetharaman B et al (2025) The Piteous Price of Progress: the environmental and health costs of modern society. In: Prakash C, Kesari KK, Negi A (eds) Sustainable Development Goals Towards Environmental Toxicity and Green Chemistry. World Sustainability Series. Springer, Cham. https://doi.org/10.1007/978-3-031-77327-3_15
Serazetdinova YR et al (2025) Rhizobia as complex biofertilizers for wheat: biological nitrogen fixation and plant growth promotion. Foods Raw Materi 14:214–227. https://doi.org/10.21603/2308-4057-2026-1-669
Shahid M et al (2016) Zebrafish biosensor for toxicant induced muscle hyperactivity. Sci Rep 6:23768. https://doi.org/10.1038/srep23768
Singh GB, Vinayak A, Mudgal G, Kesari KK (2024) Azo dye bioremediation: an interdisciplinary path to sustainable fashion. Environ Technol Innov 36:103832. https://doi.org/10.1016/j.eti.2024.103832
Singh A et al (2025) Unveiling the transformative power of smart cellulosic nanomaterials: revisiting potential promises to sustainable future. In: Kesari KK, Prakash C, Khalid M, Negi A (eds) Functionalized Cellulose Materials. Engineering Materials. Springer, Cham. https://doi.org/10.1007/978-3-031-76953-5_1
Singh GB, Dey S, Mudgal G (2025). Microbial contributions to a circular economy. In: Kaur S, Dhiman S, Tripathi M (eds) Microbial Metabolomics. Springer, Singapore. https://doi.org/10.1007/978-981-96-4824-5_18
Smith B (2021) Bren Smith's open-sourced 3D ocean farm model can feed a hungry planet (2021). https://bioneers.org/bren-smiths-open-sourced-3d-ocean-farm-model-can-feed-hungry-planet-ztvz1709/
Sugumaran R et al (2022) A retrospective review of global commercial seaweed production—current challenges, biosecurity and mitigation measures and prospects. Int J Environ Res Public Health 19:7087. https://doi.org/10.3390/ijerph19127087
SubFarm (2022) SubFarm – submersible aquaculture cage system. SubFarm https://subfarm.no
Sukhdhane KS, Kripa V, Divu D, Vase VK, Mojjada SK (2018) Integrated multi-trophic aquaculture systems: a solution for sustainability. Aquac Asia 22:26–29
Taljaard S, Van Niekerk L (2013) How supportive are existing national legal regimes for multi-use marine spatial planning?—the South African case. Mar Policy 38:72–79. https://doi.org/10.1016/j.marpol.2012.05.021
Thakur H, Kaur S, Singh A, Singh GB, Mudgal G (2025) Smart sustainable materials: a blueprint for a better tomorrow. In: Kesari KK, Prakash C, Khalid M, Negi A (eds) Functionalized Cellulose Materials. Engineering Materials. Springer Cham. https://doi.org/10.1007/978-3-031-76953-5_11
Tont SA, Delistraty DA (1977) Food resources of the oceans: an outline of status and potentials. Environ Conserv 4:243–252. https://doi.org/10.1017/S0376892900026084
Underwater farming: exploring the new agricultural frontier and its environmental benefits (2025). https://noticiasambientales.com/innovation/underwater-farming-exploring-the-new-agricultural-frontier-and-its-environmental-benefits/
Verma S, Negi NP, Pareek S, Mudgal G, Kumar D (2022) Auxin response factors in plant adaptation to drought and salinity stress. Physiol Plant 174:e13714. https://doi.org/10.1111/ppl.13714

Vinayak A, Mudgal G, Singh GB (2021) Environment persistent free radicals: long-lived particles. In: Kesari KK, Jha NK (eds) Free Radical Biology and Environmental Toxicity. Molecular and Integrative Toxicology. Springer Cham. https://doi.org/10.1007/978-3-030-83446-3_1
Wang X (2022) Managing land carrying capacity: key to achieving sustainable production systems for food security. Land 11:484. https://doi.org/10.3390/land11040484
Waycott B (2022) Regenerative ocean farming is trending, but can it be a successful business model? https://www.globalseafood.org/advocate/regenerative-ocean-farming-is-trending-but-can-it-be-a-successful-business-model/
Zaimis U, Ozolina S, Jurmalietis R (2021) Production of seaweed derived bioplastics. Engineering for rural development. In: Proceedings of the 20th international scientific conference (Latvia). Latvia University of Life Sciences and Technologies, pp 1692–1695. https://doi.org/10.22616/ERDev.2021.20.TF370

MIX
Papier aus verantwortungsvollen Quellen
Paper from responsible sources
FSC® C105338

If you have any concerns about our products,
you can contact us on
ProductSafety@springernature.com

In case Publisher is established outside the EU,
the EU authorized representative is:
Springer Nature Customer Service Center GmbH
Europaplatz 3, 69115 Heidelberg, Germany

Printed by Libri Plureos GmbH
in Hamburg, Germany